Optical Wireless Communication Theory and Technology

Series Editor

Xizheng Ke, School of Automation and Information Engineering, Xi'an University of Technology, Xi'an, China

The book series Optical Wireless Communication Theory and Technology aims to introduce the key technologies and applications adopted in optical wireless communication to researchers of communication engineering, optical engineering and other related majors. The individual book volumes in the series are thematic. The goal of each volume is to give readers a comprehensive overview of how the theory and technology in a certain optical wireless communication area can be known. As a collection, the series provides valuable resources to a wide audience in academia, the communication engineering research community and anyone else who are looking to expand their knowledge of optical communication.

Xizheng Ke · Chenghu Ke

Performance-Improvement Limits in Optical Wireless Coherent Communication

Science Press
Beijing

Springer

Xizheng Ke
School of Automation and Information Engineering
Xi'an University of Technology
Xi'an, Shaanxi, China

Chenghu Ke
Xi'an University
Xi'an, Shaanxi, China

ISSN 2731-5967 ISSN 2731-5975 (electronic)
Optical Wireless Communication Theory and Technology
ISBN 978-981-96-1999-3 ISBN 978-981-96-2000-5 (eBook)
https://doi.org/10.1007/978-981-96-2000-5

Jointly published with Science Press
The print edition is not for sale in China mainland. Customers from China mainland please order the print book from: Science Press.

Translation from the Chinese Simplified language edition: "无线光相干通信性能改善的极限" by Xizheng Ke and Chenghu Ke, © Science Press 2024. Published by Science Press. All Rights Reserved.

This Springer imprint is published by the registered company Springer Nature Singapore Pte Ltd.
The registered company address is: 152 Beach Road, #21-01/04 Gateway East, Singapore 189721, Singapore

Preface

Atmospheric turbulence and changes in weather conditions cause light-signal attenuation, light-intensity flicker, and spot drift, which directly affect the performance of optical wireless communication systems. Even in clear weather, atmospheric turbulence can attenuate light signals and cause light-intensity flicker, resulting in a reduced signal-to-noise ratio, and spot drift can interrupt communications.

Compared with Intensity Modulation/Direct Detection (IM/DD) optical wireless communication systems, optical wireless coherent communication has a gain of 20–23 dB. Optical wireless coherent-detection includes homodyne, heterodyne, and internal-difference detection. Coherent detection is the most effective method of suppressing the effects of atmospheric turbulence and environmental changes. This book analyzes the latest progress in several key technologies for optical wireless coherent communication.

In an analysis of optical wireless coherent-communication systems, this book focuses on the study of influence of optical wireless coherent-communication detection sensitivity, mixing efficiency, and coherent-detection limits and analyzes the coherent-detection performance of vortex beams. Chapter 1 describes the research progress in optical wireless communication in China and abroad. Chapter 2 introduces optical wireless coherent detection systems. In Chap. 3, the factors affecting the performance of wireless-optical coherent detection are introduced. In Chap. 4, the factors affecting coherent-detection sensitivity are analyzed.

In Chap. 5, the influence of the spatial-light distribution on the performance of optical wireless coherent detection is discussed. In Chap. 6, the influence of signal-wavefront distortion caused by atmospheric turbulence on the mixing efficiency is presented. Chapter 7 analyzes the wavefront-distortion characteristics of optical signals of different wavelengths, and Chap. 8 describes the decoupling control of wavefront-distortion correction with multiple correctors. Chapter 9 discusses the problem of detecting when a signal light breaks down into partially coherent light.

This book summarizes our long-standing research on optical wireless coherent communication and is part of our active exploration in this field. Doctoral student Han Maimiao and master's students Hui Yuze, Li Mengru, Xing Tian, Kang Weilong, and Wang Hairong participated in the writing of this book. We thank these graduate students for their youth and enthusiasm for the scientific research institute.

This book is the work of the author when he was an honorary professor of Xi'an University, and he would like to express his gratitude to Xi'an University for its efforts in the preparation and publication of this book.

We express our gratitude to our colleagues in this field. Our work was based on their research. We are also grateful for the contributions of those researchers not listed in the bibliography; their work enlightened the author in this field of knowledge.

Xi'an, China Xizheng Ke
June 2023 Chenghu Ke

Contents

Chapter 1
Overview of Optical Wireless Coherent Communication

1.1 Current Development Status of Optical Wireless Coherent Communication

1.1.1 Foreign Development Status

In 1880, American inventor Alexander Graham Bell invented the "optical telephone," which loaded sound signals onto a vibrating mirror for transmission, and was hailed as the beginning of contemporary optical wireless communication technology [1]. In 1960, Theodore Maiman, an American physicist, invented the world's first ruby laser, marking the entry of optical wireless communication into experimental exploration and research [2]. Professor Kao Kun proposed using optical fibers as communication media in 1966, after which, coherent optical communication developed more significantly in the field of optical fiber, compared to the field of free space.

With the development of optical-communication technology, considering that fiber-optic links cannot be established between satellites, as well as high-confidentiality requirements, optical wireless communication has once again entered the research arena and undergone unprecedented development.

At first, countries planned to apply homodyne coherent-detection technology to optical wireless communication systems using relay methods. However, because of the strict requirements for various indicators, complex equipment, and high cost of coherent detection in the 1980s, intensity modulation/direct detection technology was more fully developed and widely applied. With the improvement and maturity of various technologies, optical wireless coherent communication is gradually being developed and taking shape.

Since the 1960s, the United States was the first to research and apply optical wireless communication [3]. First, MIT's Lincoln Laboratory planned to use frequency-shift keying modulation and 220-Mbit/s code-rate coherent-detection technology to demonstrate inter-satellite laser communication. However, owing to developments in

X. Ke and C. Ke, *Performance-Improvement Limits in Optical Wireless Coherent Communication*, Optical Wireless Communication Theory and Technology,
https://doi.org/10.1007/978-981-96-2000-5_1

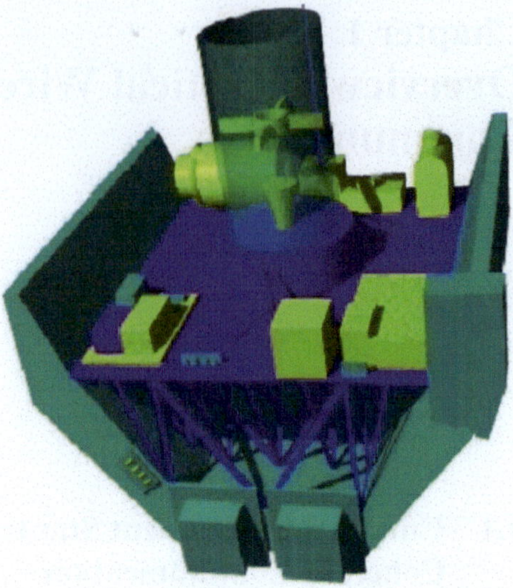

Fig. 1.1 Inter-satellite laser-communication experimental device developed by MIT's Lincoln Laboratory [3]

science and technology at the time, the demonstration experiment was only completed on the ground (Fig. 1.1).

In the early 1990s, the Jet Propulsion Laboratory of the United States National Aeronautics and Space Administration (NASA) initially planned to develop coherent optical communication, but then focused on intensity modulation/direct detection communication. At the end of the 1990s, the focus of research again turned to coherent-optical communication, exploring the detection sensitivity under different modulation modes to expand the channel capacity of inter-satellite transmissions. In 1998, the Jet Propulsion Laboratory used a laser-communication terminal test station to evaluate the performance of an optical-communication demonstration instrument [4]. The Optical Communications Telescope Laboratory was used as the ground terminal in the demonstration experiments; it was completed in 2000 [5].

In 2008, the China National Space Administration deployed and planned to implement a deep-space optical-communication plan. Figure 1.2 shows a diagram of the Lunar Laser-Communication Demonstration plan. Using wavefront-detection adaptive-correction technology and multiflutter adaptive-optics (AO) correction technology, convenient equipment was planned to communicate via lasers between the moon and the earth [6–8].

In September 2013, NASA used the Minotaur V carrier rocket to carry the Lunar Atmosphere and Dust Environment Explorer (LADEE) into space, and in October 2013, for the first time, it established a two-way communication link from the earth to the moon, with a 622-Mbit/s downlink and 20-Mbit/s uplink [9–11]. The demonstration diagram of laser communication on the moon-earth is shown in Fig. 1.3, 1.4 and 1.5.

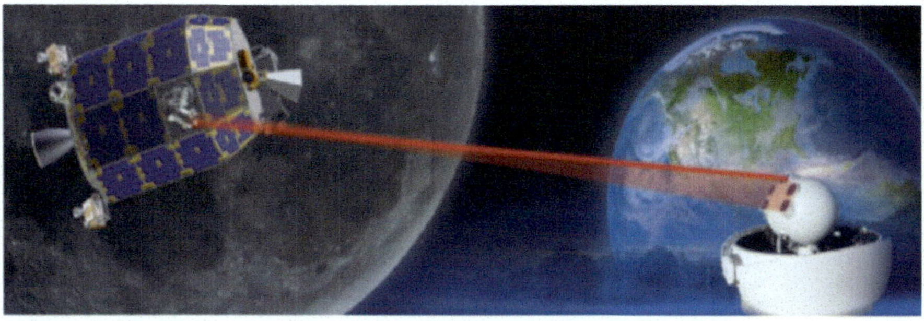

Fig. 1.2 Diagram of the Lunar Laser Communication Demonstration plan [6]

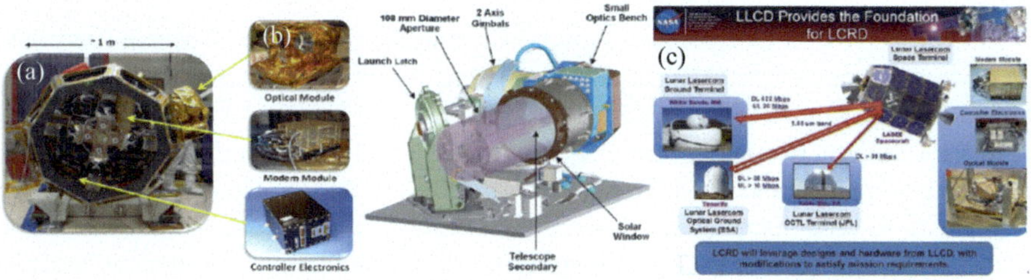

Fig. 1.3 **a** Moon–Earth Laser Communication Demonstration System. **b** Photos of satellite optical-communication terminals and their components. **c** Schematic diagram of the satellite optical-communication terminal's laser communication demonstration system [9]

Fig. 1.4 Lunar laser communication ground terminal. **a** Receiver structure model. **b** Telescope externalities structure model [9]

On June 5, 2014, the optical ground station of the Jet Propulsion Laboratory Communication Telescope Laboratory communicated via laser with the International Space Station, and transmitted a 175-MB high-definition video named "Hello, World!" at 50 Mbit/s [12, 13].

In August 2018, the China National Space Administration used the Cube Satellite to realize two-way data transmissions from orbit to ground links. The downlink transmission rate reached 200 Mbit/s, and the uplink transmission rate reached 10 Kbit/s [14].

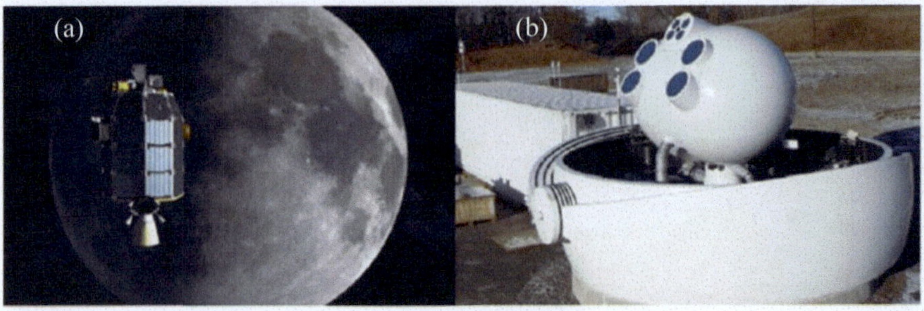

Fig. 1.5 Lunar laser communication demonstration project. **a** Lunar and **b** Earth communication terminals [11]

In 2017, NASA's Deep-Space Optical Communication project transmitted data from Mars at a rate of 250 Mbit/s, and planned to explore the Psyche asteroid located in the orbits of Mars and Jupiter in 2023 [15].

Since 1985, the European Space Agency's semiconductor-laser inter-satellite link experimental project has used the Artemis satellite and optical ground station, shown in Fig. 1.6, to conduct two-way laser link communications from the ground to the satellite. The success rate of establishing the communication link was as high as 91%, and the total communication time was as long as 78 h [16].

In 2002, the German Space Center conducted a series of experiments in the context of the TerraSAR-X satellite communication terminal program. In 2005, a communication link was established between La Palma and Tenerife, Spain. As shown in Fig. 1.7, binary phase-shift keying/homodyne coherent detection was adopted, and a 1064-nm wavelength Nd: YGA solid laser was selected. The frequency stability was less than 50 MHz per day [17].

Fig. 1.6 European laser communication experimental optical ground station [16]

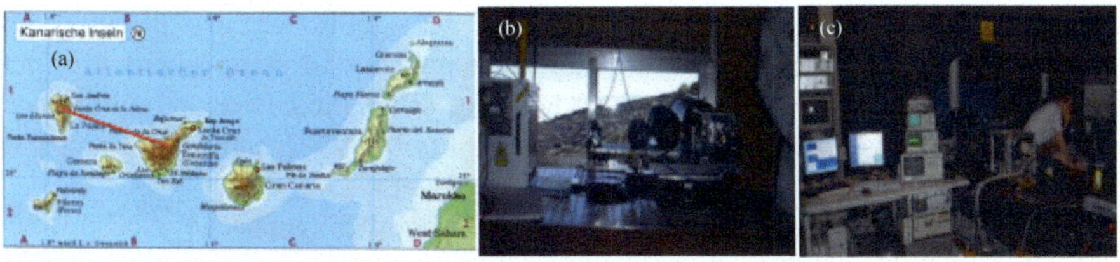

Fig. 1.7 Schematic diagram of a 142-km homodyne coherent-detection system between islands. **a** Communication link between La Palma Island and Tenerife Island. **b** Communication terminal at La Palma. **c** Communication terminal at Tenerife [17]

Experiments have shown that homodyne coherent-optical communication operates under harsh atmospheric conditions, and optical phase-locked loops can maintain stability even in high dynamic signal ranges (> 20 dB). Moreover, the link is not affected by strong background light.

In 2009, based on the above experiments, the German Space Center conducted a coherent-optical communication experiment between a satellite and a ground station in Haleakalā, Hawaii with a communication distance of about 1000 km, as shown in Fig. 1.8. This experiment did not use adaptive-optics technology for wavefront correction. Moreover, it was found that the downlink from the satellite to the ground had a higher bit-error rate than the uplink [18].

At the end of 2008, the European Space Agency began implementing the European Data Relay System (EDRS) program, which aimed to establish a 1.8-Gbit/s high-speed spatial data link between a low-orbit satellite and the ground data center through two synchronous-orbit relay satellites (EDRS-A and EDRS-C). In January 2016, the airborne optical-communication terminal EDRS-A was successfully launched into a predetermined orbit at the European Telecommunications Satellite Organization (EUTELSAT) 9B, and EDRS-C was launched in August 2019. Previously, the synchronous-orbit satellite Alphasat and low-orbit satellites Sentinel 1A and Sentinel 2A had been launched from 2013 to 2015, as pioneers of the EDRS program [19].

Fig. 1.8 Coherent optical laser communication system at Haleakalā. **a** Ground station. **b** System terminal [18]

From March to April 2016, the EUTELSAT 9B satellite successfully communicated with the European Space Agency's optical ground station, and was put into commercial use in November. As of May 18, 2018, the EDRS-A satellite had completed 10,000 laser connections with a reliability of 99.8% and transmitted 500 TB of data [19].

The European Space Agency planned to launch another synchronous-orbit relay satellite to expand the EDRS program, making it a global data-relay communication system by 2020 [20].

In 2024, the ESA plans to use deep-space laser communication technology to launch a spacecraft to solar–terrestrial Lagrange point L5 for long-term monitoring of solar variation, as shown in Fig. 1.9. Deep-space laser communication technology will be used to transmit the collected solar-image information back to Earth [21].

In 1994, Japan proposed the Engineering Test Satellite Six (ETS-VI) program, which developed a laser-communication system, and successfully achieved communication between ground stations and a satellite equipped with the system [22]. In 2000, Japan developed a bidirectional ultra-high-speed laser-communication

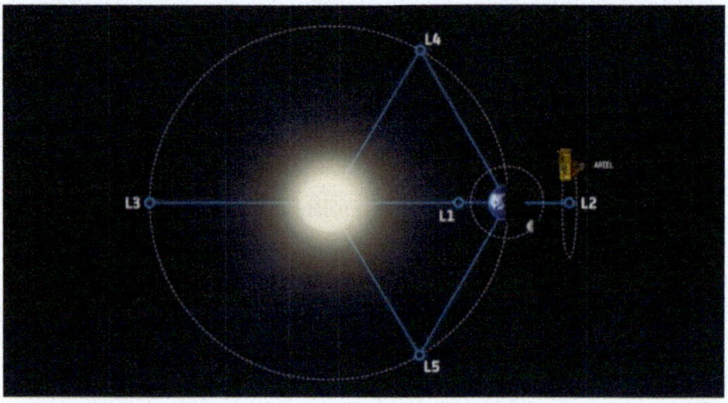

Fig. 1.9 Lagrange point between the sun and the earth [21]

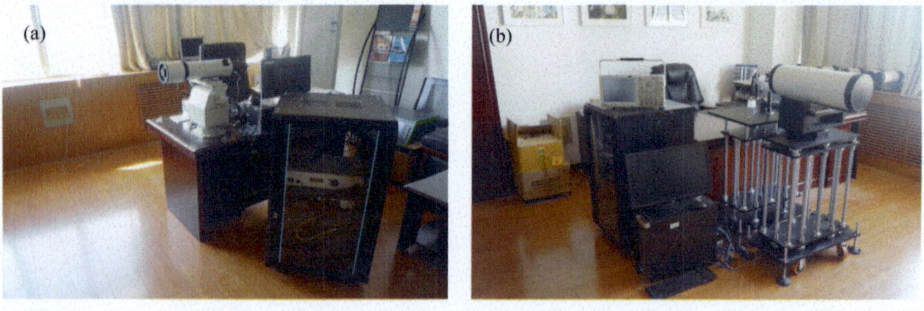

Fig. 1.10 Optical wireless coherent communication system developed by Xi'an University of Technology. **a** Transmitter; **b** receiver

terminal on the ground [23]. In 2006, the Japanese National Institute of Information and Communication Technology (NICT) conducted a bidirectional communication experiment using this satellite, with a ground station under the German Space Agency [24].

An on-board coherent-optical communication receiver independently designed by NICT was successfully developed in 2008 [25]. In 2021, two-way laser communication was conducted between ground stations and synchronous-orbit satellites at a rate of 10 Gbit/s. Related technologies will gradually be applied to space ranging, satellite networking, and other aspects [26].

The United States, Germany, and other developed Western countries have taken the lead in a series of achievements in the field of optical wireless communication. Signal modulation has gradually evolved from the initial beam-intensity modulation to subcarrier modulation and phase-amplitude modulation. Demodulation has evolved from direct detection to coherent detection. The terminals of both communication parties have developed from one-on-one to one-to-many, many-to-many, and point-to-point transmissions. Equipment and instruments are developing towards low power consumption, miniaturization, and high integration.

1.1.2 Domestic Development Status

The Changchun University of Science and Technology is mainly engaged in laser-communication experiments between static and dynamic targets in different scenarios. Two-way laser-communication experiments have been conducted and successfully demonstrated in scenarios such as ship-to-ship, airship-to-ship, and twin helicopter-to-helicopter. In 2015, an optical transceiver for laser communication between low-orbit and synchronous-orbit satellites was designed and developed. In 2016, a laser-communication optical transceiver was developed for synchronous-orbit-to-ground communication [27].

Harbin Institute of Technology is primarily engaged in research on satellite laser communications. In 2011, the "Haiyang 2" satellite realized a satellite–ground two-way laser-communication experiment, and in 2017, the Shijian-13 satellite was successfully launched from the Xichang Satellite Launch Center. It established a laser link with the ground base station and achieved stable communication [28].

The University of Electronic Science and Technology of China has conducted research in the fields of inter-satellite and satellite–ground laser communication, including inter-satellite optical-communication acquisition and tracking technology, weak-beacon optical signal-processing technology, satellite–ground coherent-optical communication technology, and high-precision alignment-information acquisition methods [29].

The Shanghai Institute of Optics and Fine Mechanics, Chinese Academy of Sciences, has conducted a series of studies on the development of coherent-detection devices for inter-satellite and satellite–ground laser communication, optical phase-locking technology, and underwater optical communication [30].

Fudan University is primarily engaged in technological research and exploration in the field of visible-beam communication. It has achieved advanced levels in modulation rates, process-device design, and other areas, such as indoor visible-beam communication, positioning, and underwater visible-beam communication [31].

Wuhan University realized laser communication between ships and coastal buildings in 2013, and 2.5-Gbit/s high-speed and stable laser communication between ships and coastal buildings in 2015 [32].

The Air Force Engineering University conducted theoretical simulation research in the fields of space laser communication, satellite–ground laser communication, unmanned-aerial-vehicle aerial laser communication, and visible-beam communication [33].

From the end of 2016 to the beginning of 2017, the high-speed satellite–ground coherent-optical communication terminal developed by the Shanghai Institute of Technical Physics of the Chinese Academy of Sciences successfully carried the Mozi satellite to complete on-orbit two-way link communication. The uplink adopted pulse-position modulation with a communication rate of 20 Mbit/s, whereas the downlink adopted binary phase-shift keying modulation with a communication rate of 5.12 Gbit/s [34].

Since 2014, the Institute of Optics and Electronics of the Chinese Academy of Sciences has developed a remote satellite–ground free-space coherent optical communication system and analyzed the improvement of the bit-error rate after using adaptive-optical correction [35]. In 2018, it jointly completed a coherent laser-communication experiment between synchronous-orbit satellites and the ground with the Fifth Academy of Aerospace Science and Technology. The laser communication terminal was launched in the second half of 2019 using the Shijian 20 satellite bus [36].

In recent years, Xi'an University of Technology has investigated laser-transmission characteristics in atmospheric turbulence [37], coding and decoding of wireless laser communication systems, optical wireless subcarrier modulation, channel estimation and equalization [38], alignment, tracking, and acquisition technology [39], coherent-detection technology and the design of optical transceiver systems [40], optical orthogonal frequency-division multiplexing modulation, and optical multiple-input multiple-output system research.

The theory and technology of optical wireless transmission in complex environments have been theoretically explored and experimentally verified. In 2016, a space laser-communication system terminal with intensity modulation/direct detection was successfully developed. It had standard Ethernet data-access capability and reliably transmitted duplex data, voice, and images [41].

Since 2017, Xi'an University of Technology has successfully conducted a optical wireless coherent communication experiment with a 10-km link near the ground, through cooperation with the China Academy of Astronautics and Space Research, with the support of the National Natural Science Foundation of China and the major industrial chain project of Shaanxi Province. It can transmit video signals in real time. By using differential binary modulation/heterodyne detection technology, the communication rate can reach 10 Gbit/s [42].

In 2021, the team developed a near-infrared-band space laser communication system using intensity modulation/direct detection with a maximum transmission rate of 10 Gbit/s. The transmitting end used image tracking, and the receiving end used two-dimensional reflector control to suppress atmospheric turbulence to rapidly align the beam. The transmitting and receiving ends used a non-coaxial working system [43].

Space laser-communication systems are developing toward large bandwidth, high reliability, and beacon-free beams, from point-to-point transmissions to point-to-multipoint transmissions. Application scenarios have evolved from initial space laser communication to ground laser communication, and even 6G has been considered for applications in space laser communication.

Space laser communication in atmospheric turbulence is also developing from traditional common line-of-sight alignment, tracking, and capture technology, with centralized control between the transmitter and receiver, to noncommon line-of-sight technology with independent control between the transmitter and receiver, gradually reducing the dependence of moving base scenarios on stable platforms.

1.2 State of the Art of Optical Wireless Coherent-Detection Technology

1.2.1 Foreign Development Status

With the increase in research in the field of optical wireless coherent detection, foreign scholars have intensely investigated coherent detection technologies. Among these, studies in the field of optical wireless coherent detection technology have used different modulation and detection methods.

In 1975, David et al. [44] derived a general expression for the signal-to-noise ratio of heterodyne detection, which indicated that the optimal local-oscillator beam distribution should be the same as the signal field distribution on the detector surface. The optimal detector radius was obtained when the signal beam on the circular detector was Airy distributed and the local-oscillator beam was uniformly distributed.

In 2007, Lee et al. [45] experimentally compared and analyzed the coherent-detection and multichannel coherent-detection performance of balanced homodyne receivers. The experimental results showed that both detection methods exceeded the signal-to-noise ratio of direct detection by 5 dB, and multichannel coherent detection had a sensitivity improvement of 2 dB–3 dB compared with coherent detection.

In 2009, Painchaud et al. [46] introduced a single-port rejection-ratio parameter to quantify the performance of a balanced detection system. The results indicated that the performance of the balanced detection system largely depended on the characteristics of the optical devices used at the front end of the balanced detector, i.e., the input port.

In 2010, Salem et al. [47] derived a mathematical model for the heterodyne detection efficiency of systems in different coherent states and conducted a simulation analysis. The results showed that the angular displacement between the signal beam and local-oscillator beam had the most significant impact on the degradation of the heterodyne-detection efficiency. Among them, partially coherent beams were less affected by angular displacement than coherent beams and it was only necessary to ensure that the coherence widths of the signal beam and local-oscillator beam were equal. Coherent detection systems can be used to detect partially coherent beams.

In 2012, Huynh et al. [48] proposed a novel coherent-receiver scheme using phase-modulation detection methods. The experimental results showed that the proposed scheme could successfully demodulate the modulated signal of a 10 Gbit/s link, further simplifying the front end of the coherent receiver.

In 2022, Kakarla et al. [49] analyzed the sensitivity of a coherent-detection system combining M-ary pulse-position modulation (M-PPM) and quadrature phase-shift keying (QPSK). The study found that the system sensitivity was best when using M-PPM + QPSK modulation, and the system sensitivity increased with the increase of the decimal number M. The system sensitivity obtained using 64-PPM + QPSK modulation was 0.6 dB higher than that of QPSK modulation.

1.2.1.1 Research on Optical Wireless Coherent Detection Technology in Atmospheric Turbulence

In 2000, Belmonte et al. [50] studied the effect of atmospheric turbulence on the optical power and system efficiency of coherent heterodyne receivers. It was found that when the atmospheric refractive-index structure constant $C_n^2 = 10^{-14}\ \mathrm{m}^{-2/3}$ and the transmission distance was 5 km, the coherent received beam power of the system was significantly reduced, and the system efficiency was reduced by approximately 5 dB.

In 2003, Belmonte [51] studied the effect of the receiving-antenna aperture parameters on the detection performance of external differences under atmospheric-turbulence conditions. Numerical simulations showed that atmospheric turbulence could disrupt the initial beam phase, thereby reducing the ability of the emitted beam to focus on the target plane. For a shorter-wavelength beam transmission, the heterodyne-detection performance obtained by receiving antennas with small and large apertures was consistent.

In 2017, Safari et al. [52] studied the effects of arrival-angle fluctuation on optical wireless coherent-detection systems under atmospheric-turbulence and receiver-vibration conditions. The simulation results showed that in the presence of arrival-angle fluctuations, coherent detection achieved a better gain than direct detection; however, it was more sensitive to arrival-angle fluctuations. When the interruption probability was 10^{-7} orders of magnitude, coherent detection achieved a gain of 31.3 dB.

In 2018, Sahota et al. [53] proposed a novel homodyne detection system to overcome the loss of beam-intensity flicker and compared it with direct detection, using the error rate as an evaluation indicator. The results indicated that with moderate beam-intensity scintillation, the error rate of direct detection was 10^{-1} orders of magnitude, whereas the error rate of the new zero-difference detection system could be effectively reduced to 10^{-5} orders of magnitude.

1.2.2 Domestic Development Status

To further promote the practical application of optical wireless coherent communication, domestic scholars have explored improving the performance of optical wireless coherent-detection systems at the receiving end. Research has been conducted on optical wireless coherent detection technology using different modulation and detection methods.

In 2007, Xu et al. [54] analyzed the impact of atmospheric turbulence on the performance of 1550-nm laser heterodyne-detection systems and conducted theoretical analyses and experimental research. Through a comparative analysis of the measurement results, it was clear that understanding the impact of atmospheric turbulence on heterodyne detection systems played a guiding role in heterodyne-detection technology.

In 2011, Wang et al. [55] derived a signal-to-noise-ratio model using balanced and general coherent-detection systems and conducted simulations and experimental comparisons. The results indicated that when the beam-splitting ratio was 0.5, the output signal-to-noise ratio of the balanced detector was 19 dB higher than that of a general coherent-detection system.

In 2013, Niu et al. [56] compared the performance of optical wireless communication systems using subcarrier intensity modulation and optical wireless coherent communication systems in gamma–gamma turbulent channels. The research results indicated that, compared to the optical wireless communication system with subcarrier intensity modulation, the sensitivity of the optical wireless coherent communication system increased by 24–30 dB.

In 2016, Liu et al. [57] presented an expression for the efficiency of homodyne detection and analyzed the relationship between this efficiency and the root-mean-square value of the residual wavefront. The results showed that the atmospheric coherence length affected the homodyne detection efficiency under the conditions of a specific atmospheric Greenwood frequency and closed-loop control bandwidth. When the atmospheric coherence length was 9 cm, the homodyne detection efficiency of the AO system after correction improved by approximately 25%.

Optical wireless coherent detection technology has been used to suppress atmospheric turbulence, phase noise, and system jitter. In 2007, Tian et al. [58] proposed an algorithm using fuzzy logic to suppress the impact of the modulation-frequency jitter offset on the coherent detection process. The experimental results showed that the algorithm expanded the range of the modulation-frequency jitter offset by $\pm$ 6.2%, while ensuring that the coherent-detection accuracy was not affected.

In 2017, Chang et al. [59] derived analytical formulas for the phase-error variance and bit-error rate of a Costas-loop coherent detection system, and analyzed the effects of the laser linewidth, incident-signal optical power, phase error, and photodiode responsiveness on the system's bit-error rate. The results showed that when the phase error increased, the responsivity of the photodiode or the optical power of the incident signal decreased. Moreover, the bit-error rate of the coherent detection system using the Costas loop gradually increased.

In 2018, Zhou et al. [60] designed a 90° Costas homodyne coherent receiver for optical wireless coherent communication systems. When the optimal signal power-division ratio was 0.05, the receiver sensitivity was improved by nearly 2.65 dB, which is close to the hot-noise limit in optical wireless coherent communication systems.

In 2019, Liu et al. [61] proposed an array detector to address the speckle-effect problem in heterodyne detection and simulated and evaluated the performance of array detectors in improving the signal-to-noise ratio of coherent optical systems. Simulation results showed that this method effectively improved the signal-to-noise ratio of the heterodyne system.

In 2020, Ren et al. [62] proposed a differential signal-detection method for a dual-frequency coherent laser and analyzed the effects of the laser linewidth, amplitude ratio, and other parameters on the differential signal. Through experiments, it was found that noise from the lasers and frequency shifters had an impact on the amplitude of the differential signal. The center frequency of the differential signal underwent a beam change, and the linewidth fluctuated, owing to the instantaneous change in the laser.

In 2022, Tao et al. [63] defined atmospheric coherence time to describe the fluctuation speed of the received beam field and analyzed the trend of the influence of the coherent-detection system parameters on the atmospheric coherence time. The results showed that the atmospheric coherence time was proportional to the atmospheric coherence length, laser wavelength, and receiving aperture; it first increased and then decreased with the waist radius, and can be used as an important parameter to measure the impact of atmospheric turbulence on coherent detection.

Professor Ke Xizheng from Xi'an University of Technology has carried out systematic research in the field of optical wireless coherence detection. In 2015, Kong et al. [64] analyzed the impact of different beam-field distributions on the performance of heterodyne detection systems in atmospheric turbulence. The study showed that the different beam-field distributions differed significantly in terms of the coherent-detection efficiency, and that the beam distribution should be reasonably selected for the experiment. In 2016, Chen [65] studied the non-optical heterodyne detection (NOHD) method for subcarrier modulation and demonstrated through numerical simulations and field experiments that the system could achieve a detection gain that was approximately 20 dB higher than that of direct detection.

In 2018, Chen [66] considered the impact of channel noise on the optical wireless coherent-detection performance, established a signal-to-noise-ratio model with a common-mode rejection ratio, and built a new double-balanced detection system. In

2018, Mao [67] completed a 1-km optical wireless coherent communication experiment and analyzed the impact of beam-transmission and polarization characteristics on the detection sensitivity of the receiving end.

In 2019, Wu [68] designed a digital differential phase-shift keying demodulator using a Costas phase-locked loop and experimentally verified it. In 2019, Tan [69] developed a optical wireless heterodyne detection system and conducted 1.3-km field experiments. In the same year, Ma [70] designed a single-particle optimization algorithm for polarization control in optical wireless coherent communication systems. Six-hundred-meter field experiments showed that this scheme effectively improved the mixing efficiency of the receiver system by 64%.

In 2021, Bai [71] proposed a loose tail-ring carrier-recovery method for QPSK modulation in coherent optical communication systems. The experimental results indicated that the signal-to-noise ratio of the system output was 20 dB. In 2023, Yang et al. [72] established a mathematical model, based on the theory of coherent detection, between wavefront distortion, mixing efficiency, and mixing gain. In addition, they analyzed the improvement limits of wavefront correction on the mixing efficiency and mixing gain under different atmospheric-turbulence intensities.

1.3 Domestic and Foreign Development Status of Dual-Wavelength Adaptive-Optics Systems

Signal beams of different wavelengths are combined with a beacon beam and sent through the transmitting antenna. The beam passes through the atmospheric channel and is received at the receiving end. After the beam is incident on the deformable mirror, it is reflected through a dichroic beam splitter to separate the signal and beacon beams.

After beam shaping, the beacon beam enters the wavefront sensor and the signal beam is coupled to the receiving fiber. The wavefront sensor measures the beacon beam to obtain the wavefront-phase distortion information, and the calculation system controls the deformable mirror to generate the corresponding conjugate phase surface for real-time signal-beam correction. The process of correcting the distortion information in the entire control program is shown in Fig. 1.11. The entire system is in a real-time closed-loop state.

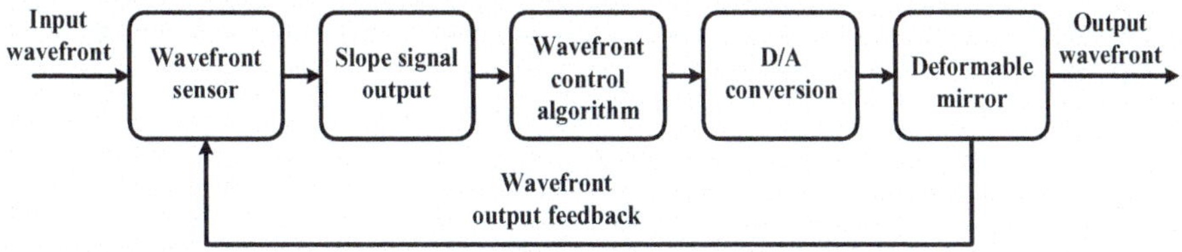

Fig. 1.11 Block diagram of an adaptive-optical correction program

Atmospheric turbulence has the greatest and most severe impact on the quality of space laser communication because it can increase the error rate and even cause random communication interruptions, resulting in a unstable laser communication link [73]. Therefore, researchers have introduced adaptive-optics technology [74] into laser-communication systems to suppress the impact of atmospheric turbulence on their communication quality [75–77].

After adaptive-optics technology was applied to the laser-communication field, researchers conducted extensive research on its practical application in uplinks, downlinks, and horizontal links (ground–ground), according to the actual environment and channel characteristics. Moreover, the control bandwidth, control algorithm, and spatial-compensation capability of the adaptive-optics system, including deformable mirrors, wavefront sensors, and other devices, have been extensively studied to improve the correction accuracy [78].

In 1996, Tyson first demonstrated the role of adaptive transmission in practice through numerical and theoretical simulations. It could compensate for signal attenuation and wavefront fluctuations in air-to-ground laser communication links, which significantly reduced the beam-intensity flicker effect and error rate [75]. The Japanese Communication Laboratory attempted to use wavefront sensor-free technology for laser-link communication between the International Space Station and the ground [76]. In 2002, the Lawrence Livermore Laboratory used adaptive-optics technology in the downlink of a slanted-transmission communication link to improve the optical-fiber coupling efficiency of the received signal from 20 to 55% [77].

Years of development on adaptive-optics systems in space laser communications gradually produced a dual-wavelength closed-loop control correction system that could effectively handle most scenarios. It had the advantages of a high correction accuracy, low engineering cost, and stable system operation.

The dual-wavelength working mechanism of dual wavelength adaptive-optics system leads to certain correction residuals, because atmospheric dispersion leads to differences in wavefront distortion of different wavelength beams. Moreover, an error exists between the wavefront phase obtained by the wavefront sensor measuring the beacon beam, and the wavefront phase of the signal beam. This error eventually causes correction residuals as the closed-loop system passes. Therefore, to improve the correction accuracy of adaptive-optical systems, researchers have conducted numerous studies and experiments on the distortion difference between beams of different wavelengths and their impact on the correction accuracy.

In 1971, Ishimaru et al. [79] produced the scintillation spectra of plane waves, spherical waves, and Gaussian beams in weak turbulence, indicating that the modulation effects of atmospheric turbulence on beams of different wavelengths and modulation degrees are different.

In 1972, Ishimaru et al. [80] calculated the cross-correlation power spectra between the amplitudes and phases of different wavelength beams, and analyzed the coherence based on the calculation results. The results were obtained based on weak turbulence and the Rytov turbulence hypothesis. The atmospheric refractive-index structure constant was indirectly measured according to the correlation coefficient relationship. The measured results were consistent with the direct measurement, which proved the correctness of the conclusions.

In 1975, Fante et al. [81] studied the effects of logarithmic amplitude-fluctuation chromatic aberrations and phase-compensated chromatic aberrations on the characteristics of phase-compensated optical links transmitted through atmospheric turbulence.

In 1982, Hogge et al. [82] analyzed the system errors of adaptive-optics systems using beacon-beam wavefront-distortion data to correct signal-beam wavefront distortions. From the perspective of geometrical optics and beam diffraction, the author described the difference between the wavefront phase distortion caused by the signal beam and that by the beacon beam during their propagation in the atmosphere. The results indicated that there were differences in the wavefront distortion between plane waves of different wavelengths, after passing through the same atmospheric channel; shorter-wavelength beams were more susceptible to turbulence, and the degree of influence was more severe.

In 1983, Winocur et al. [83] proposed to decompose the phase-distortion error caused by atmospheric turbulence between signal beams and beacon beams of different wavelengths in an adaptive-optics system with Zernike polynomials. Ground and satellite–ground communication links were analyzed using this method.

The analysis results indicated that the phase-distortion error between beams of different wavelengths in the first-order (tilted) Zernike mode caused a relatively small system-correction residual. However, with an increase in the Zernike polynomial order, the phase-difference error increased significantly and seriously affected the correction accuracy of the adaptive-optics system. The distortion phase error between beams of different wavelengths was also related to factors such as the receiving aperture, propagation-path length, and turbulence intensity.

In 1986, Ben-Yosef et al. [84] measured the intensity-scintillation correlation coefficients of two beams with different wavelengths emitted from a common optical path in the same atmospheric channel. The results indicated that, under weak turbulence conditions, if the Kolmogorov spectrum was used, the correlation between the scintillation of beam intensities of different wavelengths was higher than that predicted.

As the turbulence intensity increased, the relative proportion of small-scale turbulent vortices in the atmospheric turbulence increased, leading to a decrease in the cross-correlation coefficient between the two beams. The correlation coefficient between the two was also influenced by the internal and external turbulence scales. Finally, the numerical relationship between the scintillation of beam intensities of different wavelengths was analyzed. The ratio between the scintillation of beam intensities of different wavelengths was proportional to the 7/12th power of the corresponding wavenumber ratio.

In 2007, Li et al. [85] found that the performance of an adaptive-optics system was affected by atmospheric turbulence and was related to the wavelength and numerical aperture. They also established a glint model of the arrival angle under plane- and spherical-wave models and demonstrated the impact of the aperture and wavelength on the system-correction performance through numerical simulation.

In 2008, Devaney et al. [86] explained in detail the cause of the chromatic aberration caused by beams of different wavelengths passing through atmospheric turbulence, as well as the influence of chromatic aberration caused by atmospheric dispersion on the correction accuracy of high-order adaptive-optics systems [20].

The effect of the wavefront-distortion difference included two main aspects. First, it caused the offset measurement positions of the two wavelengths on the wavefront sensor to differ. Second, the distortion difference continuously accumulated under the influence of various factors, such as the transmission distance, atmospheric-turbulence intensity, and external turbulence scale relative to the receiving aperture. Finally, it was proven experimentally that correcting chromatic aberrations significantly improved the correction accuracy of adaptive-optics systems, but overcorrection had the opposite effect.

In 2015, Shubenkova et al. [87] established an experimental link for indoor space laser communication in an indoor atmospheric-turbulence simulation test chamber, where the signal and beacon beams shared the same optical path for emission. The communication distance of the link was 700 m, signal-beam wavelength was 1064 nm, and beacon-beam wavelength was 530 nm. At the receiving end, a Shack–Hartmann wavefront sensor was used to measure the wavefront phase distortion of the two optical signals. The measurement results are shown in Fig. 1.12.

According to Zernike polynomials and interferometric analyses, it was concluded that in weak-turbulent communication links with a transmission distance of 700 m, the wavefront phase distortion of the 530-nm signal beam was more severe than that of the 1064-nm signal beam and was approximately 35% higher than that of the signal beam.

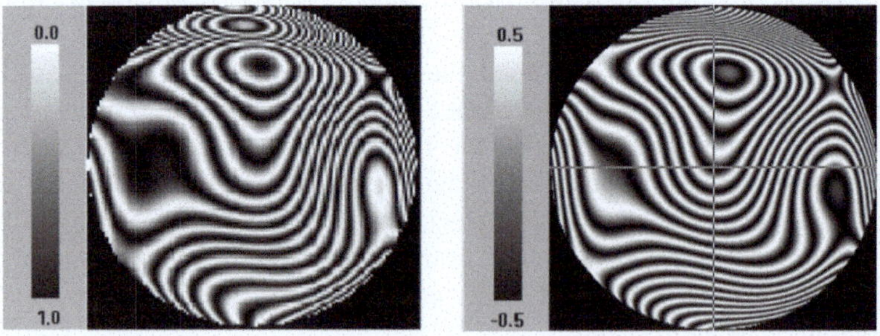

(a) Beacon phase distortion (b) Signal-beam phase distortion

Fig. 1.12 Wavefront phase distortion of signal beam and beacon beam at the receiving end

In 2020, Xu et al. [88] decomposed the pupil according to the Jones matrix to obtain the receiving wavefront phase, analyzed the polarization aberration generated by the system, and proposed an overall analysis method for the polarization difference in adaptive-optics systems. Using the proposed method, the polarization chromatic aberration caused by the surface coating of optical components in a Cassegrain optical antenna between 1300 and 589-nm beams was analyzed. The decomposition results showed that this part of the polarization chromatic aberration was composed of a defocused aberration and a small spherical aberration, with the defocused aberration reaching 1.14 rad. Moreover, the polarization chromatic aberration reduced the far-field spot of the 1300-nm optical signal from 0.8 to 0.6.

In 2020, Ke Xizheng et al. [89] deduced a formula for the overall fluctuation variance of the wavefronts of beams with different wavelengths and explained that under the same transmission conditions, the overall fluctuation variance of the wavefronts of beams with different wavelengths was different. According to the atmospheric coherence length, the variation trend of the overall fluctuation variance of the wavefront under the influence of pressure, atmospheric-turbulence intensity, and wavelength was numerically simulated. The results showed that the internal turbulence scale had a greater impact on the correlation between the two beams, whereas the external turbulence scale had a smaller impact.

References

1. Chen ZH (1993) The invention of the optical telephone, a milestone in the history of modern optical communication. Opt Commun Technol 17(5):288–289
2. Wang SL, Fang FZ (2017) High power lasers and their development. Progress in Laser and Optoelectron 54(9):090005
3. Marshalek RG, Koepf GA (1988) Comparison of optical technologies for intersatellite links in a global telecommunications network. Opt Eng 27(8):663–676
4. Jeganathan M, Portillo A, Racho CS et al (1999) Lessons learnt from the optical communications demonstrator (OCD). Free-Space Laser Commun Technol XI 3615:23–30
5. Wright MW, Valley GC (2005) Yb-Doped fiber amplifier for deep-space optical communications. J Beamwave Technol 23(3):1369–1374
6. Pan W, Hu Y (1998) Overview and current status of space laser communication research in the United States. J Univ Electron Sci Technol China 27(5):541–545
7. Kim II, Stieger R, Koontz JA et al (1998) Wireless optical transmission of fast ethernet, FDDI, ATM, and ESCON protocol data using the TerraLink laser communication system. Opt Eng 37(12):3143–3155
8. Khatri FI, Robinson BS, Semprucci MD et al (2015) Lunar laser communication demonstration operations architecture. Acta Astronaut 111:77–83
9. Beyer AD, Shaw MD, Marsili F et al (2015) Tungsten silicide superconducting nanowire single-photon test structures fabricated using optical lithography. IEEE Trans Appl Supercond 25(3):2200805
10. Čierny O, Cahoy KL (2018) On-orbit beam pointing calibration for nanosatellite laser communications. Opt Eng 58(4):041605

11. Boroson DM, Robinson BS (2014) The lunar laser communication demonstration: NASA's first step toward very high data rate support of science and exploration missions. Space Sci Rev 185:115–128
12. Stevens ML, Boroson DM (2012) A simple delay-line 4-PPM demodulator with near-optimum performance. Opt Express 20(5):5270–5280
13. Oaida BV Wu W, Erkmen BI et al (2014) Optical link design and validation testing of the optical payload for lasercomm science (OPALS) system. Free-Space Laser Communication and Atmospheric Propagation XXVI 8971 89710U
14. Rose TS, Rowen DW, LaLumondiere SD (2019) Optical communications downlink from a low-earth orbiting 1.5U CubeSat. Opt Express, 27(17):24382–24392
15. Laurent B, Camus JP, Sein E (1990) SILEX: The first European optical space communications system [J]. Acta Astronaut 22:299–303
16. Loscher A (2010) Atmospheric influence on a laser beam observed on the OICETS-ARTEMIS communication demonstration link. Atmos Meas Tech 3(5):1233–1239
17. Knopp MT, Spoerl A, Gnat M et al (2020) Towards the utilization of optical ground-to-space links for low earth orbiting spacecraft. Acta Astronaut 166:147–155
18. Gregory M, Heine F, Kämpfner H (2012) Commercial optical inter-satellite communication at high data rates. Opt Eng 51(3):031202
19. Sterr U, Dallmann D, Heine F (2016) Planning constraints of low grazing altitude GEO-LEO laser links based on in-orbit data. Opt Eng 55(11):111608
20. Calzolaio D, Curreli F, Duncan J (2020) EDRS-C—the second node of the European data relay system is in orbit. Acta Astronaut 177:537–544
21. Sodnik Z, Heese C, Arapoglou P-D et al (2018) European deep-space optical communication program. Free-Space Laser Communication and Atmospheric Propagation XXX, 105240Q
22. Chida Y, Soga H, Yamaguchi Y et al (1998) On-orbit attitude control experiments using ETS-VI by H_∞ control and two-degree-of-freedom control. Control Eng Pract 6(9):1109–1116
23. Samejima S, Tanaka M, Ohtomo I (1992) Overview of satellite on-board multibeam communications system for ETS-VI. Acta Astronaut 28:293–300
24. Noda Y, Shao Y (2002) Experimental research on optical communication on space stations. Prog Laser Optoelectron 39(1):15–19
25. Henniger H, Ludwig A, Horwath J (2010) Performance bounds of DPSK and OOK for low elevation optical LEO downlinks. Radio Engineering 19(4):589–595
26. Fujiwara Y, Mokuno M, Jono T et al (2007) Optical inter-orbit communications engineering test satellite (OICETS). Acta Astronaut 61:163–175
27. Teng YJ, Song YS, Tong SF et al (2018) Research on acquisition performance of laser communication system based on airship platform. J Opt 38(6):0606005
28. Wu F, Yu SY, Ma ZT et al (2014) Correction of aiming angle deviation and in orbit verification of satellite ground laser communication link. China Laser 41(6):0605008
29. Huang J, Deng K (2020) Several theoretical issues in the study of atmospheric coherent laser communication. J Quantum Electron 37(5):556–565
30. Xu N, Liu LR, Liu DA et al (2008) Optical phase locked loop in interstellar coherent optical communication. Prog Laser Optoelectron 45(4):25–33
31. Wang Y, Huang X, Tao L et al (2015) 4.5-Gb/s RGB-LED based WDM visible light communication system employing CAP modulation and RLS based adaptive equalization. Opt Express 23(10):13626–13633
32. Lin Y, Ai Y, Shan X (2017) Identification of electro-optical tracking systems using genetic algorithms and nonlinear resistance torque. Opt Eng 56(3):033105
33. Hao SW, Li YJ, Zhao SH et al (2021) Research on optimal power allocation for inter satellite visible light communication based on non orthogonal multiple access. China Laser 48(7):0706002
34. Si JJ, Zhu R, Zhao SW et al (2020) A fast laser wavelength locking system based on piezoelectric ceramics and Fiber Bragg grating. China Laser 47(12):1201002
35. Guo YM, Ma Y, Rao CH (2014) Optimal closed-loop bandwidth of tilt correction loop for adaptive optical systems. Acta Physica Sinica 63(6):429–433

36. Chen M, Liu C, Rui D et al (2018) Experimental results of 5-Gbps free-space coherent optical communications with adaptive optics. Opt Commun 418:115–119
37. Wang J, Ke XZ (2017) Speckle characteristics of partially coherent beams propagating in atmospheric turbulence. Infrared Laser Eng 46(7):0722003
38. Chen D, Hui J (2021) Parameter estimation of Gamma-Gamma fading channel in free space optical communication. Opt Commun 488:126830
39. Ke X, Zhang P (2020) Automatic focusing control in beaconless APT system. J Russ Laser Res 41(1):61–71
40. Ke X, Lei S (2016) Spatial light coupled into a single-mode fiber by a Maksutov-Cassegrain antenna through atmospheric turbulence. Appl Opt 55(15):3897–3902
41. Wu PF, Ke XZ, Yuan Q (2014) Development of a 10Mbit/s ~ 1Gbit/s rate adaptive wireless optical communicator. J Xi'an Univ Technol 30(4):443–447
42. Jiali W, Xizheng K (2020) Development of adaptive optical correction and polarization control modules for 10-km free-space coherent optical communications. J Mod Opt 67(3):189–195
43. Yang SJ, Ke XZ, Wu JL, Liu XG (2022) Fast alignment of wireless optical communication using two-dimensional mirrors. China Laser 49(11):1106001
44. Fink D (1975) Coherent detection signal-to-noise. Appl Opt 14(3):689–690
45. Lee W, Izadpanah H, Delfyett PJ et al (2007) Coherent pulse detection and multi-channel coherent detection based on a single balanced homodyne receiver. Opt Express 15(5):2098–2105
46. Painchaud Y, Poulin M, Morin M et al (2009) Performance of balanced detection in a coherent receiver. Opt Express 17(5):3659–3672
47. Salem M, Rolland JP (2010) Heterodyne efficiency of a detection system for partially coherent beams. J Opt Soc Am A 27(5):1111–1119
48. Huynh TN, Nguyen L, Barry LP (2012) Novel coherent self-heterodyne receiver based on phase modulation detection. Opt Express 20(6):6610–6615
49. Kakarla R, Mazur M, Schroder J et al (2022) Power efficient communications employing phase sensitive pre-amplified receiver. IEEE Photonics Technol Lett 34(1):3–6
50. Belmonte A, Rye BJ (2000) Heterodyne lidar returns in the turbulent atmosphere: performance evaluation of simulated systems. Appl Opt 39(15):2401–2411
51. Belmonte A (2003) Analyzing the efficiency of a practical heterodyne lidar in the turbulent atmosphere: telescope parameters. Opt Express 11(17):2041–2046
52. Huang SJ, Safari M (2017) Free-space optical communication impaired by angular fluctuations. IEEE Trans Wireless Commun 16(11):7475–7487
53. Sahota JK, Dhawan D (2018) Reducing the effect of scintillation in FSO system using coherent based homodyne detection. Optik 171:20–26
54. Xu J, Mao HM, Zhen SL et al (2007) Research on the degradation of spatial coherence caused by atmospheric turbulence in laser heterodyne detection. Laser Infrared 37(12):1245–1249
55. Wang CH, Gao L, Pang YJ et al (2011) Experimental study on the influence of beam splitting ratio on the signal-to-noise ratio of a 2 μm balanced coherent detection system. J Opt 31(11):1104002-1-1104002-6
56. Niu MB, Cheng J, Jonathan FH (2013) Error rate performance comparison of coherent and subcarrier intensity modulated optical wireless communications. Opt Commun Netw IEEE/OSA J 5(6):554–564
57. Liu W, Wang L, Yao KN et al (2016) Efficiency analysis of homodyne detection for a coherent lidar with adaptive optics. J Korean Phys Soc 69(12):1750–1754
58. Tian HT, Zhang CX, Jin J et al (2007) The influence of modulation frequency jitter on coherent detection and its elimination algorithm. Opt Prec Eng 15(4):604–610
59. Chang S, Song YS, Liu Y et al (2017) The performance analysis of coherent detection based on the optical Costas loop. Optik 130:1333–1338
60. Zhou HJ, Zhu ZZ, Xie WL et al (2018) Investigation of signal power splitting ratio for BPSK homodyne receiver with an optical Costas loop. Opt Eng 57(8):1–7
61. Liu Y, Zeng X, Cao C et al (2019) Modeling the heterodyne efficiency of array detector systems in the presence of target speckle. IEEE Photon J 11(4):1–9

62. Ren JY, Sun HY, Zhao YZ et al (2020) Analysis of factors influencing differential frequency signal in dual frequency laser coherent detection. Laser and Infrared 50(4):6
63. Azizigul A, Tao ZW, Liu SW et al (2022) The influence of atmospheric turbulence on the temporal coherence characteristics of the received light field. J Phys 71(23):245–252
64. Kong YX, Ke XZ, Yang Y (2015) Research on the influence of atmospheric turbulence on spatial coherent optical communication. Prog Laser and Optoelectron 52(8):080601-1–080601-7
65. Chen JN (2016) Key Technologies and experimental research on subcarrier modulation non optical heterodyne detection in wireless optical communication. Xi'an: Xi'an University of Technology
66. Chen M (2018) Research on channel noise and its impact on coherent detection performance in wireless optical coherent communication. Xi'an: Xi'an University of Technology
67. Mao YZ (2018) Research on mode degeneration and polarization characteristics in coherent optical communication. Xi'an: Xi'an University of Technology
68. Wu JL, Yang YQ, Wang Y (2019) Research on all digital DPSK demodulation technology in coherent optical communication. J Hunan Univ Sci Eng 40(10):12–17
69. Tan ZK (2019) Factors affecting the performance of heterodyne detection in wireless optical communication and experimental research. Xi'an: Xi'an University of Technology
70. Ma BB (2019) Experimental study on polarization control in coherent optical communication systems. Xi'an: Xi'an University of Technology
71. Bai JJ (2021) Research and implementation of QPSK modulation and demodulation in coherent optical communication systems. Xi'an: Xi'an University of Technology
72. Shangjun Y, Tian X, Chenghu K et al (2023) Effect of wavefront distortion on the performance of coherent detection systems: theoretical analysis and experimental research. Photonics 10(5):493
73. Tyson RK (1991) Principles of adaptive optics. Academic Press, New York
74. Kudielka KH, Hayano Y, Klaus W et al (1999) Low-order adaptive optics system for free-space lasedcom: design and performance analysis. Intl Workshop on A.O. for Industry and Medicine, Murcia, Spain, (5 Jun 1999–9 Jun 1999), 1291:1–6
75. Tyson RK (1996) Adaptive optics and ground-to-space laser communications. Appl Opt 35(19):3640–3646
76. Lukin VP, Fortes BV (2002) Phase-correction of turbulence distortions of an optical wave propagating under conditions of strong intensity fluctuations. Appl Opt 41(27):5616–5624
77. Yang HZ, Li XY, Jiang WH (2007) Application progress of Adaptive optics technology in atmospheric optical communication system. Prog Laser and Optoelectron 10:61–68
78. Wilks SC, Morris JR, Brase JM et al (2002) Modeling of adaptive optics-based free-space communications systems. Proc SPIE 48(21):121–128
79. Ishimaru A (1978) Wave propagation and scattering in random media. Academic Press, New York
80. Ishimaru A (1972) Temporal frequency spectra of multifrequency waves in turbulent atmosphere. IEEE Trans Antennas Propag 20(1):10–19
81. Fante RL (1975) Electromagnetic beam propagation in turbulent media. Proc IEEE 63(12):1669–1692
82. Hogge CB, Butts RR (1982) Effects of using different wavelengths in wave-front sensing and correction. JOSA 72(5):606–609
83. Winocur J (1983) Dual-wavelength adaptive optical systems. Appl Opt 22(23):3711–3715
84. Ben-Yosef N, Goldner E, Weitz A (1986) Two-color correlation of scintillations. Appl Opt 25(19):3486–3489
85. Rong JH, Tang ZL, Xie YM et al (2013) Topological optimization design of structures under random excitations using SQP method. Eng Struct 56(13):2098–2106
86. Devaney N, Goncharov AV, Dainty JC (2008) Chromatic effects of the atmosphere on astronomical adaptive optics. Appl Opt 47(8):1072–1081
87. Gorelaya AV, Shubenkova EV, Dmitriev DI et al (2015) Investigation of dual-wavelength laser beam propagation along the in-door atmospheric path. Optics in Atmospheric Propagation and Adaptive Systems XVIII. International Society for Optics and Photonics

88. Xu BW (2020) Influence of wavefront rotation and polarization chromatic aberration on correction ability of Adaptive optics system. Chengdu: University of the Chinese Academy of Sciences (Institute of Optics and Electronics, Chinese Academy of Sciences)
89. Ke X, Chen X (2020) Correcting wavefront distortion of dual-wavelength beams due to atmospheric turbulence with a correction coefficient. Opt Photon J 10(4):64–77

Chapter 2
Optical Wireless Coherent Communication Systems

A complete optical wireless coherent communication system consists of two major components: a transmitter and a receiver. Optical wireless coherent communication must satisfy a series of conditions, such as narrow-linewidth matching, polarization-state matching, wavefront-phase matching, and a constant frequency difference between the signal beam and the local-oscillation beam.

This chapter provides a detailed analysis of the basic principles of coherent optical communication systems, describing the working principles of the signal-modulation system at the transmitter, the spatial optical-fiber coupling system at the receiver, the optical mixing system, the double-balanced detection system, the polarization control system, the beam-source frequency control system, and the signal-demodulation system.

It also introduces the key technologies of each subsystem that influence the performance of coherent optical communication. The performance of each subsystem is experimentally verified. This chapter lays the foundation for the subsequent development of optical wireless coherent communication for long-range transmission in free space.

2.1 Overview

Coherence is the property of correlating physical quantities at different locations or times as a wave propagates. The coherence of a beam is when two or more trains of beam waves overlap or cancel each other when they meet in space, resulting in a new beam wave. Coherent-optical modulation uses a variety of modulation techniques, such as intensity, frequency, phase, and polarization state. Coherent detection is based on the interferometric superposition phase extinction of the beam. A local-oscillation beam is used at the receiving end to amplify the received signal beam, making the

X. Ke and C. Ke, *Performance-Improvement Limits in Optical Wireless Coherent Communication*, Optical Wireless Communication Theory and Technology,
https://doi.org/10.1007/978-981-96-2000-5_2

interference-output intermediate frequency (IF) signal carry the source information [1, 2].

Compared with intensity-modulated/direct-detection optical communication systems, coherent optical communication systems can achieve a gain in detection sensitivity of approximately 20 dB (zero-difference detection can reach approximately 23 dB), allow transmission over longer distances at the same transmission power, are more conducive to suppressing atmospheric-channel fading, and are more suitable for detecting weak signals [3].

2.2 Coherent-Optical Communication-System Components

A coherent-optical communication system using phase modulated/coherent detection is shown in Fig. 2.1. The source information at the transmitting end is driven and amplified, then loaded onto the electro-optical modulator by external modulation to complete the electro-optical conversion. It is transmitted in free space by the transmitting optical antenna after collimation.

The receiving optical antenna converges the beam and collimates the output, which is coupled into a single-mode fiber and connected to a 90° optical mixer, where the signal beam and the local-oscillation beam are mixed. The double-balanced detection system converts the optical signal into an electrical signal.

The polarization control system matches the polarization state of the signal beam with that of the signal-beam input port of the mixer by correcting the polarization state of the signal beam. The frequency control system stabilizes the frequency of the IF signal by correcting the wavelength of the local-oscillation beam. The demodulation system demodulates and decodes the IF signal to finally complete the transmission from the source to the host.

Let the beam field E_s of the signal beam input to the electro-optical modulator be expressed as

$$E_s(t) = A_s \cdot e^{i(\omega_s t+\varphi_0)}. \tag{2.1}$$

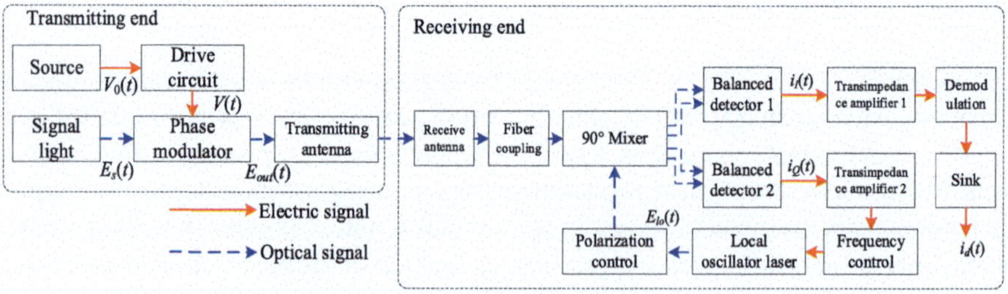

Fig. 2.1 Schematic diagram of a coherent optical communication system

In Eq. (2.1), A_s, ω_s, and φ_0 are the amplitude, angular frequency, and initial phase of the signal beam field, respectively. The signal source $V_0(t)$ acts on the phase electro-optical modulator after driving and amplifying the circuit. The optical field expression of the modulator-output optical signal E_{out} is

$$E_{out}(t) = E_s \cdot e^{i\Delta\varphi} = A_s \cdot e^{i\left(\omega_s t + \varphi_0 + \frac{\pi \cdot V_0(t) \cdot k}{V_\pi}\right)}. \tag{2.2}$$

In Eq. (2.2), k is the driver amplification, half-wave voltage $V_\pi = \lambda G / {n_0}^3 \gamma_{33} \Gamma L$, λ is the wavelength, G is the electrode spacing, ${n_0}^3$ is the effective refractive index of the crystal, γ_{33} is the crystal photoelectric coefficient, Γ is the photoelectric-overlap field integral, and L is the electrode length [4]. Let the optical field E_{lo} of the local-oscillation beam be expressed as

$$E_{lo}(t) = A_{lo} \cdot e^{i(\omega_{lo} t + \varphi_{lo})}. \tag{2.3}$$

In Eq. (2.3), A_{lo}, ω_{lo}, and φ_{lo} are the amplitude, angular frequency, and initial phase of the local-oscillation beam, respectively. The 90° mixer is a passive device with two fiber-optic input ports and four fiber-optic output ports; thus, the phases of the output four-way optical signals differ by 90°, respectively [5].

$$E_{0^\circ}(t) = A_s \cdot e^{i(\omega_s t + \varphi_0 + \Delta\varphi)} + A_{lo} \cdot e^{i(\omega_{lo} t + \varphi_{lo})} \tag{2.4}$$

$$E_{180^\circ}(t) = A_s \cdot e^{i(\omega_s t + \varphi_0 + \Delta\varphi)} - A_{lo} \cdot e^{i(\omega_{lo} t + \varphi_{lo})} \tag{2.5}$$

$$E_{90^\circ}(t) = A_s \cdot e^{i(\omega_s t + \varphi_0 + \Delta\varphi)} + i \cdot A_{lo} \cdot e^{i(\omega_{lo} t + \varphi_{lo})} \tag{2.6}$$

$$E_{270^\circ}(t) = A_s \cdot e^{i(\omega_s t + \varphi_0 + \Delta\varphi)} - i \cdot A_{lo} \cdot e^{i(\omega_{lo} t + \varphi_{lo})}. \tag{2.7}$$

In Eqs. (2.4) to (2.7), $\Delta\varphi$ is the modulated phase. Any two-phase difference-of-180° optical signals, E_{0° and E_{180°, E_{90° and E_{270°, are paired on the balanced detector. The two photodiodes inside the balanced detector are detected separately and the output signal is processed differently, using the product and the difference. The frequency term $\omega_s + \omega_{lo}$ is far beyond the detector bandwidth and cannot be detected. The difference frequency term $\omega_s - \omega_{lo}$ is calculated; then, by transimpedance-amplifier amplification, the 90° output phase difference of the two IF electrical signals $i_I(t)$ and $i_Q(t)$ is as follows:

$$i_I(t) = 2 \cdot \beta \cdot \alpha \cdot A_s \cdot A_{lo} \cdot \cos\left((\omega_s - \omega_{lo})t + \varphi_0 + \pi \frac{k \cdot V_0(t)}{V_\pi} + \varphi_{lo}\right) \tag{2.8}$$

$$i_Q(t) = 2 \cdot \beta \cdot \alpha \cdot A_s \cdot A_{lo} \cdot \sin\left((\omega_s - \omega_{lo})t + \varphi_0 + \pi \frac{k \cdot V_0(t)}{V_\pi} + \varphi_{lo}\right). \tag{2.9}$$

In Eqs. (2.8) and (2.9), α is the photoelectric-conversion coefficient of the balanced detector and β is the amplification of the transimpedance amplifier. A digital signal-processing technique is used to demodulate either signal $i_I(t)$ or $i_Q(t)$ to obtain the demodulated baseband signal $i_d(t)$, as follows:

$$i_d(t) = 2 \cdot \beta \cdot \alpha \cdot A_s \cdot A_{lo} \cdot \cos\left(\frac{k \cdot V_0(t) \cdot \pi}{V_\pi} + \varphi_0 + \varphi_{lo}\right). \tag{2.10}$$

Thus, the source information $V_0(t)$ is transmitted to the host to become information $i_d(t)$.

2.2.1 *Signal Modulation*

Lithium niobate ($LiNbO_3$) crystals have excellent piezoelectric, electro-optical, nonlinear optical, and acoustic surface-wave properties [6]. They have been widely used in optical modulators, optical waveguide substrates, optical isolators, and narrowband filters, and have broad application prospects in photonic mass memory and optical integration [7]. $LiNbO_3$ crystals have good beam transmission, and the waveguide-fabrication processes are more mature, without the effect of natural birefringence.

Under the action of an external electric field, the refractive index of the crystal changes, and $LiNbO_3$ has the lowest transmission loss compared with other polymers, which can greatly reduce the effect of chirp signals. In addition, owing to the internal structure of $LiNbO_3$ in the form of traveling-wave electrodes, more mature modulation rates have reached 40 Gbit/s [8], which can meet the increasingly urgent requirements for large-bandwidth applications.

Figure 2.2 shows the internal optical-waveguide structures of the phase, Mach–Zehnder, and in-phase-quadrature (IQ) modulators using $LiNbO_3$. The internal optical-waveguide structure of the phase modulator is shown in Fig. 2.2a. It is based on the primary Planck and secondary Kerr electro-optical effects, where a change in the applied electric field causes a change in the refractive index of the crystal, resulting in a phase jump of the output optical field from zero to π [9].

The internal transmission structure of the phase modulator is shown schematically in Fig. 2.3a, where the polarizer located at the input of the modulator acts as a starter; only the vertical vibrational component exists before the beam is incident onto the crystal, and the vibrational direction is parallel to the main induction axis of the crystal. Thus, the applied voltage only changes the phase without changing the polarization state.

The internal optical-waveguide structure of a Mach–Zehnder modulator is shown in Fig. 2.2b. It consists of two phase modulators placed in parallel to form the Mach–Zehnder interference structure. The drive is composed of simultaneous radio-frequency (RF) and direct-current (DC)-bias drives. The intensity or phase can be modulated by stabilizing the DC bias point.

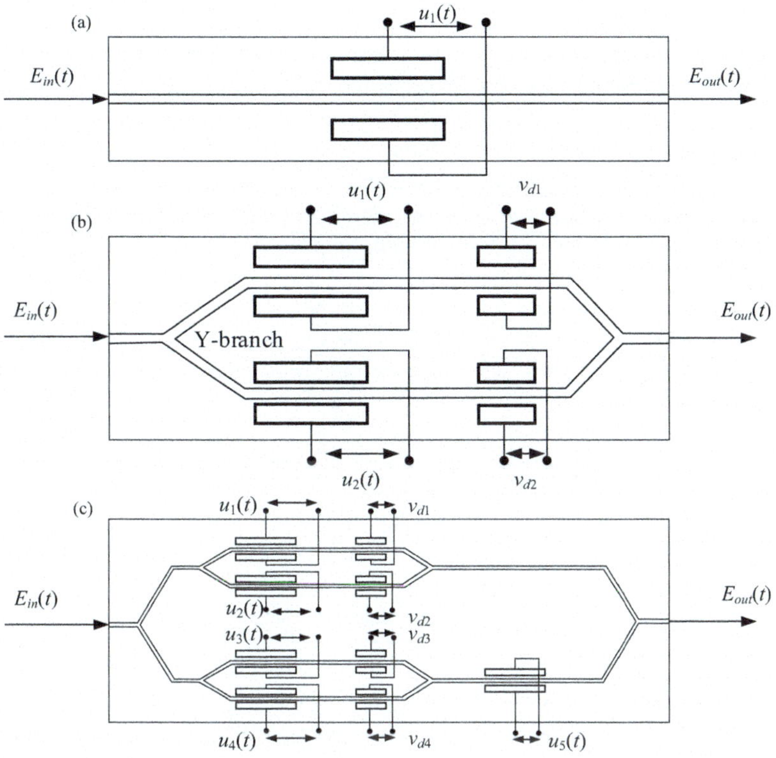

Fig. 2.2 Structure of the optical waveguides of $LiNbO_3$ modulators. **a** Phase modulator, **b** Mach–Zehnder modulator, **c** IQ modulator

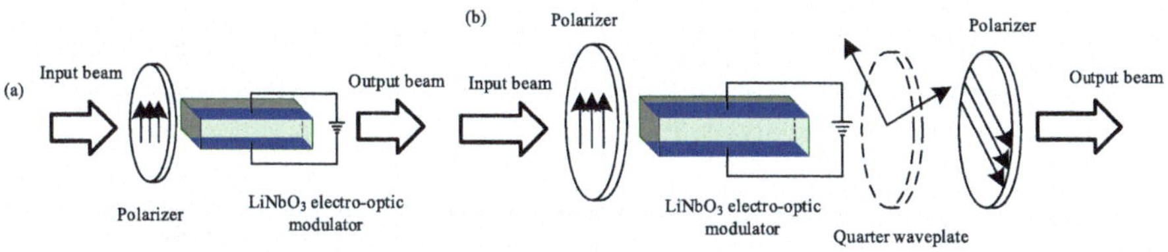

Fig. 2.3 Transmission structure of $LiNbO_3$ modulators. **a** Phase modulator, **b** Mach–Zehnder modulator

A schematic diagram of the Mach–Zehnder modulator transmission structure is shown in Fig. 2.3b. It shows that the output optical power of the modulator is proportional to the cosine square of the angle between the starter and detector by applying a voltage to the crystal located between them. The output power of the intensity modulator is maximized by adjusting the waveplate to produce a phase shift according to Malus's law.

The IQ modulator consists of two Mach–Zehnder modulators connected in parallel, with a phase modulator connected in series with one of them. The internal optical waveguide is shown in Fig. 2.2c. When the IQ modulator is fixed as a push–pull structure, the IQ modulator requires two RF drive signals and three DC-bias signals to modulate a higher-order signal. An RF driver is used to achieve phase hopping, and the bias voltage stabilizes the output signal power. By setting different combinations of RF-drive and DC-bias voltages, higher-order amplitude quadrature phase modulation and single-sideband-suppressed carrier transmission can be created.

For the electro-optical modulator with the Mach–Zehnder structure shown in Fig. 2.2b, the output-modulated signal can be expressed as in [10] by applying a voltage between the two arms, which causes the optical signal between the two arms to interfere.

$$
\begin{aligned}
E_{out}(t) &= \frac{E_{in}}{2}\left[e^{i\pi\left(\frac{u_1(t)}{V_{\pi RF1}}+\frac{v_{d1}}{V_{\pi DC1}}\right)}+e^{i\pi\left(\frac{u_2(t)}{V_{\pi RF2}}+\frac{v_{d2}}{V_{\pi DC2}}\right)}\right] \\
&= E_{in}\cos\left[\frac{\pi\left(\frac{u_1(t)}{V_{\pi 1}}+\frac{v_{d1}}{V_{\pi DC1}}\right)-\pi\left(\frac{u_2(t)}{V_{\pi 2}}+\frac{v_{d2}}{V_{\pi DC2}}\right)}{2}\right]e^{i\frac{\pi\left(\frac{u_1(t)}{V_{\pi 1}}+\frac{v_{d1}}{V_{\pi DC1}}\right)-\pi\left(\frac{u_2(t)}{V_{\pi 2}}+\frac{v_{d2}}{V_{\pi DC2}}\right)}{2}}.
\end{aligned}
\tag{2.11}
$$

In Eq. (2.11), the input RF signal voltages of the two arms are $u_1(t)$ and $u_2(t)$ and the DC voltage signals are v_{d1} and v_{d2}. Each applied voltage produces the following phase shift:

$$
\varphi_1=\frac{\pi u_1(t)}{V_{\pi 1}},\ \varphi_2=\frac{\pi u_2(t)}{V_{\pi 2}},\ \varphi_{d1}(t)=\frac{\pi v_{d1}}{V_{\pi\,DC1}},\ \text{and } \varphi_{d2}(t)=\frac{\pi v_{d2}}{V_{\pi\,DC2}}. \tag{2.12}
$$

In Eq. (2.12), $V_{\pi 1}$, $V_{\pi DC1}$, $V_{\pi 2}$, and $V_{\pi DC2}$ are the voltages required to phase shift the optical signals in Fig. 2.2b on the upper and lower arms of the waveguide by 180°. For convenience of analysis, it can be considered that, in general [11],

$$
V_\pi = V_{\pi 1} = V_{\pi 2} = V_{\pi DC1} = V_{\pi DC2}. \tag{2.13}
$$

At this point, the output optical signal $E_{out}(t)$ can be expressed as

$$
\begin{aligned}
E_{out}(t) &= \frac{E_{in}}{2}\left[e^{i\pi\left(\frac{u_1(t)}{V_\pi}+\frac{v_{d1}}{V_\pi}\right)}+e^{i\pi\left(\frac{u_2(t)}{V_\pi}+\frac{v_{d2}}{V_\pi}\right)}\right] \\
E_{in}&\cos\left(\frac{\pi(u_1(t)-u_2(t))+\pi(v_{d1}-v_{d2})}{2V_\pi}\right)e^{i\frac{\pi(u_1(t)+u_2(t))+\pi(v_{d1}-v_{d2})}{2V_\pi}}.
\end{aligned}
\tag{2.14}
$$

For the Mach–Zehnder modulator in the push–pull operation shown in Fig. 2.2b, there is only one RF drive input port on the modulator. Let the RF drive signal input to the modulator be $V_{in}(t)$, which is divided by the internal voltage to obtain $u_1(t)$

and $u_2(t)$, at which point $V_{in}(t) = 2u_1(t) = -2u_2(t)$. A DC voltage is used to set the bias voltage V_{bias} between the two arms of the modulator. Let $V_{bias} = v_{d2} - v_{d1}$; then, the modulator output signal is

$$E_{out}(t) = E_{in}\cos\left[\frac{\pi}{2V_\pi}(V_m(t) - V_{bias})\right]e^{i\frac{\pi V_{bias}}{2V_\pi}}. \tag{2.15}$$

The power of the output optical signal is

$$P_{out}(t) = |E_{out}(t)|^2 = \frac{1}{2}P_m\left[1 + \cos\left(\frac{\pi}{V_\pi}(V_m(t) - V_{bias})\right)\right]. \tag{2.16}$$

The DC bias curve of the Mach–Zehnder modulator and the corresponding DC bias point can be obtained by varying the bias voltage V_{bias} between the two arms when the RF drive signal $V_m(t) = 0$, based on Eq. (2.16), as shown in Fig. 2.4 [12]. For different modulation formats, different optimal bias points should be set for Mach–Zehnder modulators. For example, when the intensity is modulated, it should be controlled by setting the DC bias at the Quad + point, where there is an optimal linear correspondence between the DC bias voltage and the optical power. When modulating the phase, the optimal operating bias point for the Mach–Zehnder modulator is the null point.

However, external conditions, such as the cutting process of the crystal inside the Mach–Zehnder modulator, temperature-drift characteristics of the material, accuracy of the processing process, uneven force, effect of the applied electric field, and temperature changes can cause the bias point of the modulator to drift when operating, thus causing the modulated signal quality to deteriorate. In practice, the bias point of the Mach–Zehnder modulator should be controlled using appropriate techniques.

In general, the bias point can be monitored by measuring the modulator's output optical power. The bias voltage can be adjusted using feedback control when the

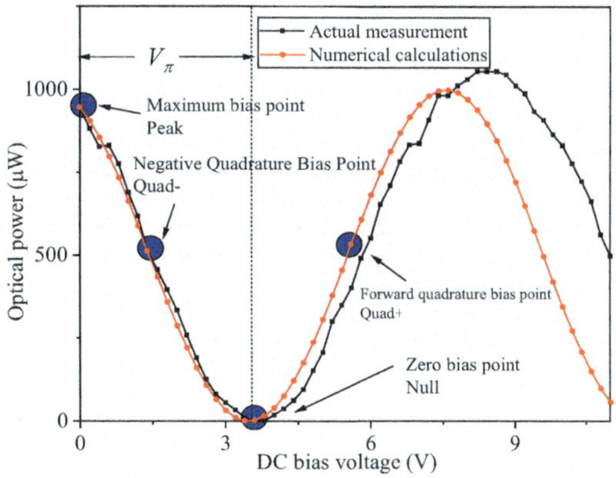

Fig. 2.4 Direct-current bias curve of a Mach–Zehnder modulator

bias point drifts, thereby effectively stabilizing the modulator at the optimum operating bias point. X-cut Y-transmitting $LiNbO_3$ optical waveguides fabricated using plasmonic-exchange technology have a slower DC-bias point drift than Z-cut Y-transmitting $LiNbO_3$ optical waveguides. The operation can typically be stabilized in a short time without automatic control.

2.2.2 Spatial Optical-Fiber Coupling

After the signal beam has been transmitted through free space, the space beam at the receiving end must be coupled into a single-mode fiber using fiber-coupling technology to meet the subsequent requirements of fiber-optic mixers and balanced detectors for fiber-optic transmission. The coupling efficiency is an important parameter for measuring the spatial beam–fiber coupling index. The higher the coupling efficiency, the stronger the fiber output power at the receiving end, and the more obvious the improvement in the coherent optical communication-system mixing gain.

Figure 2.5 shows a schematic diagram of the spatial beam–fiber coupling. The coupling efficiency is defined as the ratio of the optical power transmitted by the fiber to the spatial optical power located at the end face of the fiber. The coupling efficiency can typically be calculated using either geometrical optics or mode-field analysis, which considers the fluctuating nature of the beam and can lead to a more complete solution.

Using the mode-field analysis method in polar coordinates, the coupling efficiency is defined as [13]

$$\eta = \frac{\left| \iint U_i^*(r) U_f(r) r dr d\theta \right|^2}{\iint U_i^*(r) U_i(r) r dr d\theta \cdot \iint U_f^*(r) U_f(r) r dr d\theta}. \tag{2.17}$$

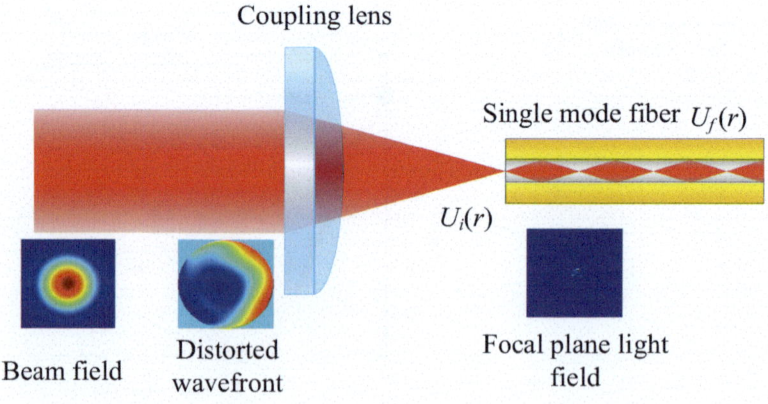

Fig. 2.5 Schematic diagram of optical-fiber coupling

In Eq. (2.17), $U_i(r)$ is the distribution of the optical field in the focal plane after the beam has been converged by the lens, $U_f(r)$ is the distribution of the single-mode fiber electromagnetic field over the fiber cross-section, and * denotes the conjugate operation. Considering the mode field of the coupled fiber and the fiber mode-field matching problem of the fiber mixer, the space beam must be coupled directly to a single-mode fiber. The single-mode fiber optical field can be expressed as [14]

$$U_f(r) = \sqrt{\frac{2}{\pi}} \frac{1}{W_m} \exp\left[-\left(\frac{r}{W_m}\right)^2\right]. \tag{2.18}$$

In Eq. (2.18), r is the radial distance in the cross section of the fiber and W_m is the radius of the mode field of the single-mode fiber. For the commonly used Corning SMF-28e + fiber with a wavelength of $\lambda = 1550$ nm, a typical value is 5.25 μm. Considering the limitations of the aperture diameter D_A of the receiving lens, the optical field of a plane wave located in the focal plane of the lens using the Fresnel diffraction formula can be written as [15]

$$U_i(r) = \frac{1}{i\lambda f} \exp(i\kappa f) \exp\left(\frac{i\kappa r^2}{2f}\right) \pi \left(\frac{D_A}{2}\right) \left[\frac{2J_1\left(\frac{\pi D_A r}{\lambda f}\right)}{\frac{\pi D_A r}{\lambda f}}\right]. \tag{2.19}$$

In Eq. (2.19), f is the coupling-lens focal length, $J_1(\cdot)$ is the first-order first-class Bessel function, and $\kappa = 2\pi/\lambda$ represents the wavenumber. Substituting Eqs. (2.18) and (2.19) into Eq. (2.17), the coupling efficiency can be expressed as [16], after simplification:

$$\eta = \frac{4\left|\int_0^{\infty} \exp(-i\kappa f) \exp\left(\frac{-i\kappa r^2}{2f}\right) J_1\left(\frac{\pi D_A r}{\lambda f}\right) \exp\left[-\left(\frac{r}{W_m}\right)^2\right] dr\right|^2}{W_m^2 \int_0^{\infty} \frac{\left[J_1\left(\frac{\pi D_A r}{\lambda f}\right)\right]^2}{r} dr}. \tag{2.20}$$

Setting $\lambda = 1550$ nm, $W_m = 5.25$ μm, and $D_A = 25.4$ mm, numerical simulations were performed based on Eq. (2.20) and the results are shown in Fig. 2.6. The maximum extreme value of the coupling efficiency exists when the relative aperture diameter $D_A/f = 0.21$ and $\eta = 81.45\%$. In practical engineering applications, for a 1" (25.4 mm) diameter lens, an aspherical planoconvex lens with $f = 125$ mm can be selected for coupling.

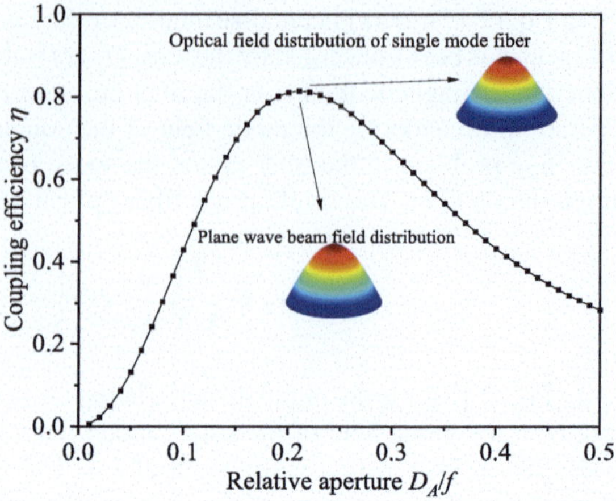

Fig. 2.6 Change curve of the coupling efficiency versus relative aperture D_A/f

2.2.3 *Optical Mixing*

An optical mixer is a passive device that mixes the signal beam and local-oscillation beam. Each output signal has a fixed phase difference and contains both signal-beam and local-oscillation-beam components [17]. Depending on the fixed phase difference between the output signals of the mixer, optical mixers can be divided into 90° mixers and 180° mixers. Depending on the coupling method of the input and output signals of the mixer, mixers can be divided into spatial optical mixers and fiber-optic optical mixers.

Ninety-degree fiber-optic mixers play a crucial role in the fields of both wireless optical coherent communication and fiber-optic optical coherent communication because of their compact external structure, high integration, and mature technology. The 90° fiber-optic mixer plays a crucial role in both optical wireless coherent communication and fiber-optic optical coherent communication [18, 19]. A 90° fiber-optic mixer with a polarization-diversity multiplexing (PDM) function expands the channel capacity by converting the polarization state of the signal beam and the local-oscillation beam and mixing them in the entire system [20].

We let the signal beam be the beam vector and the x-axis angle of the θ line polarization beam; the Jones vector is expressed as $S = [S_1, S_2]^T$. The local vibration beam is the beam vector and the x-axis angle of 0° is the polarization beam; the Jones vector is expressed as $Lo = [Lo_1, 0]^T$, where ^{T} represents the matrix transposition. After the signal beam passes through the polarization-splitting prism, the two split beams are, respectively,

$$\left\{\begin{bmatrix} S_1 \\ S_2 \end{bmatrix}\right\}^2 = \left\{\begin{bmatrix} S_1 \\ 0 \end{bmatrix}\right\}^2 + \left\{\begin{bmatrix} 0 \\ S_2 \end{bmatrix}\right\}^2. \tag{2.21}$$

After passing through the beam-splitting prism, the two halves of the beam are

$$\left\{\begin{bmatrix} Lo_1 \\ 0 \end{bmatrix}\right\}^2 = \left\{\begin{bmatrix} \frac{1}{\sqrt{2}}Lo_1 \\ 0 \end{bmatrix}\right\}^2 + \left\{\begin{bmatrix} \frac{1}{\sqrt{2}}Lo_1 \\ 0 \end{bmatrix}\right\}^2. \tag{2.22}$$

The Jones matrix entering x-polarization (x-pol) mixing is the first term of the sum of the squares of Eqs. (2.21) and (2.22). The Jones matrix entering y-pol mixing is the second term of the sum of the squares of Eqs. (2.21) and (2.22).

The waveplate consists of two transmissive axes (fast and slow) that are perpendicular to each other, and the phase difference between them is denoted as δ. The different types of waveplates are related to the phase difference, as shown in Table 2.1.

For x-pol mixing, after the signal beam has passed through the quarter-wave plate, the Jones matrix can be expressed as

$$\cos\frac{\delta}{2}\begin{bmatrix} 1 - i\tan\frac{\delta}{2}\cos 2\theta & -i\tan\frac{\delta}{2}\sin 2\theta \\ -i\tan\frac{\delta}{2}\sin 2\theta & 1 + i\tan\frac{\delta}{2}\cos 2\theta \end{bmatrix}\Bigg|_{\substack{\delta=\frac{\pi}{2} \\ \theta=\frac{\pi}{4}}} \cdot \begin{bmatrix} S_1 \\ 0 \end{bmatrix} = \begin{bmatrix} \frac{1}{\sqrt{2}}S_1 \\ \frac{1}{\sqrt{2}}S_1 \cdot e^{-i\frac{\pi}{2}} \end{bmatrix}. \tag{2.23}$$

After passing through the one-half waveplate, the Jones matrix of this oscillating beam can be expressed as

$$\begin{bmatrix} \cos^2\theta + \sin^2\theta \cdot e^{i\delta} & \sin\theta\cos\theta - \sin\theta\cos\theta \cdot e^{i\delta} \\ \sin\theta\cos\theta - \sin\theta\cos\theta \cdot e^{i\delta} & \sin^2\theta + \cos^2\theta \cdot e^{i\delta} \end{bmatrix}\Bigg|_{\substack{\delta=\pi \\ \theta=\frac{\pi}{8}}} \cdot \begin{bmatrix} \frac{1}{\sqrt{2}}Lo \\ 0 \end{bmatrix} = \begin{bmatrix} \frac{1}{\sqrt{2}}Lo \\ \frac{1}{\sqrt{2}}Lo \end{bmatrix}. \tag{2.24}$$

According to the optical-path diagram in Fig. 2.7, the signal beam can be decomposed into two components after passing through the beam-splitting prism:

$$\begin{aligned} \begin{bmatrix} \frac{1}{\sqrt{2}}S_1 \\ \frac{1}{\sqrt{2}}S_1 \cdot e^{-i\frac{\pi}{2}} \end{bmatrix}^2 &= \left\{\begin{bmatrix} 1 & 0 \\ 0 & 1 \end{bmatrix}_{\text{project}} \cdot \begin{bmatrix} \frac{1}{2}S_1 \\ \frac{1}{2}S_1 \cdot e^{-i\frac{\pi}{2}} \end{bmatrix}\right\}^2 \\ &+ \left\{\begin{bmatrix} e^{-i\frac{\pi}{2}} & 0 \\ 0 & e^{-i\frac{\pi}{2}} \end{bmatrix}_{\text{reflect}} \cdot \begin{bmatrix} \frac{1}{2}S_1 \\ \frac{1}{2}S_1 \cdot e^{-i\frac{\pi}{2}} \end{bmatrix}\right\}^2 \\ &= \begin{bmatrix} \frac{1}{2}S_1 \\ \frac{1}{2}S_1 \cdot e^{-i\frac{\pi}{2}} \end{bmatrix}^2 + \begin{bmatrix} \frac{1}{2}S_1 \cdot e^{-i\frac{\pi}{2}} \\ \frac{1}{2}S_1 \cdot e^{-i\pi} \end{bmatrix}^2. \end{aligned} \tag{2.25}$$

Table 2.1 Relationship between different types of waveplate and the phase difference

Waveplate type	Quarter waveplate	Half waveplate	Full waveplate
Phase difference	$\pi/2$	π	2π

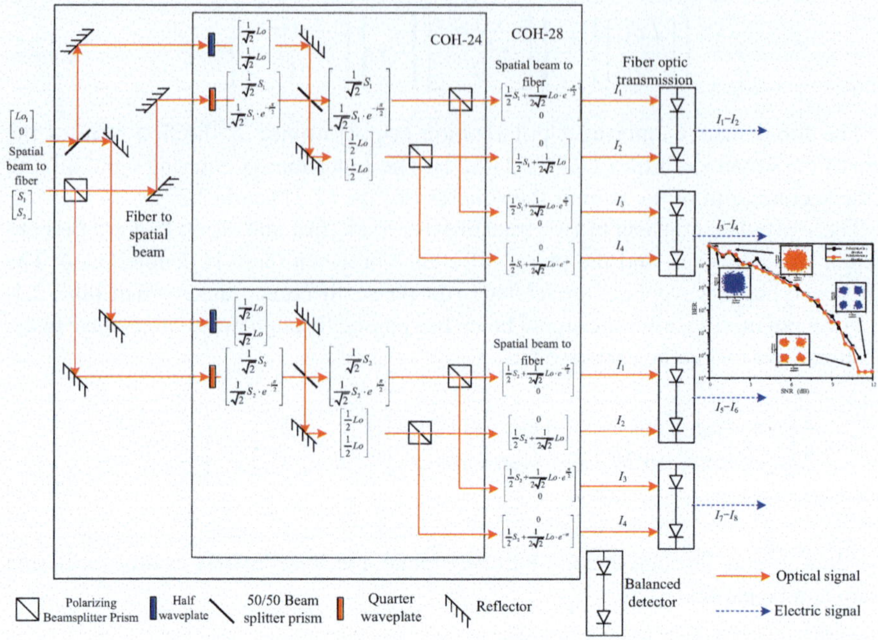

Fig. 2.7 Schematic diagram of the polarization multiplexing optical-fiber hybrid internal structure. The constellation diagram represents the data constellation-point distribution polarization in the x and y directions of the balanced-detector output electrical signal, when the signal-to-noise ratios are 1 dB and 10 dB, respectively

After passing through the spectroscopic prism, the local oscillation beam can also be decomposed into two beams:

$$\begin{bmatrix}\frac{1}{2}Lo\\ \frac{1}{2}Lo\end{bmatrix}^2=\left\{\begin{bmatrix}1&0\\0&1\end{bmatrix}_{\text{project}}\cdot\begin{bmatrix}\frac{1}{2\sqrt{2}}Lo\\ \frac{1}{2\sqrt{2}}Lo\end{bmatrix}\right\}^2+\left\{\begin{bmatrix}e^{-i\frac{\pi}{2}}&0\\0&e^{-i\frac{\pi}{2}}\end{bmatrix}_{\text{reflect}}\cdot\begin{bmatrix}\frac{1}{2\sqrt{2}}Lo\\ \frac{1}{2\sqrt{2}}Lo\end{bmatrix}\right\}^2$$
$$=\begin{bmatrix}\frac{1}{2\sqrt{2}}Lo\\ \frac{1}{2\sqrt{2}}Lo\end{bmatrix}^2+\begin{bmatrix}\frac{1}{2\sqrt{2}}Lo\cdot e^{-i\frac{\pi}{2}}\\ \frac{1}{2\sqrt{2}}Lo\cdot e^{-i\frac{\pi}{2}}\end{bmatrix}^2. \tag{2.26}$$

After the phase-delay slice, the four outputs in the x-pol direction are

$$E_1=\begin{bmatrix}\frac{1}{2}S_1+\frac{1}{2\sqrt{2}}Lo\cdot e^{-i\frac{\pi}{2}}\\ 0\end{bmatrix} \tag{2.27}$$

$$E_2=\begin{bmatrix}0\\ \frac{1}{2}S_1+\frac{1}{2\sqrt{2}}Lo\end{bmatrix}=\begin{bmatrix}e^{-i\frac{\pi}{2}}&0\\0&e^{-i\frac{\pi}{2}}\end{bmatrix}\begin{bmatrix}0\\ \frac{1}{2}S_1\cdot e^{-i\frac{\pi}{2}}+\frac{1}{2\sqrt{2}}Lo\cdot e^{-i\frac{\pi}{2}}\end{bmatrix} \tag{2.28}$$

$$E_3 = \begin{bmatrix} \frac{1}{2}S_1 + \frac{1}{2\sqrt{2}}Lo \cdot e^{i\frac{\pi}{2}} \\ 0 \end{bmatrix} = \begin{bmatrix} e^{i\frac{\pi}{2}} & 0 \\ 0 & e^{i\frac{\pi}{2}} \end{bmatrix} \begin{bmatrix} \frac{1}{2}S_1 \cdot e^{-i\frac{\pi}{2}} + \frac{1}{2\sqrt{2}}Lo \\ 0 \end{bmatrix} \tag{2.29}$$

$$E_4 = \begin{bmatrix} 0 \\ \frac{1}{2}S_1 + \frac{1}{2\sqrt{2}}Lo \cdot e^{-i\pi} \end{bmatrix} = \begin{bmatrix} e^{-i\pi} & 0 \\ 0 & e^{-i\pi} \end{bmatrix} \begin{bmatrix} 0 \\ \frac{1}{2}S_1 \cdot e^{i\pi} + \frac{1}{2\sqrt{2}}Lo \end{bmatrix} \tag{2.30}$$

Let the photoconversion factor be α; then, the output signals of the four photodetectors located in the x-pol direction are expressed as

$$I_1 = |E_1|^2 = \alpha\left[\frac{1}{4}|S_1|^2 + \frac{1}{8}|Lo|^2 + \frac{1}{2\sqrt{2}}S_1 \cdot Lo \cdot \cos\left[(\omega_{S_1} - \omega_{Lo})t + \Delta\varphi(t) - \frac{\pi}{2}\right]\right] \tag{2.31}$$

$$I_2 = |E_2|^2 = \alpha\left[\frac{1}{4}|S_1|^2 + \frac{1}{8}|Lo|^2 - \frac{1}{2\sqrt{2}}S_1 \cdot Lo \cdot \cos\left[(\omega_{S_1} - \omega_{Lo})t + \Delta\varphi(t) - \frac{\pi}{2}\right]\right] \tag{2.32}$$

$$I_3 = |E_3|^2 = \alpha\left[\frac{1}{4}|S_1|^2 + \frac{1}{8}|Lo|^2 + \frac{1}{2\sqrt{2}}S_1 \cdot Lo \cdot \cos\left[(\omega_{S_1} - \omega_{Lo})t + \Delta\varphi(t)\right]\right] \tag{2.33}$$

$$I_4 = |E_4|^2 = \alpha\left[\frac{1}{4}|S_1|^2 + \frac{1}{8}|Lo|^2 - \frac{1}{2\sqrt{2}}S_1 \cdot Lo \cdot \cos\left[(\omega_{S_1} - \omega_{Lo})t + \Delta\varphi(t)\right]\right]. \tag{2.34}$$

Two identical photodetectors form a balanced detector. The currents output from the two photodetectors in the balanced detector are differenced separately to obtain the output current of the double-balanced detector system for x-pol.

$$BPD_1 = I_1 - I_2 = \frac{1}{\sqrt{2}}S_1 \cdot \alpha \cdot Lo \cdot \sin\left[(\omega_{S_1} - \omega_{Lo})t + \Delta\varphi(t)\right] \tag{2.35}$$

$$BPD_2 = I_3 - I_4 = \frac{1}{\sqrt{2}}S_1 \cdot \alpha \cdot Lo \cdot \cos\left[(\omega_{S_1} - \omega_{Lo})t + \Delta\varphi(t)\right]. \tag{2.36}$$

Based on the above calculation, the output current of the double-balanced detection system for y-pol is obtained by deriving the beam in the y-pol direction. Note that the S-polarized beam can produce a half-wave loss, whereas the P-polarized beam does not produce a half-wave loss.

$$BPD_3 = I_5 - I_6 = \frac{1}{\sqrt{2}}S_2 \cdot \alpha \cdot Lo \cdot \sin\left[(\omega_{S_2} - \omega_{Lo})t + \Delta\varphi(t)\right] \tag{2.37}$$

$$BPD_4 = I_7 - I_8 = \frac{1}{\sqrt{2}}S_2 \cdot \alpha \cdot Lo \cdot \cos\left[(\omega_{S_2} - \omega_{Lo})t + \Delta\varphi(t)\right]. \tag{2.38}$$

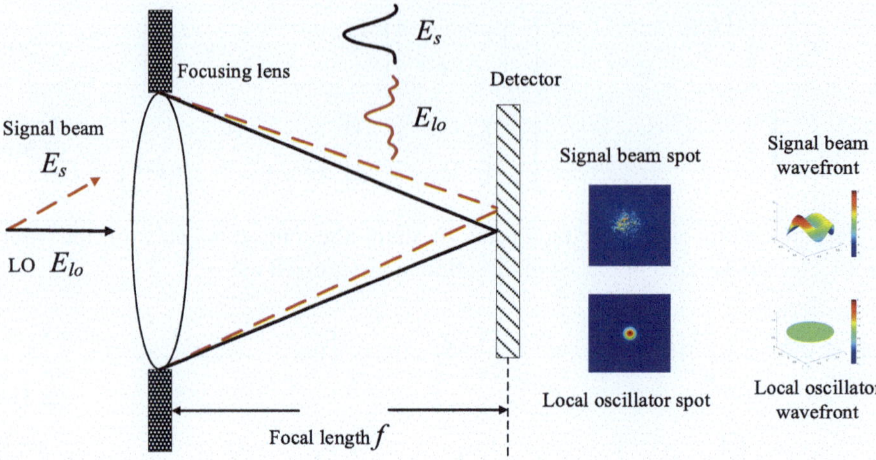

Fig. 2.8 Diagram of coherent detection in a coherent optical communication system

Figure 2.8 shows the schematic diagram of coherent optical communication systems for coherent detection. The mixed signal and local-oscillation beams form interference on the detector surface, thus completing the transmission of the signal. The mixing efficiency is an important indicator of coherent optical communication. It ensures that the wavelength, polarization, phase, spot size, beam-field distribution, and a series of other conditions are matched to obtain a high mixing efficiency.

Coherent optical communication can be divided into zero-difference detection and outlier detection, based on the frequency difference between the signal beam and local-oscillation beam [21]. When the frequency of the signal beam is equal to the frequency of the local-oscillation beam, $\omega_s = \omega_{lo}$, it is called zero-difference detection. The mixing efficiency η_{homodyne} can be expressed as [22]

$$\eta_{\text{homodyne}} = \frac{\left[\int_0^{2\pi}\int_0^R A_s A_{lo}\cos(\varphi_s - \varphi_{lo}) r dr d\varphi\right]^2}{\int_0^{2\pi}\int_0^R |E_s|^2 r dr d\varphi \cdot \int_0^{2\pi}\int_0^R |E_{lo}|^2 r dr d\varphi}. \tag{2.39}$$

In Eq. (2.39), R is the effective mixing radius of the signal beam and the local-oscillation beam.

When the frequency of the signal beam is not equal to the frequency of the fundamental beam, i.e., $\omega_s \neq \omega_{lo}$, it is called outlier detection. The mixing efficiency $\eta_{\text{heterodyne}}$ is expressed as [23]

$$\eta_{\text{heterodyne}} = \frac{\left[\int_0^{2\pi}\int_0^R A_s A_{lo}\cos(\varphi_s - \varphi_{lo}) r dr d\varphi\right]^2 + \left[\int_0^{2\pi}\int_0^R A_s A_{lo}\sin(\varphi_s - \varphi_{lo}) r dr d\varphi\right]^2}{\int_0^{2\pi}\int_0^R |E_s|^2 r dr d\varphi \cdot \int_0^{2\pi}\int_0^R |E_{lo}|^2 r dr d\varphi}. \tag{2.40}$$

2.2.4 Double-Balanced Detection

Two identical balanced detectors in a coherent optical communication system constitute a double-balanced detection system. Figure 2.9 shows the internal structure of the double-balanced detection system using two balanced detectors. Photodetectors D_1 and D_2 are connected in series. The pull-up resistor R_2 and the pull-down capacitor C_2 located in photodetector D_1 cause the photocurrent generated by photodetector D_1 to form a pressure difference, while photodetectors D_1 and D_2 form a series structure. The two are then amplified by a differential operation to output the electrical signal.

When considering the noise introduced by the photodetector inside the balanced detector, Eqs. (2.31) and (2.32) can be written as

$$I_1 = \alpha\left[\frac{1}{4}|S_1|^2 + \frac{1}{8}|Lo|^2 + \frac{1}{2\sqrt{2}}S_1 \cdot Lo \cdot \cos\left[(\omega_{S_1} - \omega_{Lo})t + \Delta\varphi(t) - \frac{\pi}{2}\right]\right] + n_1(t) \tag{2.41}$$

$$I_2 = \alpha\left[\frac{1}{4}|S_1|^2 + \frac{1}{8}|Lo|^2 - \frac{1}{2\sqrt{2}}S_1 \cdot Lo \cdot \cos\left[(\omega_{S_1} - \omega_{Lo})t + \Delta\varphi(t) - \frac{\pi}{2}\right]\right] + n_2(t). \tag{2.42}$$

In Eqs. (2.41) and (2.42), the first two terms are the DC components of the signal beam and the local-oscillation beam, respectively. The third term is the differential frequency term, and the fourth term is the relative-intensity noise introduced by the local-oscillation beam. Equations (2.41) and (2.42) show the difference processing;

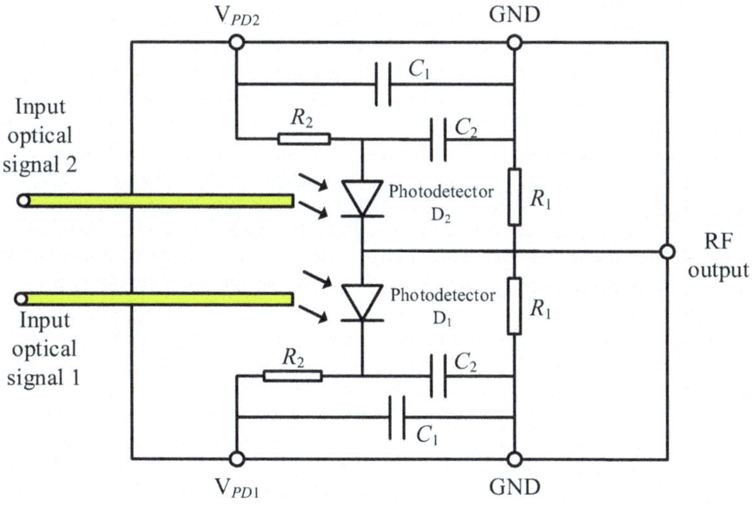

Fig. 2.9 Balanced-detector internal structure diagram

the balanced-detector output current can be expressed as

$$BPD_1 = I_1 - I_2 = \frac{1}{\sqrt{2}} S_1 \cdot \alpha \cdot Lo \cdot \sin\left[\left(\omega_{S_1} - \omega_{Lo}\right)t + \Delta\varphi(t)\right] + \Delta n(t). \quad (2.43)$$

In Eq. (2.43), $\Delta n(t) = n_1(t) - n_2(t)$, which is the residual relative intensity noise (RIN). Ideally, the differential operation can effectively suppress the relative intensity noise; the IF signal in Eq. (2.43) will have double the amplitude of the AC component of Eq. (2.41) and Eq. (2.42), which can significantly improve the signal-to-noise ratio.

Considering the effects of scattered particle noise (caused by electron-emission inhomogeneity), relative-intensity noise (normalized average power noise in the optical-power rise and fall of the laser), and thermal noise (electron Brownian motion in the passive device) introduced into the photodetector by the local-oscillation beam, the following photodetection methods can be used: direct detection, outlier detection, and zero-difference detection, with signal-to-noise ratios SNR_S, SNR_B, and SNR_H [24], respectively. In direct detection, a photodiode is balanced in the detector to directly detect the optical signal after mixing the output.

$$SNR_S = \frac{\frac{1}{8}\alpha \cdot \beta^2 \cdot R \cdot P_S \cdot P_{LO}}{2\beta^2 \cdot e[\alpha(P_S + P_{LO}) + I_d]R \cdot B + R_{IN} \cdot B(\alpha \cdot P_{LO})^2 + \frac{4k_B TB}{R}} \quad (2.44)$$

$$SNR_B = \frac{\frac{1}{4}\alpha \cdot \beta^2 \cdot R \cdot P_S \cdot P_{LO}}{2\beta^2 \cdot e[\alpha(P_S + P_{LO}) + I_d]R \cdot B + R_{IN} \cdot B[(1-k)\alpha P_{LO}]^2 + \frac{4k_B TB}{R}} \quad (2.45)$$

$$SNR_H = \frac{\frac{1}{2}\alpha \cdot \beta^2 \cdot R \cdot P_S \cdot P_{LO}}{2\beta^2 \cdot e[\alpha(P_S + P_{LO}) + I_d]R \cdot B + R_{IN} \cdot B[(1-k)\alpha P_{LO}]^2 + \frac{4k_B TB}{R}}. \quad (2.46)$$

In Eqs. (2.44)–(2.46), the three summation terms in the denominator of the signal-to-noise ratio denote the scattered particle noise, relative intensity noise, and thermal noise, respectively; $\alpha = \eta e/hv$ is the detector responsivity; η is the detector quantum efficiency; e is the electron-charge constant; h is the Planck constant; v is the optical frequency; $c = \lambda v$ is the speed of the beam; and λ is the wavelength. β is the amplifier-circuit gain factor, R is the output-load impedance, P_S is the signal optical power, P_{LO} is the fundamental optical power, I_d is the detector dark current, B is the detector effective bandwidth, k is the detector-response agreement, R_{IN} is the laser RIN factor, k_B is the Boltzmann constant, and T is the absolute temperature.

When SNR = 0 dB, i.e., the power of the signal is equal to the power of the noise, the value of the power detected by the detector is the noise-equivalent power. The minimum power required by the signal beam is called the detection sensitivity of the coherent optical communication system (i.e., the coherent detection limit). The system bit-error rate (BER) is obtained using various methods [25].

$$BER_S = \frac{1}{2} erfc\left(\frac{SNR_S}{4}\right) \tag{2.47}$$

$$BER_B = \frac{1}{2} erfc(SNR_B) \tag{2.48}$$

$$BER_H = \frac{1}{2} erfc(SNR_H). \tag{2.49}$$

2.2.5 *Polarization Control*

The polarization of an optical wave is the shift in the vibration direction of the electric vector relative to the vertical axial position in the transmission direction. Beam waves are electromagnetic waves, and the polarization of the beam is the polarization of the electromagnetic wave. The polarization of the signal beam and the polarization of the incident port of the optical mixer are aligned to match.

The polarization state of the laser beam changes randomly after an atmospheric transmission and cannot be effectively matched to the polarization state at the input port of the mixer; therefore, the polarization state of the beam signal must be adjusted and controlled before entering the mixer [26, 27].

The working principle of the polarization-preserving fiber squeeze-type polarization controller is shown in Fig. 2.10. Two optical axes, called fast and slow axes, exist in the perpendicular fiber direction. By applying a force of magnitude F in the slow-axis direction, the phase time delay between them changes the polarization state. The squeeze force F applied to the cross-section of the polarization-preserving fiber and the birefringent phase difference γ are related as follows [28]:

$$\gamma = \frac{KF}{d}. \tag{2.50}$$

In Eq. (2.50), d is the core diameter and $K = 9.5 \times 10^{-5}$ rad/m is a constant. When the fiber is subjected to pressure F at an angle θ in the x-axis direction, the corresponding Jones matrix can be expressed as [29]

$$J(\theta, \gamma) = \begin{bmatrix} \cos\left(\frac{\gamma}{2}\right) + i\cos(2\theta)\sin\left(\frac{\gamma}{2}\right) & i\sin(2\theta)\sin\left(\frac{\gamma}{2}\right) \\ i\sin(2\theta)\sin\left(\frac{\gamma}{2}\right) & \cos\left(\frac{\gamma}{2}\right) - i\cos(2\theta)\sin\left(\frac{\gamma}{2}\right) \end{bmatrix}. \tag{2.51}$$

Figure 2.11 illustrates the internal structure of the polarization controller, which consists of three fiber-optic squeezers, each independent of the others and each at 45° to the others.

As shown in Fig. 2.12, by increasing the pressure on the first (F_1) or third (F_3) fiber squeezer, the polarization state rotates clockwise at the intersection of the YOZ plane and the Poincaré sphere. By increasing the second (F_2), the polarization state

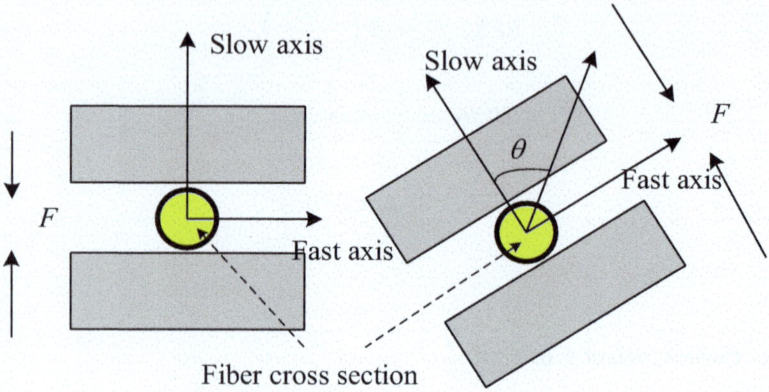

Fig. 2.10 Relationship between the pressure direction and the fast and slow axes

rotates clockwise at the intersection of the XOZ plane and the Poincaré sphere. Thus, by separately setting the reference values of F_1, F_2, and F_3 and making any combination, any change in the polarization state can be achieved, thus spreading over the entire Poincaré sphere [30].

Let the input and output polarization-state Jones vectors of the polarization controller be E_{in} and E_{out}, respectively. Then, the model of the signal beam field through the polarization controller can be expressed as

$$E_{out} = J(0, \gamma_3)J\left(\frac{\pi}{4}, \gamma_2\right)J(0, \gamma_1)E_{in}, \tag{2.52}$$

such that the squeezing principle of the polarization controller in Eq. (2.52) can be expressed using the Mueller matrix [31]:

$$M = \begin{bmatrix} \cos^2\left(\frac{\gamma}{2}\right) + \sin^2\left(\frac{\gamma}{2}\right)\cos(4\theta) & \sin^2\left(\frac{\gamma}{2}\right)\sin(4\theta) & -\sin(\gamma)\sin(2\theta) \\ \sin^2\left(\frac{\gamma}{2}\right)\sin(4\theta) & \cos^2\left(\frac{\gamma}{2}\right) - \sin^2\left(\frac{\gamma}{2}\right)\cos(4\theta) & \sin(\gamma)\cos(2\theta) \\ \sin(\gamma)\sin(2\theta) & -\sin(\gamma)\cos(2\theta) & \cos(\gamma) \end{bmatrix}. \tag{2.53}$$

In Eq. (2.53), γ is the birefringent phase difference formed by the force on the fiber extrusion, and θ is the angle between the force on the fiber and the x-axis. The Mueller matrices for 0° and 45° fiber extruders are, respectively,

$$M(0, \gamma) = \begin{bmatrix} 1 & 0 & 0 \\ 0 & \cos(\gamma) & \sin(\gamma) \\ 0 & -\sin(\gamma) & \cos(\gamma) \end{bmatrix} M\left(\frac{\pi}{4}, \gamma\right) = \begin{bmatrix} \cos(\gamma) & 0 & -\sin(\gamma) \\ 0 & 1 & 0 \\ \sin(\gamma) & 0 & \cos(\gamma) \end{bmatrix}. \tag{2.54}$$

Stokes vectors S_{in} of the input signal beam and S_{out} of the output signal beam are related as follows:

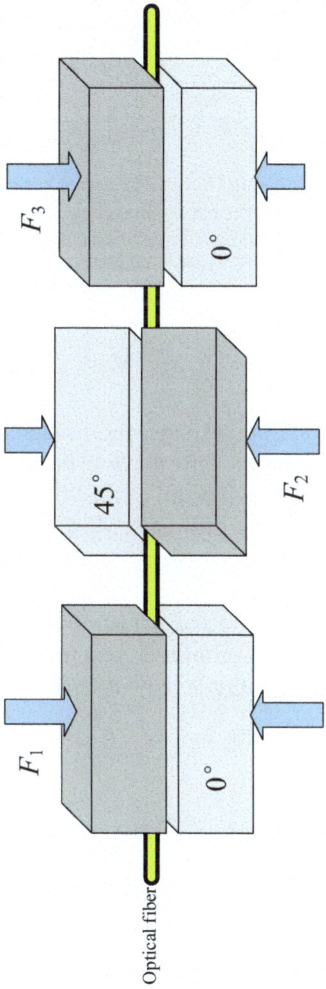

Fig. 2.11 Polarization-controller internal structure diagram

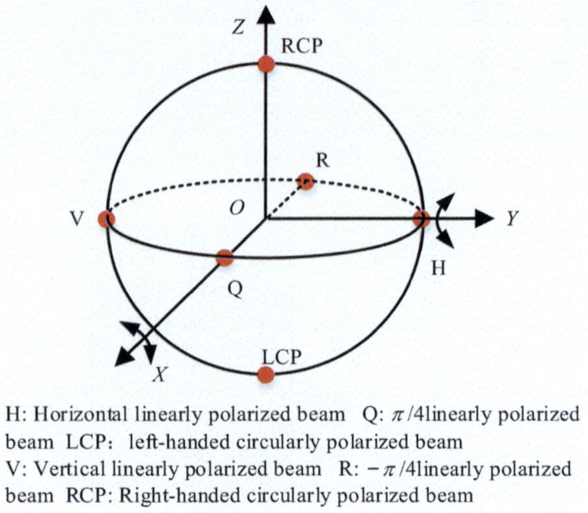

Fig. 2.12 Polarization-state changes on the Poincaré sphere

$$S_{out} = M(0, \gamma_3)M\left(\frac{\pi}{4}, \gamma_2\right)M(0, \gamma_1)S_{in}. \quad (2.55)$$

In coherent optical communication systems, the fundamental beam is a stable polarized beam, and the polarization state only needs to be manually adjusted to align with the incident port of the fundamental beam of the 90° optical mixer. Changes in the polarization state of the signal beam can directly affect the IF signal characteristics of the balanced-detector output. The polarization control structure using the IF signal as feedback is shown in Fig. 2.13 [32].

As shown in Fig. 2.13, the change in the polarization state of the signal beam can be reflected in the Q-way IF-signal amplitude output from Balanced Detector 2. The data processor uses the signal amplitude or peak-to-peak value as the evaluation

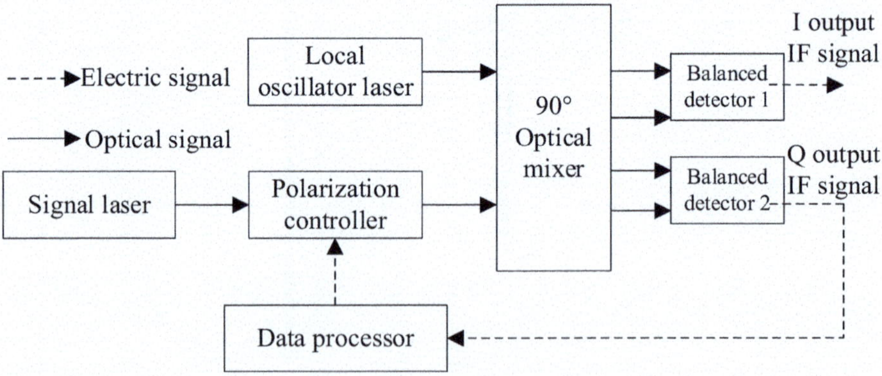

Fig. 2.13 Polarization control scheme with the intermediate-frequency signal as feedback

function to control the polarization controller by extracting the acquisition and using an intelligent optimization algorithm for iterative operations. Taking the particle-swarm algorithm as an example, the velocity and position iterative expressions are [32], respectively,

$$v_i^d(t+1) = w \cdot v_i^d(t) + c_1 \cdot r_1\left[p_{best}(t) - x_i^d(t)\right] + c_2 \cdot r_2\left[g_{best}(t) - x_i^d(t)\right] \quad (2.56)$$

$$x_i^d(t+1) = x_i^d(t) + v_i^d(t+1). \quad (2.57)$$

In Eqs. (2.56) to (2.57), t is the number of iterations of the algorithm, d is the dimension of the solution space, and i is the number of particles ($i = 1, 2, \ldots, N$). ${v_i}^d$ denotes the velocity of the ith particle in the dth dimension, ${x_i}^{\text{d}}$ denotes the corresponding position, w is the inertia weight, and c_1 and c_2 are the cognitive and social parameters, respectively. p_{best} denotes the position of the ith particle corresponding to the individual optimum obtained in the traversal, g_{best} denotes the global optimum obtained by all particles in the traversal process, and r_1 and r_2 denote two mutually independent random numbers following a (0, 1) uniform distribution.

The particles are equated to the applied voltage of the polarization controller, thereby controlling the signal-beam polarization state. This method can effectively reduce the coupling loss between devices, compared with schemes that use scramblers and polarization beam splitters [32].

2.2.6 *Beam-Source Frequency Control*

Optical wireless coherent communication for IF-signal demodulation processing requires the IF-signal frequency to be relatively stable. The signal beam and local-oscillation laser are affected by external vibrations, temperature changes, heat-dissipation conditions, and other factors. Moreover, both produce simultaneous random variations, causing the frequency difference to also show simultaneous random variations. The inverse of the ratio of the average value of the laser frequency f_{mea} to the frequency variation of $\Delta f(\tau)$ over time is called ${S_v}^{-1}(\tau)$. This represents the frequency stability of the laser,

$$S_v^{-1}(\tau) = \frac{\Delta f(\tau)}{f_{mea}}. \quad (2.58)$$

Considering that the frequency of the IF signal varies with time, two IF signals i_{I} and i_{Q} can be expressed as [33]

$$i_I(t) = 2 \cdot \beta \cdot \alpha \cdot A_s \cdot A_{lo} \cdot \cos\left[[\omega_s - \omega_{lo} \pm [\Delta\omega_s(t) + \Delta\omega_{lo}(t)]]t + \varphi_0 + \pi\frac{V(t)}{V_\pi} + \varphi_{lo}\right] \quad (2.59)$$

$$i_Q(t) = 2 \cdot \beta \cdot \alpha \cdot A_s \cdot A_{lo} \cdot \sin\left[[\omega_s - \omega_{lo} \pm [\Delta\omega_s(t) + \Delta\omega_{lo}(t)]]t + \varphi_0 + \pi \frac{V(t)}{V_\pi} + \varphi_{lo} \right]. \tag{2.60}$$

In Eqs. (2.59) and (2.60), the source frequencies of both the signal beam and the local-oscillation beam produce random variations over time, i.e., $\Delta\omega_s(t)$ and $\Delta\omega_{lo}(t)$. The inherent difference-frequency terms between the signal beam and the local-oscillation beam, $\omega_s - \omega_{lo}$, and the frequency-drift error term, $\Delta\omega_s(t) + \Delta\omega_{lo}(t)$, generated by the signal beam and the local-oscillator beam, are in the same order of magnitude. $\Delta\omega_s(t) + \Delta\omega_{lo}(t)$ will drift randomly over a range of time, resulting in a carrier-frequency value, $\omega_s - \omega_{lo} \pm [\Delta\omega_s(t) + \Delta\omega_{lo}(t)]$, over time.

The random frequency drift of the IF signal then increases the difficulty of subsequent demodulation processing and requires automatic frequency control to facilitate subsequent demodulation processing.

Laser frequencies are typically regulated using temperature-compensation and piezoelectric control methods. Temperature compensation is a coarse adjustment in laser-frequency regulation, with a large adjustment range, slow response rate, and low adjustment sensitivity; it is not suitable for practical coherent optical communication systems. The piezoelectric control method does fine adjustments in laser-frequency regulation systems, where the piezoelectric ceramic is a device capable of converting mechanical and electrical energy to each other.

The shape of the piezoelectric ceramic in the laser cavity changes, owing to the changes in the external voltage, thus changing the length of the cavity within the laser as well as the output wavelength. Owing to its fast response rate and high regulation sensitivity, piezoelectric control is used in coherent optical communication systems to regulate the frequency.

Figure 2.14 shows the frequency control system of coherent optical communication using an external frequency discriminator. Using the frequency value of the IF signal as feedback, the frequency discriminator extracts the frequency information and converts it into an electrical signal. It then uses an embedded system to convert from analog to digital. After the verdict and digital-to-analog conversion amplification, the reference voltage is adjusted up and down to change the cavity length to drive the internal piezoelectric ceramic of the local-oscillation laser [34]. Thus, the frequency is controlled by varying the output wavelength by adjusting the reference voltage up or down [35].

In an actual system, both the signal laser and the local-oscillator laser are selected as narrow-linewidth lasers with a KHz linewidth to enable them to produce beat-frequency interference [36]; the local-oscillator laser also has a voltage-controlled tunable function. In this system, the bandwidth of the balanced receiver is 200 MHz, and the required IF-signal frequency value is assumed to be stable at 100 MHz, corresponding to a wavelength difference of 0.007 nm. To maintain a constant frequency difference between the signal beam and the local-oscillation beam, the voltage applied to the voltage-controlled oscillator of the local-oscillation laser is V_{base}:

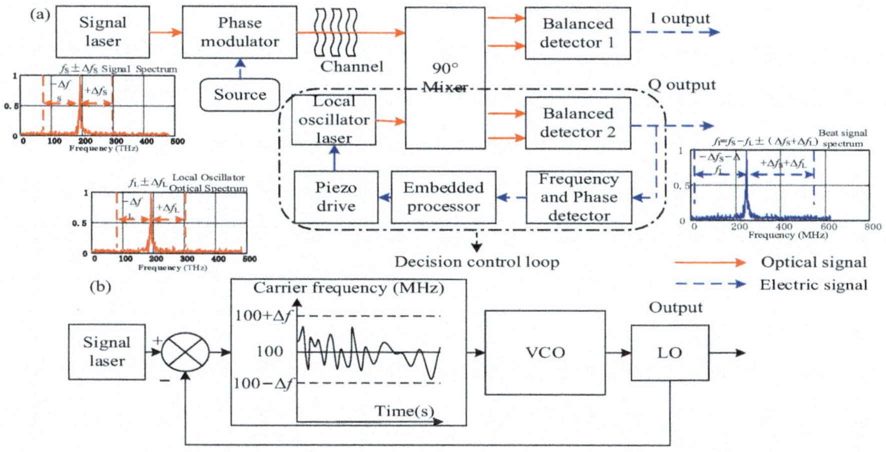

Fig. 2.14 Frequency control system. **a** Structure diagram, **b** schematic diagram

$$f_{lo} - f_s = V_{base} \cdot \rho = 0.007 \text{ nm}. \tag{2.61}$$

In Eq. (2.61), f_s is the signal optical frequency, f_{lo} is the local-oscillation optical frequency, and ρ is the voltage-controlled oscillation factor. The frequency f_{IF} of the IF signal after stabilization by regulation can be expressed as

$$\begin{aligned} f_{IF} &= f_s - \left[f_{lo} + \rho(V_{base} + \Delta V)\right] \pm (\Delta f_s + \Delta f_{lo}) \\ &= f_s - \left[f_{lo} + \rho \cdot k \cdot \left(\tfrac{V_{base}}{k} + \Delta V_{STM32}\right)\right] \pm (\Delta f_s + \Delta f_{lo}). \end{aligned} \tag{2.62}$$

In Eq. (2.62), k is the amplification of the piezoelectric-ceramic drive-voltage amplifier; it determines the adjustable range of the frequency compensation. ΔV is the output-voltage variation, ΔV_{STM32} is the embedded-processor output-voltage variation, Δf_s is the signal optical frequency drift, and Δf_{lo} is the local-oscillation optical frequency drift. The value of f_{IF} can be made relatively constant by changing ΔV_{STM32} to compensate for $\Delta f_s + \Delta f_{lo}$, such that the local-oscillator laser tracks the signal laser frequency.

2.2.7 *Signal Demodulation*

In a coherent optical communication demodulation system, the received signal can only be coherently demodulated if the locally generated carrier frequency and phase are synchronized with the frequency and phase of the received IF signal. Figure 2.15 shows the implementation of a Fourier transform–based carrier-frequency estimation algorithm to estimate the frequency of the IF signal [37].

A squared transformation of $i_I(t)$ in Eq. (2.8) yields

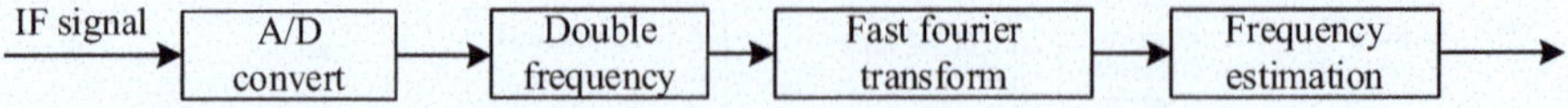

Fig. 2.15 Frequency-estimation schematic diagram

$$i_I^2(t) = 2(\beta \cdot \eta \cdot A_s \cdot A_{lo})^2 \cdot \cos\left(2(\omega_s - \omega_{lo})t + 2\varphi_0 + 2\pi \frac{V(t)}{V_\pi} + 2\varphi_{lo} + 1\right). \tag{2.63}$$

The two-fold frequency $2(\omega_s - \omega_{lo})$ information is obtained by performing a Fourier transform on $i_I{}^2(t)$.

After obtaining the two-fold frequency information of the carrier frequency, the coherent carrier of the signal can be extracted using a phase-locked loop. The phase-locked loop is a feedback-control system that enables phase tracking of the input signal, thus synchronizing the input signal with the output signal.

The phase-locked loop consists of three main components: a phase discriminator, loop filter, and voltage-controlled oscillator. The frequency and phase information of the carrier signal are recovered by the coherent demodulation of a differential binary phase-shift keyed IF signal using a Costas loop, as shown in Fig. 2.16, where the input-tuned signal is set to suppress the carrier's double-sideband signal.

$$u(t) = m(t)\cos(\omega_c t + \varphi) \tag{2.64}$$

In Eq. (2.64), $\omega_c = \omega_s - \omega_{lo}$ and $\varphi = \varphi_0 + \pi \bullet V(t) / V_\pi + \varphi_{lo}$. The phase jump caused by $V(t)$ can be expressed in terms of $m(t)$. $m(t)$ is the source signal and $m(t) = \pm 1$. The output of the voltage-controlled oscillator is

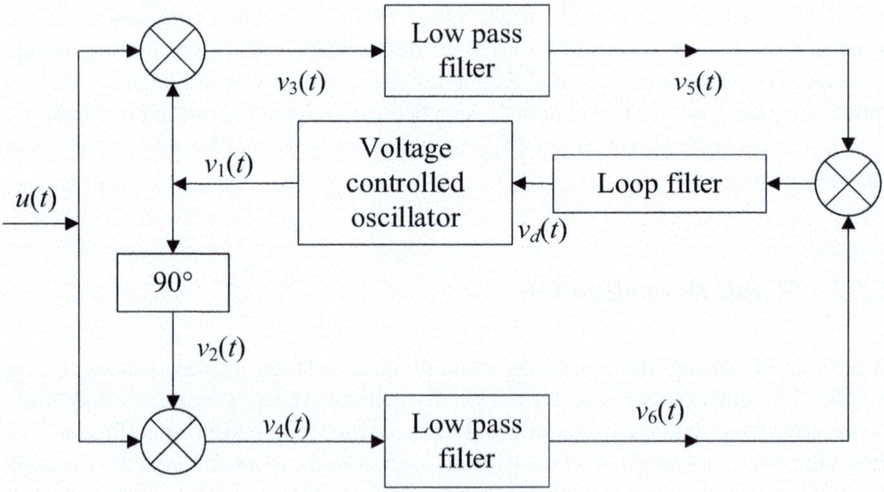

Fig. 2.16 Costas phase-locked loop schematic diagram

$$v_1(t) = \cos\left(\omega_c t + \varphi'\right) \tag{2.65}$$

$$v_2(t) = \sin\left(\omega_c t + \varphi'\right). \tag{2.66}$$

Then, the phase difference between the local carrier and the input carrier is

$$\Delta\varphi = \varphi' - \varphi. \tag{2.67}$$

The output after the multiplier is

$$v_3(t) = 0.5m(t)\left[\cos\Delta\varphi + \cos\left(2\omega_c t + \varphi + \varphi'\right)\right] \tag{2.68}$$

$$v_4(t) = 0.5m(t)\left[\sin\Delta\varphi + \sin\left(2\omega_c t + \varphi + \varphi'\right)\right]. \tag{2.69}$$

After a low-pass filter,

$$v_5(t) = 0.5m(t)\cos\Delta\varphi \tag{2.70}$$

$$v_6(t) = 0.5m(t)\sin\Delta\varphi. \tag{2.71}$$

Finally, the error signal $v_d(t)$ is obtained by passing through the loop filter:

$$v_d(t) = 0.125\sin 2\Delta\varphi. \tag{2.72}$$

When the loop stabilizes, $\Delta\varphi = 0$. Solving Eq. (2.70) yields $v_5(t) = 0.5m(t)$, confirming that the output of the in-phase branch is the coherently demodulated baseband signal.

2.3 Key Factors Affecting the Performance of Coherent Optical Communications

2.3.1 Effect of the Modulation Drive Characteristics on the Constellation Diagram and BER

Using Monte Carlo simulation, according to Eq. (2.2) and Fig. 2.1, when the modulation method uses binary phase-shift keying, different voltage values $V(t) = 2.5$ V, 3.5 V, and 4.5 V are used to drive the modulator. The constellation diagram of the electrical signal via IF is obtained as shown in Fig. 2.17.

As the drive-signal voltage increases, the constellation points in Fig. 2.17 drift in a circular direction proportional to the drive voltage. That is, when the drive voltage

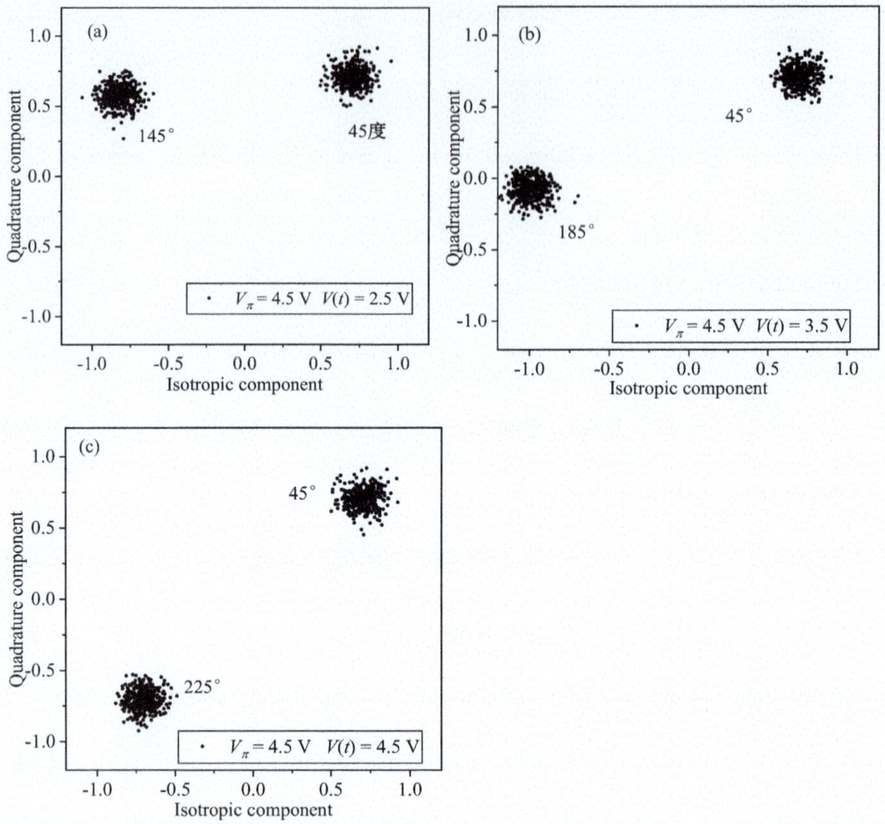

Fig. 2.17 Signal-constellation diagram of a phase modulator driven by different amplitude voltages. **a** $V(t) = 2.5$ V, **b** $V(t) = 3.5$ V, **c** $V(t) = 4.5$ V

is $V(t) = 4.5$ V, the two phase points produce an angular difference of 180° in the constellation diagram. As the drive voltage decreases, the angular difference between the two phase points also decreases proportionally.

Using Monte Carlo simulation, the BER trend of the coherent-optical communication system is calculated with the modulator drive voltage under different SNR scenarios, based on Eqs. (2.2) and (2.10), as shown in Fig. 2.18. When the drive voltage $V(t)$ is equal to the half-wave voltage V_π, the BER of the system is very small, and increases as the drive voltage increases or decreases.

When the drive voltage was much higher than the half-wave voltage, the $LiNbO_3$ crystal inside the modulator was permanently damaged. This indicates that optimum modulation is achieved when the modulator drive circuit outputs a voltage equal to the modulator's half-wave voltage. Driving voltages that are too high or too low affect the system's communication performance.

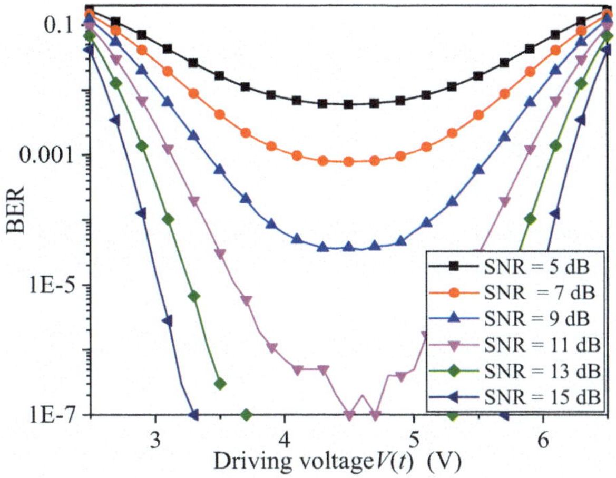

Fig. 2.18 Influence of different amplitude driving voltages on the system bit-error rate

When considering the amplitude response of the modulating driver circuit, the frequency response should also be considered, and a suitable passband should be selected to enable distortion-free signal amplification.

Let the unit-shock response of the modulation drive circuit in Fig. 2.1 be $h(t)$, and take a first-order RC filter as an example; the equivalent circuit is shown in Fig. 2.19. When the filter is a low-pass equivalent circuit, the time-domain expressions for the continuous and discrete types can be written as

$$V(t) = V_0(t) - RC\frac{dV(t)}{dt} \tag{2.73}$$

$$V(n) = \frac{V_0(n) + \frac{RC}{T_s}V(n-1)}{1 + \frac{RC}{T_s}}. \tag{2.74}$$

In Eqs. (2.73)–(2.74), R, C, and T_s represent the resistance, capacitance, and sampling interval, respectively, of the first-order low-pass filter. Take the low-pass filter circuit in Fig. 2.19a as an example. Setting $R = 100\ \Omega$, $C = 7.5$ nF, and $T_s =$

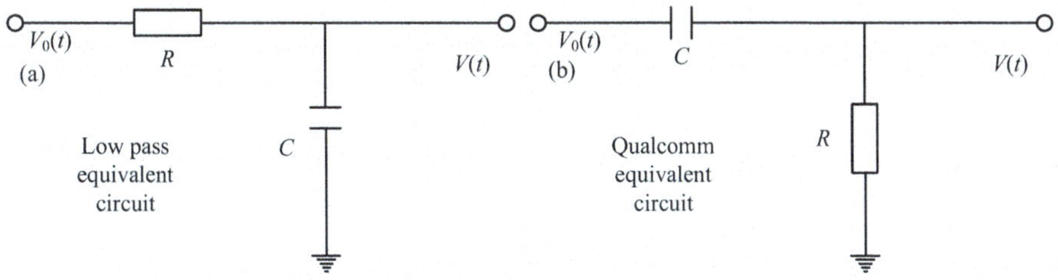

Fig. 2.19 Equivalent circuit with first-order filtering-characteristic amplifier. **a** Low-pass, **b** high-pass

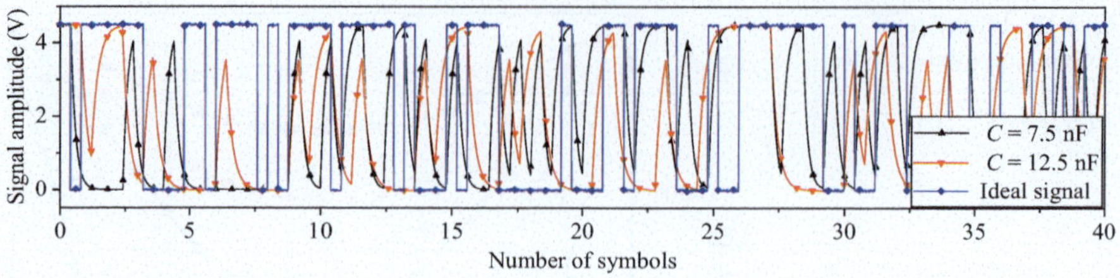

Fig. 2.20 Phase-modulator driving signal amplified by low-pass filtering characteristics

50 ns, as well as $R = 100\ \Omega$, $C = 12.5$ nF, and $T_s = 50$ ns, the output waveform of the drive signal after low-pass filtering and amplification is obtained according to Eq. (2.73), as shown in Fig. 2.20. From Fig. 2.20, it can be seen that the high and low levels of the square-wave signal cannot be sustained within a code cycle, and the squareness of the waveform is obviously distorted.

Taking the high-pass filter circuit shown in Fig. 2.18b as an example, when the response frequency of the high-pass filter is greater than the frequency of the input signal, its continuous and discrete time-domain expressions are as follows:

$$V(t) = C\frac{d[V_0(t) - V(t)]}{dt} \cdot R \tag{2.75}$$

$$V(n) = [V_0(n) - V_0(n-1) + V(n-1)] \cdot \frac{RC}{RC + T_s}. \tag{2.76}$$

Setting $R = 100\ \Omega$, $C = 50$ nF, and $T_s = 50$ ns, as well as $R = 100\ \Omega$, $C = 100$ nF, and $T_s = 50$ ns, the output waveform of the drive signal after high-pass filtering and amplification is shown in Fig. 2.21. At this point, the output waveform could not maintain the full level value within one code cycle. As the number of zeros and ones in the code element increased, the level dropped more obviously, and the amplified signal produced a significant DC drift characteristic.

Monte Carlo simulations were used to calculate the constellation-diagram variation of the output drive-signal waveform when the response frequency of the drive circuit was below and above the actual input signal, as shown in Fig. 2.22.

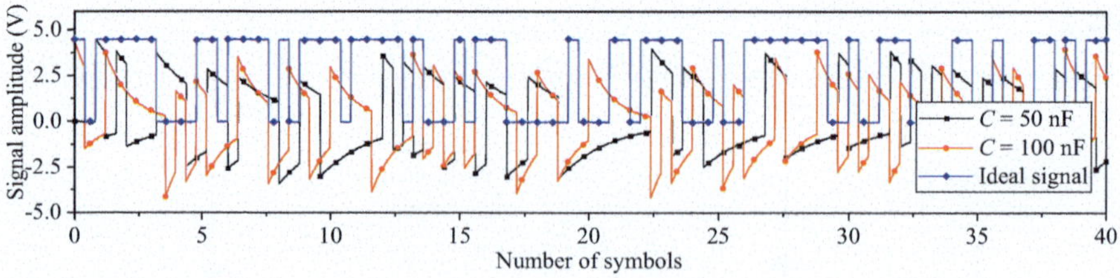

Fig. 2.21 Phase-modulator driving signal amplified by high-pass filtering characteristics

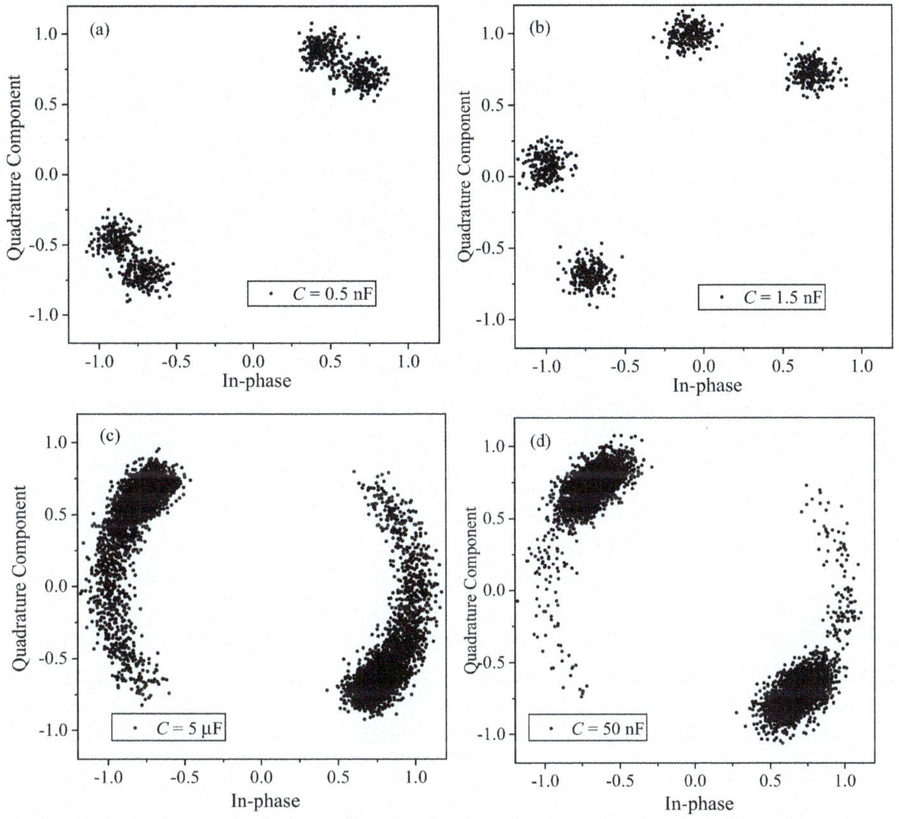

Fig. 2.22 Signal constellation of a voltage-driven phase modulator under different filtering conditions. **a** Low-pass (LP) $C = 0.5$ nF, **b** LP $C = 1.5$ nF, **c** high-pass (HP) $C = 5$ μF, **d** HP $C = 50$ nF

When the signal frequency is greater than the response frequency of the driver circuit, the constellation points of the signal exhibit a circumferential copy spread at the modulation position. The rotation angle of the circumferential copy spread becomes more pronounced as the passband cutoff frequency of the low-pass circuit decreases. This is due to the irregularity of the driving waveform at the rising and falling edges, which leads to the copy spread of the constellation diagram and results in code-element misinterpretation at the receiver.

When the signal frequency is less than the frequency response of the driver circuit, the constellation diagram shows a circumferential "tailing" of the signal at the position of the modulation point, with most of the modulation points located at the tail end of the "tailing." As the passband cut-off frequency of the high-pass circuit increases, the "tailing" phenomenon becomes more pronounced. The drive signal after the high-pass circuit produces a DC component drift that causes the constellation diagram to

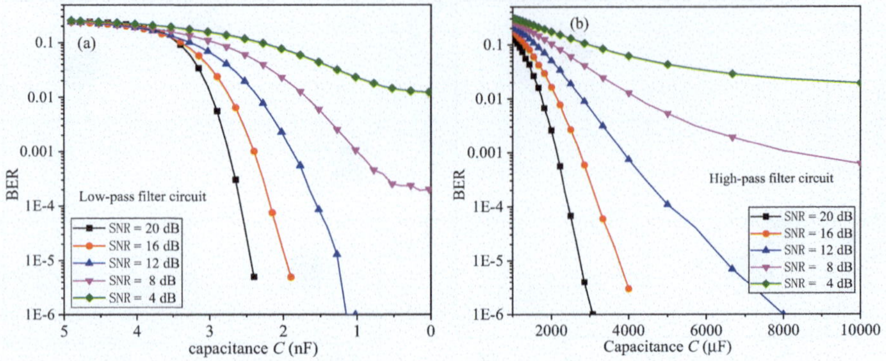

Fig. 2.23 Influence of the driving circuit passband characteristics on the system bit-error rate. **a** Low-pass characteristic, **b** high-pass characteristic

rotate. The relative drive voltage is insufficient to cause a non-uniform distribution of the "tailing" constellation points.

Monte Carlo simulations were used to calculate the variation in the BER of the system with capacitance C based on Eq. (2.10) for each of the two cases, with different passband characteristics of the driver circuit. As shown in Fig. 2.23, the best-quality output drive signal is achieved when the capacitance value is reduced in the low-pass circuit or increased in the high-pass circuit, such that the drive circuit has a passband that matches the input signal, and the drive voltage is equal to the half-wave voltage.

2.3.2 *Effect of Changes in the Signal-Beam Polarization State on the Amplitude of the iF Signal*

We assume that the polarization state of the input signal beam is (0.5, − 0.49, 0.71). Figure 2.24 shows the corresponding normalized amplitude change of the IF signal, when keeping the third waveplate in the polarization controller stationary and changing the first and second waveplates, as well as keeping the first waveplate stationary and changing the second and third waveplates.

As shown in Fig. 2.24, there are multiple locally optimal solutions for the amplitude of the IF signal; therefore, the three waveplate phase variations γ_1, γ_2, and γ_3 and the normalized amplitude of the IF signal (AIF) can be expressed using the following multi-input cost function:

$$\max A_{IF} = f(\gamma_1, \gamma_2, \gamma_3). \tag{2.77}$$

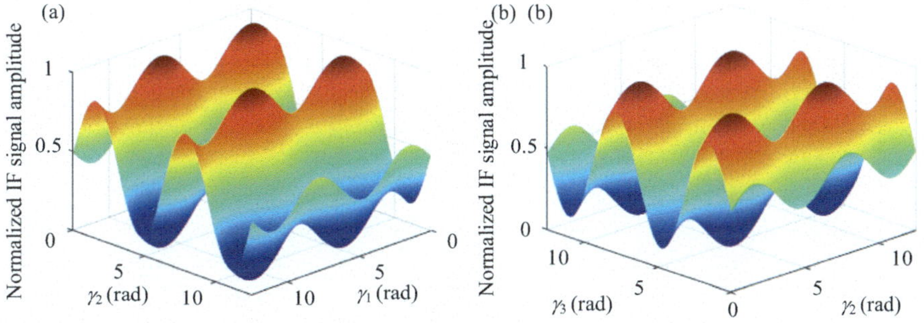

Fig. 2.24 Relationship between the phase change of the polarization controller waveplate and the amplitude of the IF signal. **a** First and second waveplates, **b** second and third waveplates

2.3.3 Effect of a Wavefront Mismatch Between the Signal Beam and Local-Oscillation Beam on the Mixing Efficiency

Based on the definition of the mixing efficiency, Eqs. (2.39) and (2.40), it can be seen that the mixing efficiency of a coherent optical communication system is significantly affected when the phase of the signal beam does not match the phase of the local-oscillation beam, i.e., $\varphi_s \neq \varphi_{lo}$.

The wavefront phase can usually be expanded using a set of orthogonal Zernike basis functions; based on Eqs. (2.39) and (2.40), the following values are set: electron power $e = 1.6 \times 10^{-19}$ C, quantum efficiency $\eta = 0.6$, Planck's constant $h = 6.63 \times 10^{-34}$ J·s, signal-beam wavelength $\lambda = 1550$ nm, and detector bandwidth $B = 40$ GHz. The Zernike coefficient is in the range of 0 μm–10 μm, detector responsivity $R = 0.6$ A/W, optical frequency $v = 1.94 \times 10^{14}$ Hz, and signal-beam amplitude $A_{lo} = 10^3 A_s$. The wavefront phase in the Zernike mode of orders 1, 3, 5, 7, and 9 is used to calculate the relationship between the Zernike coefficient and the system mixing efficiency and signal-to-noise ratio. The effect of the change in the Zernike coefficient on the mixing efficiency of the system is shown in Fig. 2.25.

As can be seen from the results in Fig. 2.25, as the value of the Zernike coefficient increases at each order, there is no significant difference in the trend of the mixing efficiency, regardless of whether zero-difference or outlier detection is used.

This is because the integral value of the cosine term of the wavefront phase difference between the signal beam and the local-oscillation beam is much larger than the integral value of the sine term. The second term of the numerator in Eq. (2.40) can be omitted when the wavefront aberration is relatively small. The mixing efficiency decreases as the distortion increases for a given Zernike order. The third-order defocus term has the most significant effect compared to the other orders, and the same number of higher-order aberrations has less effect on the mixing efficiency than lower-order aberrations.

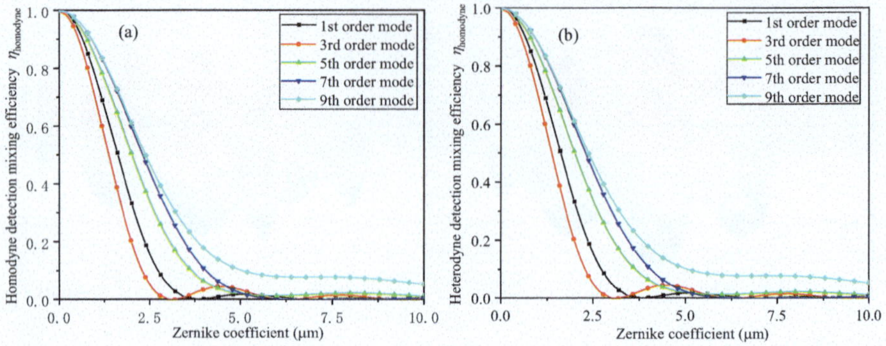

Fig. 2.25 Influence of the Zernike-coefficient change on the mixing efficiency. **a** Homodyne detection, **b** heterodyne detection

2.3.4 Effect of the Random Drift of the IF Signal Frequency on the Constellation Diagram and BER

The IF signal in Eq. (2.59), which is affected by the frequency drift, is demodulated to obtain the baseband signal, as in Eq. (2.78):

$$i_d(t) = \beta \cdot \alpha \cdot A_s \cdot A_{lo} \cdot \cos\left[[\Delta\omega_s(t) + \Delta\omega_{lo}(t)]t + \varphi_0 + \pi \frac{V(t)}{V_\pi} + \varphi_{lo} \right]. \quad (2.78)$$

Figure 2.26 shows the corresponding constellation diagram, showing that the drift of the carrier frequency in the constellation diagram has a circular direction with respect to the position of the modulation point. The larger the frequency-drift range, the more obvious the dispersion of the constellation diagram.

For a coherent optical communication system using quadratic phase-shift keying, the signal constellation at the receiver is obtained by taking different values of the frequency drift, as shown in Fig. 2.27. The random drift of the IF signal-frequency pair in a quadratic phase-shift keying system will also cause the constellation to disperse circumferentially around the modulation point.

Taking different carrier-frequency drift values, $\Delta\omega_s(t) + \Delta\omega_{lo}(t)$, the BER of the BPSK and QPSK systems were calculated, and the results are shown in Fig. 2.28. For the same signal-to-noise ratio, the frequency drift caused the BER of the system to increase with increasing frequency-drift value. The phase difference between the modulation points of the QPSK system is smaller than that of the BPSK system. Moreover, the same amount of frequency-drift error has a more serious impact on the QPSK system.

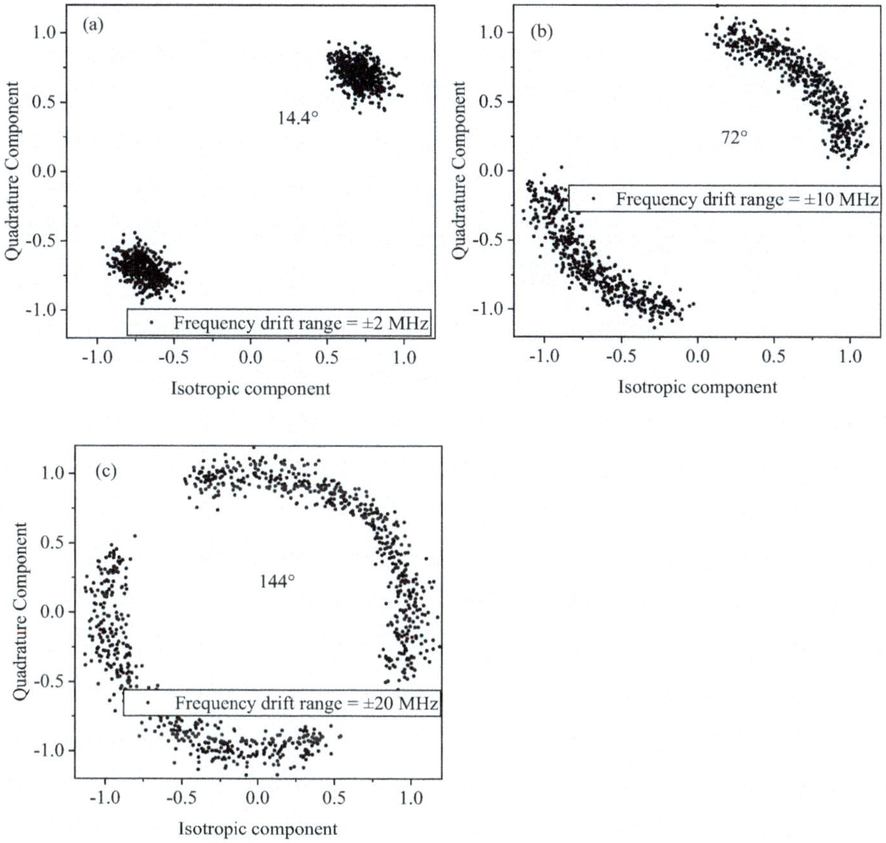

Fig. 2.26 BPSK system with SNR = 20 dB constellation. **a** ± 2 MHz, **b** ± 10 MHz, **c** ± 20 MHz

2.3.5 Effect of the Signal-Transmission Bandwidth and Signal Optical Power on the Detection Sensitivity and BER

Based on Eqs. (2.44) to (2.49), we use the following values: photodetector quantum efficiency $\eta = 0.8$, electron-charge constant $e = 1.6 \times 10^{-19}$ C, Planck constant $h = 6.63 \times 10^{-34}$ J·s, optical frequency $v = 193.4$ THz, wavelength $\lambda = 1550$ nm, and detector responsiveness $\alpha = 0.9982$ A/W. The internal gain coefficient of the amplifier circuit $\beta = 0.4$, output load impedance $R = 50\ \Omega$, signal optical power $P_S = 10$ nW, fundamental optical power $P_{LO} = 5$ mW, detector dark current $I_d = 5$ nA, detector effective bandwidth $B = 50$ GHz, detector response agreement $k = 0.975$, laser RIN-noise figure $R_{IN} = -135$ dB/Hz, Boltzmann constant $k_B = 1.38 \times 10^{-23}$ J/K, and absolute temperature $T = 290$ K. The effects of the detector bandwidth on

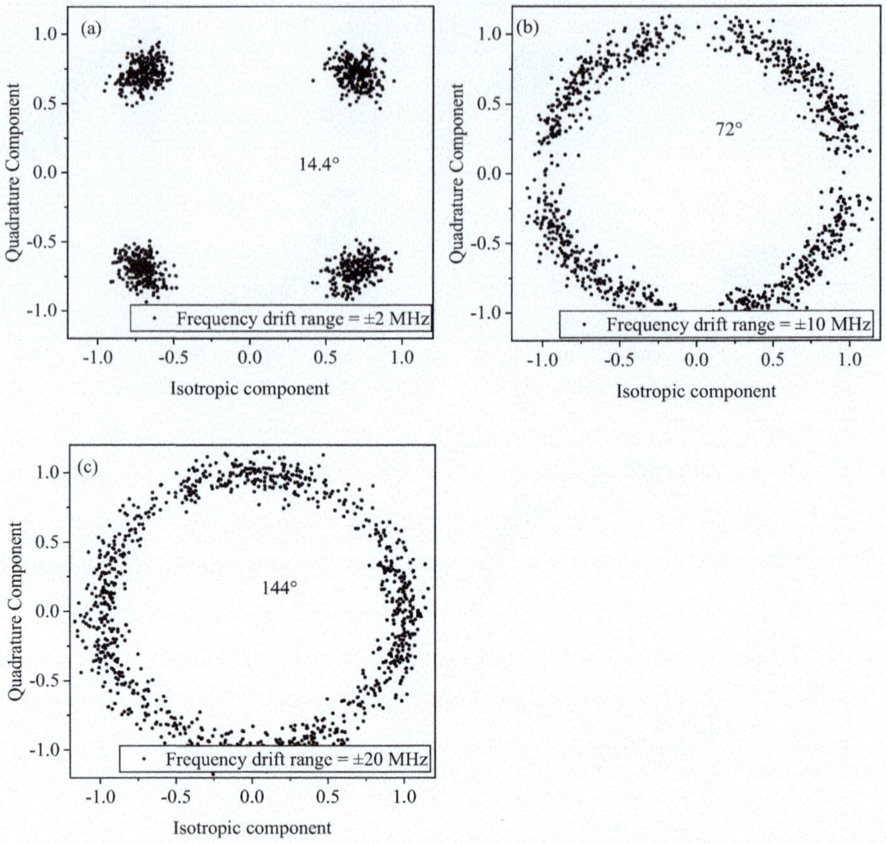

Fig. 2.27 QPSK system with SNR = 20 dB constellation. **a** ± 2 MHz, **b** ± 10 MHz, **c** ± 20 MHz

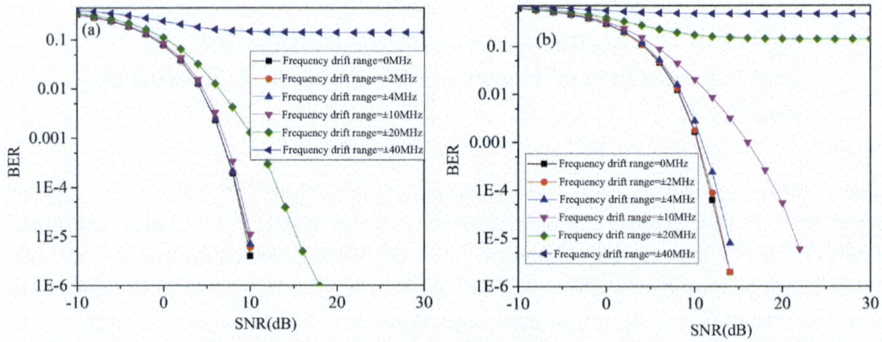

Fig. 2.28 Bit-error rate curve caused by carrier-frequency drift. **a** BPSK, **b** QPSK

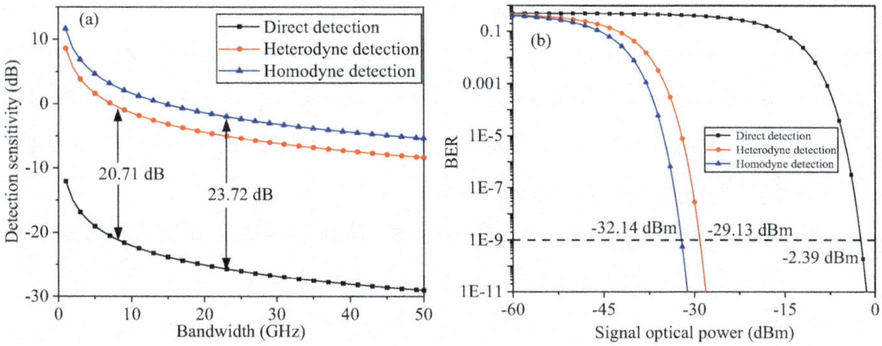

Fig. 2.29 Different modulation and detection methods. **a** Influence of the detector bandwidth on the detection sensitivity, **b** influence of the signal optical power on the BER

the detection sensitivity and the signal optical power on the BER were obtained for different modulation and detection methods.

From the calculation results in Fig. 2.29, it can be seen that the detection sensitivity of the system tends to decrease as the bandwidth of the detector increases. In addition, the detection sensitivity of outlier detection has a gain of 20.71 dB compared to direct detection, and the detection sensitivity of zero-difference detection has a gain of 23.72 dB compared to direct detection.

When the BER of the communication system is 10^{-9}, the signal optical power requires approximately 600 μW (−2.39 dBm) when using direct detection, whereas when using coherent detection, the optical power only requires approximately 1 μW (−32.14 dBm). This shows that coherent detection requires less received optical power for the same order of magnitude of BER tolerance and that the advantages of zero-difference detection over outlier detection are more obvious.

References

1. Kaushal H, Kaddoum G (2017) Optical communication in space: challenges and mitigation techniques. IEEE Commun Surveys and Tutor 19(1):57–96
2. Faruk MS, Savory SJ (2017) Digital signal processing for coherent transceivers employing multilevel formats. J lightwave Tech 35(5):1125–1141
3. Ghassemlooy Z, Arnon S, Uysal M et al (2015) Emerging optical wireless communications-advances and challenges. IEEE J Sel Areas Commun 33(9):1738–1749
4. Hui KW, Chiang KS, Wu B et al (1998) Electrode optimization for high-speed traveling-wave integratedoptic modulators. J Lightwave Technol 16(2):232–238
5. Wan L, Zhou Y, Liu L et al (2013) Realization of a free-space 2×4 90 degrees optical hybrid based on the birefringence and electro-optic effects of crystals. J Opt 15(3):035402
6. Peterson GE, Ballman AA, Lenzo PV et al (1964) Electro-optic properties of $LiNbO_3$. Appl Phys Lett 5(3):62–64
7. Burns WK, Howerton MM, Moeller RP et al (1998) Broad-band reflection traveling-wave $LiNbO_3$ modulator. IEEE Photonics Technol Lett 10(6):805–806

8. J Kondo A Kondo K Aoki et al (2002) 40-Gb/s X-Cut $LiNbO_3$ Optical modulator with Two-Step back-slot structure. J Lightwave Technol 20(12):2110–2114
9. JingYuan L, Jin Z, Heng Y (2019) Study on the influence of Lithium Niobate Phase Modulator on the performance of coherent optical communication system. J Hunan Univ Sci Eng 40(10):18–22
10. Li Y, Zhang Y, Huang Y (2013) Any bias point control technique for Mach-Zehnder modulator. IEEE Photonics Technol Lett 25(24):2412–2415
11. Chong Zheng H, Hong Zuo L, Quan S et al (2017) Stable bias control technique for any-point locking in mach-zehnder modulator. Acta Photonica Sinica, 46(10):1023002
12. Leeb WR, Winzer PJ, Kudielka KH (1998) Aperture dependence of the mixing efficiency, the signal-to-noise ratio, and the speckle number in coherent Lidar receiver. Appl Opt 37(15):3143–3148
13. Xiaofeng S, Xiang Z, Jinfa W et al (1986) Mode-field radius investigation on mono-mode optical fibers. J China Inst Commun 7(3):49–54
14. Xizheng K, Jing L, Sichen L (2018) Automatic alignment method and realization of space optical coupling. Infrared Laser Eng 47(1):103009–0103009(7)
15. Ke D, Zhong WB, Xu W et al (2007) Technology of coupling beam into single-mode fiber in free space optical communication system. J Univ Electron Sci Technol China 36(5):889–891
16. Noé R (2005) PLL-free synchronous QPSK polarization multiplex/diversity receiver concept with digital I&Q baseband processing. IEEE Photonics Technol Lett 17(4):887–889
17. Han Y, Li G (2005) Coherent optical communication using polarization multiple-input-multiple-output. Opt Express 13(19):7527–7534
18. Kikuchi K (2011) Performance analyses of polarization demultiplexing based on constant-modulus algorithm in digital coherent optical receivers. Opt Express 19(10):9868–9880
19. Lin Z, Yang Y, Xiang Q, Gu C, Yao Y (2021) Polarization demultiplexing algorithm for probabilistically shaped signals in coherent optical communication. Acta Optica Sinica, 41(6):0606002 (in Chinese)
20. Tan Z, Ke X (2018) Analysis of a heterodyne detection system affected by irradiance and phase fluctuations in slant atmospheric turbulence. Appl Opt 57(32):9596–9603
21. YunYun W, Er Hu C, Yu Z et al (2012) Experiment on adaptive optics to improve the performance of FSO. Opt Commun Technol 36(4):15–18
22. Ke X, Tan Z (2018) Effect of angle-of-arrival fluctuation on heterodyne detection in slant atmospheric turbulence. Appl Opt 57(5):1083–1090
23. Hongyang L, Yange Z, Yong A, Yonghong D, Jing C (2014) Design and implementation of balance detector used in coherent optical communication system. Laser and Optoelectronics Prog 51(7):070601 (in chinese)
24. Yuen HP, Chan VWS (1983) Noise in homodyne and heterodyne detection. Opt Lett 8(3):177–179
25. Shen B, Wang P, Polson R et al (2014) Ultra-high-efficiency metamaterial polarizer. Optica 1(5):356–360
26. Tingting M, Shoufeng T, Hang N, Xiaoyang J (2017) Effect of polarization property of signal light on mixing efficiency of space coherent detection. Laser and Optoelectronics Progress 54(2):020604 (in Chinese)
27. Tie L, Zheng KX et al (2012) Polarization control in the coherent optical detection system. Inf Laser Eng 41(11):3069–3074
28. Xi-Zheng K, Jiao W (2015) Comparison of polarization property of partially coherent beam between propagating along an uplink path and a downlink path in atmospheric turbulence. Acta Physica Sinica 64(22) 148–155
29. Zhang J, Ding S, Zhai H et al (2014) Theoretical and experimental studies of polarization fluctuations over atmospheric turbulent channels for wireless optical communication systems. Opt Express 22(26):32482–32488
30. Liu J, Li J, Bai C, Xu Y, Qian J, Wang Y (2020) Theory and method of fourier transform hyperspectral Mueller matrix imaging. Acta Optica Sinica 40(7):0711004 (in Chinese)

31. Bingbin M, Xizheng K, Ying Z (2019) Polarization control and control algorithm of beams in coherent optical communication system. Chinese J Lasers, 46(1):0106002 (in Chinese)
32. Zhenkun T, Xizheng K (2017) Mixing efficiency in coherent detection system. Laser & Optoelectronics Progress, 54(10):100605 (in Chinese)
33. Guoyu L, Kang Y, Sumei J, Yan L, Mingsheng L (2015) Linear tunable fiber ring laser based on closed-loop piezoelectric ceramics. Acta Optica Sinica 35(6):0614003 (in Chinese)
34. Yang S, Ke X (2018) Carrier frequency stability control in coherent optical communication. Laser & Optoelectronics Progress 55(4):040601 (in Chinese)
35. Xiu KY, Zheng KX, Yuan Y (2017) Influence of laser linewidth on the performance of space coherent optical communication system. Chinese J Sci Inst 38(7):1668–1674
36. Jiali W, Yaqi Y, Yi W (2019) Research on all-digital DPSK demodulation technology in coherent optical communication. J Hunan Univ Sci Eng 40(10):12–17
37. Haitao C, Huajun Y, Tuohui L et al (2011) Effect of fiber offset on spatial light-single mode fiber coupling efficiency. Laser and Infrared 41(1):75–78

Chapter 3
Factors Affecting the Sensitivity of Coherent Detection

In 1955, Forrester et al. used two beams of light with different wavelengths to interfere with and generate a new intermediate-frequency signal [1, 2], introducing coherent detection into the research arena.

Coherent-detection methods can be divided into heterodyne, homodyne, and intradyne detection. When the angular frequency of the signal-light center is the same as that of the local-oscillator (LO) light center, it is called homodyne detection. When the central angular frequency of the signal light differs from that of the LO light, it is called heterodyne detection. The signal light and local-oscillator light are interfered and mixed by a 90° mixer to obtain two orthogonal phase signals, and then the two signals are detected, which is called internal-difference detection.

The sensitivity of coherent detection refers to the limit of what a detector can detect under a certain signal-to-noise ratio or bit-error rate. Theoretically, the detection limit can reach one photon; however, in practical situations, the detection limit cannot be reached, owing to atmospheric turbulence, thermal noise in the detector, low mixing efficiency, and low detector sensitivity. Compared with a direct-detection system, the detection sensitivity of a coherent optical communication system can have a gain of approximately 20 dB [3]. It has many modulation methods [4], excellent selectivity [5], a large channel capacity [6], and other characteristics that can greatly improve the performance of optical wireless communication systems.

In subsequent research, domestic and foreign scholars analyzed the influence of thermal noise, shot noise, wavefront distortion, polarization misalignment, spot-size deviation, optical-axis deflection, coupling efficiency, detector performance, pre-optical amplification, and other aspects of the coherent-detection system on the coherent-detection limit. After years of research, it has become possible to use wavefront-correction systems and beam-polarization control technology. Numerous methods, such as fiber coupling and beam control, reducing noise interference, improving detector performance, and adding preamplifiers, have been used to improve the sensitivity of the detection system.

X. Ke and C. Ke, *Performance-Improvement Limits in Optical Wireless Coherent Communication*, Optical Wireless Communication Theory and Technology,
https://doi.org/10.1007/978-981-96-2000-5_3

The level of the coherent-detection limit directly affects the performance of a coherent communication system. As the communication distance increases, the signal intensity that can be detected by the receiving end weakens, and the sensitivity requirements for coherent detection increase. Therefore, it is necessary to conduct a detailed analysis of the factors affecting the coherent-detection limit, which is essential for improving the performance of coherent-detection systems.

This chapter introduces the research progress in coherent detection at home and abroad, as well as the research on detection limits by domestic and foreign scholars. The factors that affect the detection limit are analyzed, and the factors that domestic and foreign scholars have determined suppress these effects are summarized. Finally, the development direction of coherent-detection limits is summarized and discussed.

3.1 Direct and Coherent Detection

3.1.1 Direct Detection

Direct detection refers to directly reflecting incident signal light onto a photosensitive surface; the photodetector then outputs the corresponding current, based on the intensity of the signal light. As shown in Fig. 3.1, when the signal light reaches the receiving end, the aperture diaphragm and focusing lens converge the light signal onto the photosensitive surface of the photodetector, and the photodetector converts the light signal into a voltage or current signal. The direct-detection method has high reliability, easy implementation, and low cost; however, its detection sensitivity is too low to meet the high sensitivity requirements of long-distance optical wireless communication [9].

3.1.2 Coherent Detection

3.1.2.1 Coherent-Detection Principle

Coherent detection considers the interference between the local oscillator light and the signal light, using the changes in the light and dark fringes generated by the interference, as well as the intensity and displacement of the interference fringes, to determine the coherence and phase difference between the two coherent light beams. This principle is illustrated in Fig. 3.2.

The signal light $E_s(r)$ interferes with the local oscillator light $E_l(r)$ on the optical mixer and the interfered light wave is incident on the balanced detector. The balanced detector can detect the interference fringes, which form a differential signal. Coherent detection can be divided into zero-difference, internal-difference, and external-difference.

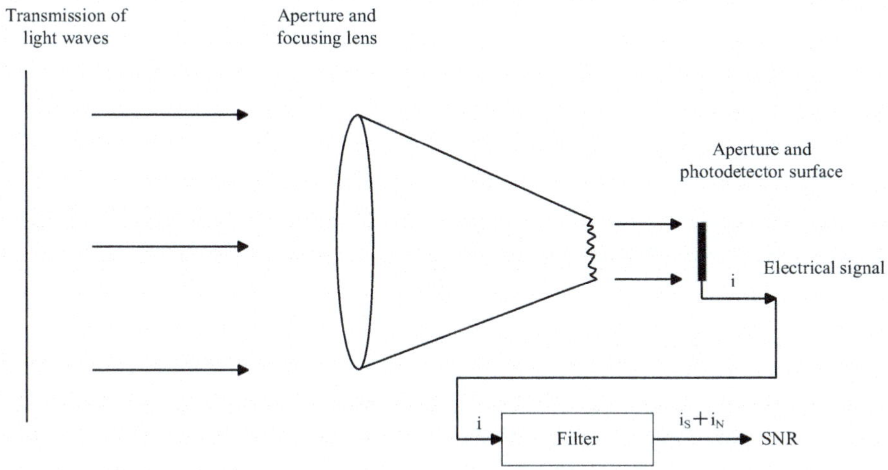

Fig. 3.1 Direct-detection system diagram model [9]

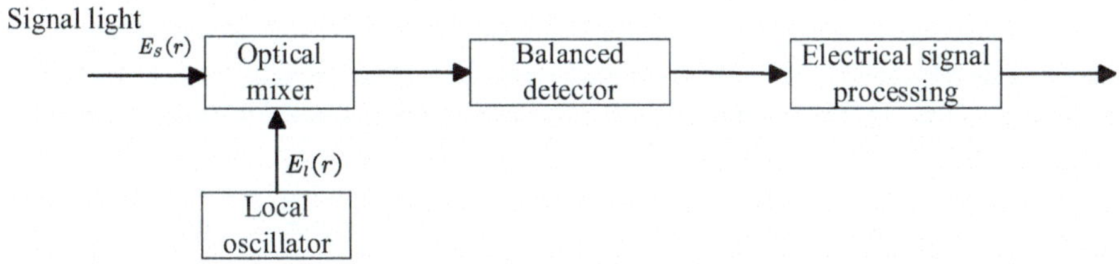

Fig. 3.2 Principle block diagram of a coherent-detection system [3]

Optical mixers are generally composed of fiber couplers and nonlinear optical crystals. A balanced detector detects weak optical signals and converts them into electrical signals, from which, the electrical signal-processing module can recover the baseband signal after electrical signal processing.

3.1.2.2 Zero-Difference Detection

When the frequencies of the local-oscillator and signal lights are the same, it is called homodyne detection. The principle block diagram of zero-difference detection is shown in Fig. 3.3. The signal light and local-oscillator light become electrical signals after being demodulated through an optical phase-locked loop. The electrical signal can be obtained as a baseband signal after passing through a low-pass filter, and the original signal can be obtained from the decision output.

The optical phase-locked loop can stabilize the frequency and phase of the local oscillator and signal light [10, 11]; however, it increases the complexity of the receiver. The advantage of homodyne detection is high sensitivity.

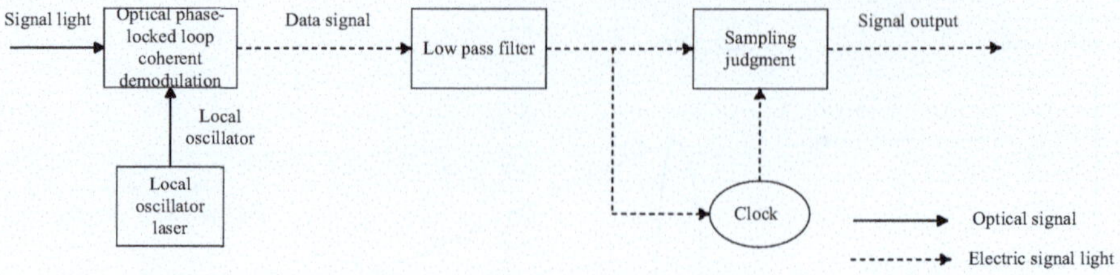

Fig. 3.3 Demodulation-principle block diagram of a zero-difference detection system

3.1.2.3 Heterodyne Detection

When the frequencies of the local-oscillator light and signal light are not equal, the coherent-detection method is called heterodyne detection. It can be divided into synchronous and asynchronous demodulation, based on the demodulation method at the receiving end.

A schematic diagram of synchronous demodulation is shown in Fig. 3.4a. The signal light and local-oscillator light form an intermediate signal after passing through the optical mixer. The signal then passes through the photodetector and becomes an electrical signal. After carrier-frequency recovery, it is multiplied by the intermediate-frequency signal. The signal output of the multiplier is then passed through a low-pass filter to obtain the baseband signal.

The principle of asynchronous demodulation is shown in Fig. 3.4b. After passing through a mixer and photodetector, the signal light and local-oscillator light pass directly through a bandpass filter and then undergo envelope detection and other processing to obtain the baseband signal directly. Although this method is relatively simple, it has strict requirements for the phase and frequency of the local-oscillator and signal lights.

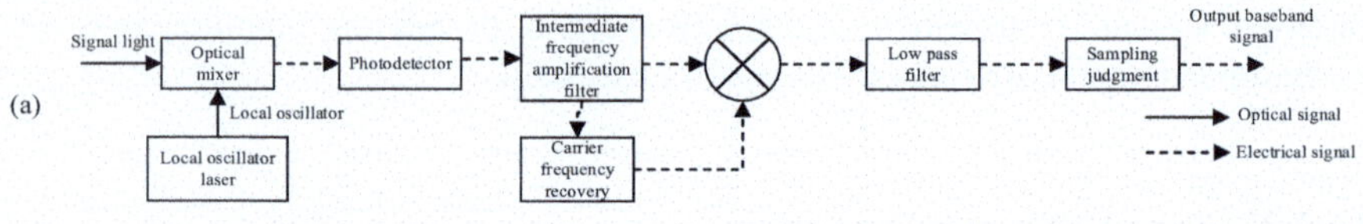

(a) Synchronous demodulation

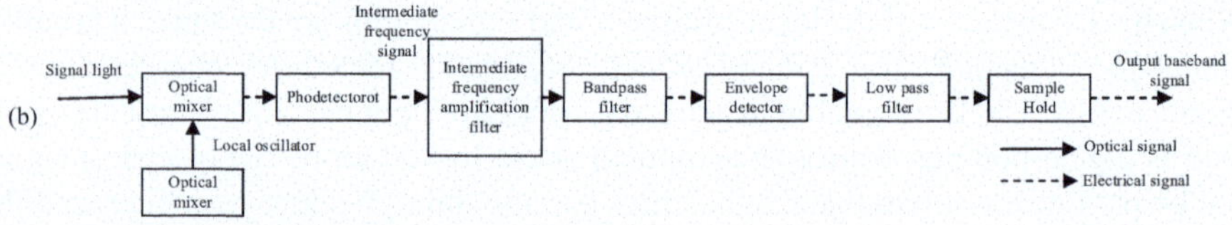

(b) Asynchronous demodulation

Fig. 3.4 Schematic diagram of heterodyne detection

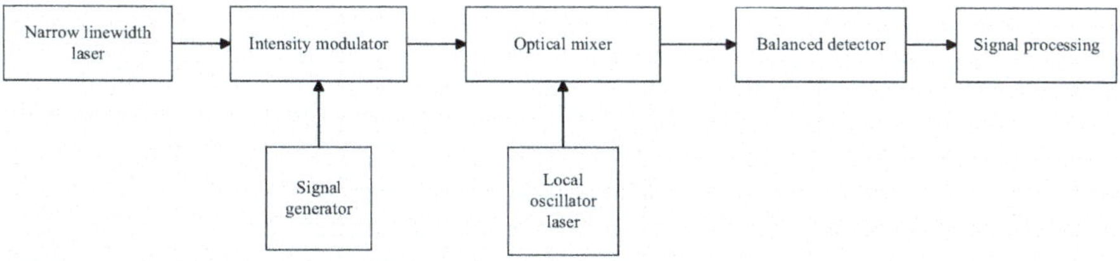

Fig. 3.5 Basic structure of an internal-difference detection system

3.1.2.4 Internal-Difference Detection

The structural diagram of an internal-difference detection system is shown in Fig. 3.5. The transmitting end is mainly composed of signal generators, laser sources, and modulators. At the receiving end, the signal light and local-oscillator light pass through a mixer to obtain an intermediate-frequency signal. The intermediate-frequency signal passes through a balanced detector and becomes an electrical signal, which is processed by the receiving end to obtain the original signal.

3.1.3 Signal-to-Noise Ratio and Sensitivity of Coherent-Detection Systems

In coherent-detection systems, the detection sensitivity refers to the limit of what a detector can detect at a certain signal-to-noise ratio (SNR) or bit-error rate (BER). In theory, the detection limit of an analyzer can reach one photon; however, in reality, the detection limit cannot be reached because of atmospheric turbulence, thermal detector noise, low mixing efficiency of the mixer, and low detector sensitivity. This section analyzes the relationship between the sensitivity and signal-to-noise ratio of coherent-detection systems.

The signal-to-noise ratio of a photodetector is defined as follows [3]:

$$SNR_{IF} = \frac{\langle i_{IF}^2 \rangle}{\langle i_N^2 \rangle} = \frac{\eta \left| \int_S E_l \cdot E_s^* ds \right|^2}{h\nu \Delta f_{IF} \cdot \int_S A_l^2 ds} = \frac{\eta \left| \int_S E_l \cdot E_s^* ds \right|^2}{h\nu \Delta f_{IF} \cdot \int_S E_l \cdot E_l^* ds} \tag{3.1}$$

where <· > means taking the time average, i_{IF} is the intermediate-frequency current, i_{N} is the noise current, η is the quantum efficiency of the ideal photosensitive surface, A_L is the amplitude of the local-oscillator light, E_l is the local-oscillator light electric field, E_s is the signal photoelectric field, and Δf_{IF} is the effective noise bandwidth.

When the mixing efficiency reaches the ideal value, the signal-to-noise ratio can be expressed as [3]

$$SNR_{IF} = SNR_0 = \frac{\eta P_S}{h\nu \Delta f} = \frac{2\eta P_S}{h\nu B}, \tag{3.2}$$

where η is the quantum efficiency, h is the Planck constant, P_S is the signal optical power, B is twice the effective noise bandwidth, and v is the carrier frequency.

From Eq. (3.2), it can be seen that when the SNR is 1, $NEP = P_S = h\nu\Delta f/\eta$. This variable is called the coherent-detection sensitivity. When the quantum efficiency of the mixer reaches 1, $\Delta f = 1$ Hz; the ultimate sensitivity of coherent detection can be obtained as one photon. However, in practical coherent detection, such high sensitivity cannot be achieved. After years of research, the detection sensitivity limit has reached the nanowatt level [3].

3.2 Foreign Research Progress

Compared to direct detection, the sensitivity of coherent detection can be significantly improved, while improving the performance of coherent communication systems. This has led many foreign scientists to invest in this field. They have analyzed the impact of coherent-detection limits from various aspects, such as wavefront distortion, detector sensitivity, intensity noise caused by the local-oscillator light, spot size, optical-axis deflection, coupling efficiency, and polarization mismatch. Methods such as wavefront correction, polarization control, high-sensitivity detectors, and searching for the optimal local-oscillator power, have been designed to continuously improve the detection sensitivity and approach the coherent-detection limit.

In 1966, Teich et al. [12] measurement of the signal-to-noise ratio and minimum detectable power for heterodyne detection of scattered radiation at 10.6 μm has been made. Using photoconductive Ge: Cu as the detector, the minimum detection power observed at a 70-kHz frequency and 270-Kbit/s bandwidth was 3.5×10^{-14} W. This corresponds to an efficiency of 1.3 at a bandwidth of 1 MHz. The minimum detectable power of 10^{-19} W is within 10 times that of the theoretical photon counter.

In 1972, Popoola et al. [13] developed an experimental device that could compensate for the influence of turbulence in real time. The experimental results showed that the device could effectively compensate for the wavefront distortion caused by atmospheric turbulence in a 300-m horizontal link.

In 1975, Cohen et al. [14] conducted a thorough study on the effects of a series of factors, such as spatial-angle mismatch, spot size, and detector performance, on the mixing efficiency of a mixer. The research results showed that when the signal light had an Airy distribution, the spatial-angle mismatch had a relatively small impact on the mixing efficiency.

In 1983, Abbas et al. [15] calculated the signal-to-noise ratio of homodyne / heterodyne receivers for two detectors. The results showed that excessively intense noise in a local oscillator could be eliminated without reducing the performance of the receiver. Later, it was experimentally proven that using two matched detectors could achieve near-quantum-limit performance.

In 1987, Tanaka et al. [16] analyzed the impact of the vector-tilt angle and offset error of the signal light on the mixing efficiency. The results showed that when the vector-tilt angle was less than 10^{-4} degrees and the offset reached one-fifth of the incident aperture, the mixing efficiency exceeded 60%.

In 1988, Calvani et al. [17] developed primitive coherent-transmission technology with polarization modulation and (dual-channel) differential heterodyne detection. In a preliminary experiment, the detection-sensitivity limit reached 4 dB and showed high immunity to laser phase noise.

In 1992, Tanaka et al. [18] studied the mixing efficiency of a detection system in which the local-oscillator and signal lights were partially coherent Gaussian beams. A is the ratio of the minimum $1/e$ radius of the signal light to the minimum $1/e$ radius of the local oscillator. B is the ratio of the input aperture to the detector aperture. The results showed that A and B were inversely proportional.

In 1998, Bar David et al. [19] used mathematical theory to demonstrate the superior performance of coherent dual-light detection over single-diode detection and established the basic principle of local-oscillator intensity. They proposed modulation and detection formats that could reduce additional noise by using the influence of phase noise.

In 1999, Wang et al. [20] proposed a novel anti-reception heterodyne-detection scheme for a monolithic integrated four-segment coherent transceiver. The proposed anti-reception heterodyne-detection scheme further simplified the design of integrated coherent transceivers and promoted the use of a single-terminal device for duplex transmissions.

In 2000, Sato et al. [21] studied and designed a phase-drift suppression method (PDSM) that suppressed the influence of phase drift and improved the stability of heterodyne beat signals, and demonstrated its application.

In 2001, Gol'tsman et al. [22] developed a new type of high-bandwidth laser detector. The experimental results showed that the detector could achieve a long-distance optical-coupling transmission with a coupling efficiency of 38%. The measurement bandwidth reached 17.5 Mbit/s, and the detection sensitivity reached 0.52 μW.

In 2002, Ricklin et al. [23] demonstrated a light-coupling technology that used integrated circuits to control a deformable mirror for wavefront correction. The results showed that, using the new coupling technology, the fiber-coupling efficiency reached 75%.

In 2003, Sobolewski et al. [24] proposed a new type of ultrafast single-photon detector, which was made of NbN films with 3.5-nm and 10-nm thicknesses. For photons at 1550 nm, the detection probability was 3.5%. The response time and jitter of the detector were 100 ps and 35 ps, respectively, and were limited by the acquisition system. At the optimal bias voltage, the dark count was less than 0.01 per second. In terms of the counting rate, jitter, and dark counting, NbN single-photon detectors were significantly superior to semiconductor detectors.

In 2004, Fernandez et al. [25] conducted an experiment on a coherent optical communication system using a 1064-nm-wavelength laser. In the experiment, the laser powers of the local-oscillator and signal lights were 50 mW and 23 mW, respectively. The heterodyne detection method was used to obtain a frequency-difference signal of 6 MHz, output from the mixer.

In 2005, Yang et al. [26] studied a radial–shear interference system with spatial phase modulation that could be used to detect wavefront distortions and control the beam quality. This interferometer was used to test the wavefront distortion of laser pulses with a 1064-nm near-infrared wavelength and 10-ns pulse time width. The software displayed the results of the pulse wavefront quality and energy distribution. The measured beam diameter reached 150 mm, and the root-mean-square test accuracy exceeded 1/15 wavelengths.

In 2007, Dauler et al. [27] proposed a multi-element superconducting-nanowire single-photon detector that consisted of multiple independently biased superconducting-nanowire single-photon detector elements that formed a continuous active region. Two superconducting-nanowire single-photon detectors with two components were experimentally produced and tested. There was no measurable crosstalk between the devices, and the relative timing jitter was less than 50 ps, which was four times the maximum counting rate of a single superconducting-nanowire single-photon detector with the same active area. Compared with single-element superconducting nanowire single-photon detectors, multi-element superconducting-nanowire single-photon detectors have larger effective areas and higher speeds.

In 2007, Park et al. [28] developed a waveguide photodetector with a hybrid waveguide structure, consisting of AlGaInAs quantum wells on a silicon waveguide. The light in the hybrid waveguide was absorbed by the AlGaInAs quantum wells under a reverse bias. The photodetector had a fiber-coupling responsivity of 0.31 A/W with an internal quantum efficiency of 90% over the 1.5 μm wavelength range.

In 2009, Xu et al. [29] developed a novel homodyne-detection structure that performed orthogonal and sequential in-phase measurements of the signal light. The results showed that the weaker the equivalent substate was, the closer the detection limit was to the standard quantum limit.

In 2010, Gregory et al. [30], in view of the impact of wavefront distortion on the detection sensitivity, installed a coherent laser-communication terminal on the ground, and used the terminal to establish a 5.635-Gbit/s communication link with a satellite. Experimental results showed that the adaptive-optics technology used in the link effectively suppressed the wavefront distortion of the signal light.

In 2010, Salem et al. [31] derived an analytical expression for heterodyne efficiency when two partially coherent Gaussian–Schell-model beams were coherently mixed on the surface of a photodetector. They provided numerical examples of the heterodyne efficiency changing with the misalignment angle, detector radius, and overlapping beam parameters. The experimental results showed that partially coherent beams were more susceptible to a decrease in the heterodyne efficiency than coherent beams.

In 2013, Gan et al. [32] developed a new type of graphene photodetector, which had the advantages of high responsiveness, high speed, and a wide spectral bandwidth. Using a metal-doped graphene junction coupled with the waveguide evanescent mode, the detector achieved a light response of more than 0.1 AW^{-1}, and a uniform response between 1450 and 1590 nm. Under zero-bias operation, it demonstrated a response rate of more than 20 GHz and a 12-Gbit/s optical data link.

In 2014, Boroson et al. [33] implemented a communication link between the ground and the moon with an information transmission rate of 622 Mbit/s. The project combined ground-terminal array transceiver technology with superconducting-nanowire single-photon detection to solve the problems of light-intensity flickering and low receiving power, and improved the detection sensitivity.

In 2015, Li et al. [34] introduced the power spectrum of atmospheric turbulence to study the effects of the atmosphere on heterodyne systems. Based on the obtained results, the relationships between the heterodyne efficiency and the detector diameter, misalignment angle, and beam parameters under different turbulence conditions were shown.

In 2016, Wang et al. [35] studied the variation in the heterodyne efficiency with the misalignment angle, detector diameter, turbulence conditions, and overlapping beam parameters, based on the Tatarskii spectral model. The results indicated that the variation in the heterodyne efficiency depended on the initial beam polarization. For larger detector diameters, a turbulent atmosphere significantly reduced the heterodyne efficiency.

In 2017, Yang et al. [36] analyzed and illustrated the effect of polarization aberrations on the output polarization state using polar and Pauli–Zernike decomposition algorithms. Based on the modified heterodyne efficiency, a model of an off-axis optical system with a periscope scanner was established and used to discuss the variation in heterodyne efficiency. The results showed that the improved heterodyne efficiency not only comprehensively described the coherent-detection system but could also be used to evaluate and minimize the polarization aberration of the optical system.

In 2018, Chen et al. [37] developed optical-fiber coupling equipment suitable for long-distance communication, which was used to compensate for random angular errors and residual aberrations. The experimental results showed that the coupling efficiency reached 73% and the mean-squared error reached 1.1% after using this device. The sensitivity of the coherent-detection system was approximately 20 dB higher than that of the direct-detection system.

In 2020, Maiti et al. [38] studied a MoTe-based photodetector that exhibited a strong optical response in the 1550-nm wavelength band, with a responsivity of up to 0.5 AW^{-1} and a bandgap modulation of 0.2 eV for a non-planar waveguide structure. In contrast to the graphene photodetector, which relies on a bandless structure, the dark current of the photodetector was reduced by approximately 100 times, and the noise equivalent power reached 90 pW $Hz^{-0.5}$.

In 2022, Ji et al. [39] proposed and demonstrated a scheme based on complementary polarization detection using three 90° optical mixers for self-coherent homodyne detection. The proposed complementary polarization-diversity coherent receiver utilized remotely transmitted local-oscillator signals without optical polarization control. It had basic stability for any input polarization state of the receiver's remote local-oscillator signal.

The above literature shows that, in coherent detection, the intensity noise [18, 19, 34] caused by the local oscillator, atmospheric turbulence [12], polarization supporting role [36], spot-size deviation and optical-axis deflection [14], coupling efficiency [22], and detector aperture size [16, 31] affect the detection sensitivity. However, through unremitting efforts, foreign scholars can reduce the impact of wavefront distortion through wavefront correction [13, 21, 23, 26, 30], reduce the impact of polarization misalignment through polarization control [35, 39], improve the coupling efficiency [22], improve the performance of photodetectors [24, 25, 27–29, 32, 33, 38], and use other methods to improve the detection sensitivity, thus approaching the limit of coherent detection.

3.3 Domestic Research Progress

In the 1980s, researchers gradually realized the importance of coherent detection in communication systems and investigated the influence of a series of factors on the detection sensitivity, such as the coupling efficiency, system thermal noise, wavefront distortion, polarization's supporting role, coupling efficiency, and atmospheric turbulence.

In 1990, Da et al. [40] developed a heterodyne atmospheric laser communication system that used a CO_2 laser as the laser source, and laid a solid foundation for future research on coherent optical communication systems.

In 2002, Huang et al. [41] developed a new type of photodetector to address the issue of the low reflectivity of Bragg reflectors. The detector exhibited high sensitivity and high-speed characteristics with an absorption-layer thickness of 0.2 μm. At a 1583-nm wavelength, the peak quantum efficiency reached 80%.

In 2003, Wang et al. [42] studied the factors that affected heterodyne efficiency and found that a spatial-angle mismatch was one of the main factors. Local-oscillator phase plane bending and non-collimation effects also affected the heterodyne efficiency.

In 2007, Jin et al. [43] studied the statistical characteristics of Gaussian white noise and laser phase noise in homodyne detection systems. They derived the probability distribution of a two-dimensional random-variable function that took two types of noise as independent variables. Based on this, the detection error caused by the generated noise was calculated and analyzed. The results indicated that it was necessary to maintain a low phase-noise level of the laser to ensure the detection accuracy of the tracking system, and to ensure high requirements for the accuracy of the phase control and tracking of the optical phase-locked loop.

In 2007, Zhao et al. [44] proposed a new receiver scheme that integrated heterodyne detection and optical-power tracking functions, established a distribution model of the detector-surface light field, and studied the heterodyne efficiency under optimal matching of the local and signal lights. Moreover, the heterodyne efficiency under incomplete matching was analyzed.

In 2008, Li et al. [45] proposed a novel adaptive-compensation method for atmospheric turbulence. The compensation effect of this method was insignificant when the relative aperture was large; for example, when $D/r_0 = 6$, the compensated intensity could only reach approximately 50% of the diffraction limit. However, when

$D/r_0 = 2$, the compensation effect reached 90%. The experimental results indicated that when the relative aperture was the same, the compensation difficulty increased with an increase in the turbulence intensity.

In 2008, Cheng et al. [46] studied an optical microwave subcarrier coherent-detection system, and simulated and analyzed its performance. The results showed that the system was less affected by atmospheric noise and atmospheric attenuation, the receiver sensitivity was high, the information-transmission quality was high, and the average bit-error rate was 4.98×10^{-9}.

In 2010, Zhao et al. [47] investigated an external-differential step demodulation method. The research results showed that the mixing efficiency was inversely proportional to the loss of the supporting-role angle. When the local-oscillator light had a thin surface bending, the mixing efficiency was directly proportional to the radius of curvature.

In 2011, Liu et al. [48] investigated the impact of different amplitude distributions of the signal light and local-oscillator light on the mixing efficiency. The experimental results showed that the Airy and Gaussian distributions achieved the maximum heterodyne efficiency, whereas the heterodyne efficiency achieved by Gaussian plane waves was 82%.

In 2011, Han et al. [49] studied the ability of adaptive-optics to compensate for the errors in space optical communication systems, based on the gamma–gamma distribution of atmospheric turbulence. The experimental results showed that adaptive-optical error-compensation technology significantly improved the detection sensitivity, and even low-order adaptive-optical compensation achieved a good compensation effect.

In 2012, Chen et al. [50] proposed a highly sensitive detector that reduced the measurement errors caused by fluctuations in the detector-light intensity and noise carried by the signal itself. The experimental results showed that the detector had advantages, such as a high sensitivity, wide operating bandwidth, and high signal-to-noise ratio, which met the research requirements of terahertz time-domain spectroscopy systems.

In 2013, Zhang et al. [51] studied the factors that affected the heterodyne efficiency, based on a binary phase-shift keying (BPSK) homodyne coherent-detection system. Among these, the collimation mismatch and beam-waist radius significantly affected the heterodyne efficiency. The experimental results showed that an optimal beam-waist radius could be found to maximize the mixing efficiency.

In 2013, Ma et al. [52] analyzed the relationship between the polarization state of the signal light and the mixing efficiency. The results showed that circularly polarized light was more stable than linearly polarized light, and when both the signal light and local-oscillator light were circularly polarized signals, the mixing efficiency of the mixer reached its maximum.

In 2013, Ke et al. [53] investigated an extruded-fiber polarization controller. The system control algorithm adopted a simulated-annealing algorithm, and an improved algorithm was adopted to automatically control the polarization. The control process adopted a blind search method. The simulation results showed that this method was very suitable for polarization control in coherent optical communications.

In 2013, Xiao et al. [54] studied the coupling conditions, systems, and modes of optical fibers, and analyzed the ideal coupling between spatial light and optical fibers. Three common types of fiber amplifiers were compared. An erbium-doped fiber amplifier, which had the best performance, was used as the front-end optical amplifier for the receiver. Subsequently, the sensitivity and signal-to-noise ratio of the coherent-detection system were analyzed.

In 2014, Hu et al. [55] developed a new type of photodetector to meet a series of requirements for coherent-light measurement systems, such as high sensitivity, wide response bandwidth, and high stability. The test results indicated that the detector could achieve a remote optical-coupling transmission and that its detection sensitivity could reach 0.52 μW. The coupling efficiency exceeded 38% and the measured bandwidth reached 17.5 Mbit/s.

In 2014, Luo et al. [56] developed an experimental platform for coherent-optical communication systems and conducted a series of experiments on the platform. The experimental results showed that when the outdoor communication distance was 600 m and the sensitivity was -17.6 dBm, the received bit-error rate was 10^{-6}. When the indoor communication distance was 6 m, the sensitivity increased by approximately 7 dB, compared to that outdoors.

In 2015, Li et al. [57] studied the mixing efficiency of a coherent-detection system using a Gaussian beam model. The research results showed that the mixing efficiency was highest when the waist radius of the signal light was small, and the waist radius of the local-oscillator light was half that of the detector radius.

In 2016, Ke et al. [58] studied the effects of the local-oscillator power and detector parameters on the system signal-to-noise ratio. The experimental results showed that the characteristic parameters of the detector affected the signal-to-noise ratio and optimal local-oscillator power. When the polarization angles of the local oscillator and signal light were larger, the optimal local-oscillator power was smaller and the signal-to-noise ratio was larger.

In 2017, China Southern Airlines et al. [59] investigated the impact of various factors on the mixing efficiency, including the deviation of the spot size and axis deflection. They studied the error-fluctuation range of the spot-size deviation and optical-axis deflection when the mixing efficiency was greater than 10%.

In 2019, Ke et al. [60] analyzed the relationship between the intermediate-frequency signal, optical-polarization state, and mixing efficiency, and developed a single-particle optimization algorithm for coherent optical communication. The experimental results showed that the mixing efficiency increased with the increase of the amplitude of the intermediate-frequency signal in a closed-loop system. The coherent-detection system could control the polarization when the amplitude of the intermediate-frequency signal was 0.001.

In 2020, Tao et al. [61] investigated the design scheme and preparation process of series superconducting-nanowire detectors (SNDs) and developed a preparation process for superconducting-nanowire single-photon detectors with high-saturation detection efficiency. For the 1550-nm laser communication band, a six-pixel NbN superconducting-nanowire detector connected by six nanowires in series was developed. The test results confirmed that the six-pixel SND achieved a system detection efficiency of up to 72.1% (dark count of 1 Hz), while achieving a resolution of one to six incident photons.

In 2021, Cui et al. [62] developed a new type of balanced photodetector, which consisted of a hybrid cross-resistance amplifier and a balanced photodiode chip. The experiment used a 1550-nm laser testing system, and the results showed that under certain conditions, the sensitivity limit of the receiver could reach -61 dBm.

In 2021, Cheng et al. [63] analyzed the relationship between the wavefront distortion and mixing efficiency, as well as the periodic impact of wavefront distortion on the system error rate, under different modulation methods. Based on the impact of wavefront distortion on the system performance, a large-amplitude wavefront-distortion residual-correction algorithm was proposed and achieved good results.

In 2022, Ke et al. [64] designed a fiber coupler with a five-degrees-of-freedom coupling device based on piezoelectric ceramics, and combined it with a variable-gain random parallel gradient-descent algorithm to determine the optimal alignment posture of a spatial-light single-mode fiber. The experimental results showed that the five-degrees-of-freedom fiber coupler effectively corrected different alignment errors, and the coupling efficiency between the spatial light and single-mode fiber reached 53.2% in a closed-loop system.

In 2022, Fan et al. [65] studied the influence of the top-gate voltage on the interface-coupling effect of graphene silicon photodetectors in response to their problems of limited light-absorption capacity and short carrier-recombination life. The experimental results showed that when the top-gate voltage was -3 V and the light-source wavelength was 635 nm, the detector's response could reach 1.1×10^4 A/W. When the light-source wavelength was 1550 nm, the detector's response reached 15 A/W. By increasing the top-gate voltage, the detector response could be increased by five to eight times.

In 2022, Lan et al. [66] designed a highly sensitive self-driving photodetector based on a graphene/silicon nanowire-array heterostructure. This detector greatly improved the photoelectric detection performance by increasing the effective light area and enhancing the absorption of the heterostructure. The experimental results indicated that when the incident-light intensity was 90 μW, and the wavelength was 810 nm, the detector response could reach 0.56 A/W.

In 2023, Xi et al. [67] adopted an easily etched and small-divergence Au/SiO_2 optical cavity, which not only improved the efficiency of nanowire light absorption, but also optimized the processing accuracy of the self-aligned chip outer contour, improving the optical-coupling efficiency of the self-aligned superconducting-nanowire single-photon detector. The experiment showed that the detector could achieve a maximum detection efficiency of 82% at a temperature of 2.2 K and a laser wavelength of 1310 nm. When the wavelength was 1200 nm–1600 nm, the system detection efficiency reached over 65%.

In 2023, Zhu et al. [68] proposed a transistor structure based on the high electron mobility of AlGaN/GaN and successfully developed a ZnO-nanowire sensitive gate photodetector. It was intended to address the problems of low solar-cell efficiency, low light absorption, and low photocurrent of traditional AlGaN/GaN high-electron-mobility transistor photodetectors using ZnO films. Compared with the conventional AlGaN/GaN transistor structure with high electron mobility, the peak responsivity of the new detector was increased by about 2.85 times, the response time was shortened to 10 ms, and the recovery time was shortened to 250 ms, greatly improving the detector performance.

The above literature shows that the local oscillator [44, 58], atmospheric turbulence [56], polarization-loss supporting role [42, 47], spot-size deviation and optical-axis deflection [59], beam-waist radius [51, 57], etc. all affect the detection sensitivity. Therefore, increasing the wavefront correction [45, 49, 63], polarization control [52, 53, 60], and preamplifier [54], improving the coupling efficiency [64] and detector performance [41, 50, 55, 61, 62, 65–68], and optimizing detection systems [46] have been proposed to improve the detection sensitivity and approach the detection limits.

3.4 Factors Affecting the Coherent-Detection Limit

In a coherent-detection system, the following factors affect the detection sensitivity:

1. Thermal noise and shot noise,
2. Wavefront distortion of the signal light,
3. Polarization mismatch between the signal light and local-oscillator light,
4. Spot-size deviation and optical-axis deflection,
5. Coupling efficiency,
6. Detector performance, and
7. Pre-optical amplification.

The first factor is the error caused by noise, which directly affects the detection sensitivity. The second, third, fourth, and fifth errors further affect the sensitivity of the detection system by affecting the mixing efficiency of the mixer. The performance of the photodetector directly affects the sensitivity of the detection system, whereas the role of the preamplifier is to amplify the weak signal light, thereby increasing the mixing efficiency. This section elaborates on the limitations of coherent detection, based on the seven aspects mentioned above.

3.4.1 Thermal Noise and Shot Noise

This section analyzes and demonstrates the impact of thermal noise and shot noise on the coherent-detection sensitivity. The thermal noise of a system is caused by the irregular movement of electrons in the conductor, and the shot noise is caused by random fluctuations of the photocurrent.

The total noise of the system can be described using the following equation [3]:

$$N_{P1} = \frac{2e^2\eta_{PC}}{h\nu}P_s\Delta f_{IF}R_L + \frac{2e^2\eta_{PC}}{h\nu}P_l\Delta f_{IF}R_L + \frac{2e^2\eta_{PC}}{h\nu}P_b\Delta f_{IF}R_L, \tag{3.3}$$

where e is the amount of electron charge, h is Planck's constant, η_{PC} is the solar-cell efficiency, ν is the carrier frequency, P_s is the signal optical power, P_l is the local-oscillator optical power, Δf_{IF} is the effective noise bandwidth, P_b is the background radiation power, and R_L is the effective resistance of the photodetector.

The thermal noise of the system can be expressed as follows [3]:

$$N_{P2} = 4kT_b\Delta f_{IF}, \tag{3.4}$$

where N_{P2} is the thermal noise of the system, T_b is the operating temperature, k is the Boltzmann constant, and Δf_{IF} is the effective noise bandwidth.

Considering the thermal noise caused by the load resistance of the optical mixer and detector, the detection sensitivity is as follows:

$$\text{NEP} = \frac{\left[\frac{e^2\eta_{PC}}{h\nu}(P_l + P_b)\right]\Delta f_{IF}R_L + 2kT_b\Delta f_{IF}}{\left[\left(\frac{e\eta_{PC}}{h\nu}\right)^2 P_l R_L - \frac{e^2\eta_{PC}}{h\nu}\Delta f_{IF}R_L\right]} \tag{3.5}$$

where e is the amount of electron charge, η_{PC} is the quantum efficiency, h is Planck's constant, ν is the carrier frequency, P_l is the local-oscillator power, P_b is the background radiation power, R_L is the effective resistance of the photodetector, k is the Boltzmann constant, and Δf_{IF} is the effective noise bandwidth.

Research has shown that when the quantum efficiency of the detector varies, the detection sensitivity is related to the local-oscillator power as follows. Below the quantum-noise limit, the detection sensitivity decreases as the local-oscillator power increases. When the power of the local oscillator remains constant, the detection sensitivity decreases as the quantum efficiency of the detector increases.

3.4.2 Wavefront Distortion of the Signal Light

Wavefront distortion is caused by an unequal optical path from the light source to each point on the wavefront; the magnitude of this optical-path difference quantifies the degree of wavefront distortion. Owing to the influence of atmospheric turbulence, a signal light transmitted in the atmosphere can follow unequal optical paths, causing the receiving end's light-wave equiphase plane to become a randomly undulating surface, resulting in different angles of incidence of the light rays incident on the optical receiving surface.

If the signal and LO lights are coherently mixed without any processing, the mixing efficiency is significantly reduced, owing to the existence of a spatial-mismatch supporting role, and the mixing efficiency decreases with an increase in

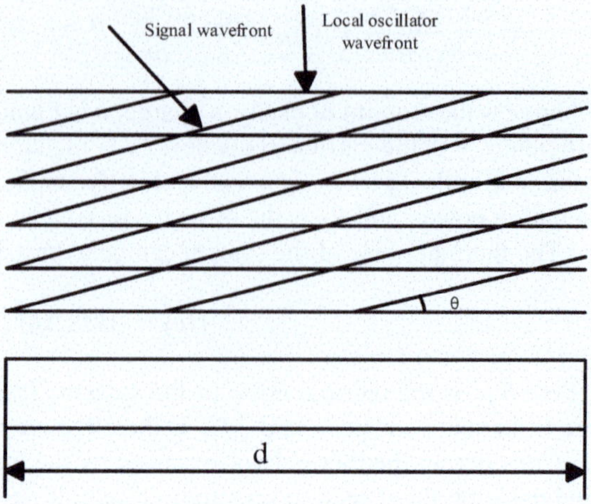

Fig. 3.6 Spatial phase relationship between the signal light and local-oscillator light [3]

the spatial mismatch supporting role. Figure 3.6 shows the relationship between the signal-light wavefront and the LO light wavefront on a photosensitive surface.

The output current of the photodetector is as follows [3]:

$$i_{IF}(t) = G\beta E_{ls}E_l\cos[(\omega_l - \omega_s)t + (\varphi_l - \varphi_s)]\frac{\sin(\omega_s d/2v_x)}{\omega_s d/2v_x} \tag{3.6}$$

where $\upsilon_x = c/\sin\theta$, G is the internal gain of the photodetector, β is the responsivity of the photodetector, and ω_s, ϕ_s and ω_l, ϕ_s are, respectively, the angular frequencies and phases of the signal light and local-oscillator light. E_{ls} is the signal photoelectric field considering wavefront distortion, d is the side length of the photosensitive surface of the detector, θ is the spatial supporting-role mismatch, E_l is the local-oscillator light field, and t is time.

In a coherent-light detection system, the mixing efficiency and difference-frequency current decrease with an increase in the spatial adaptive supporting-role angle, which directly affects the sensitivity of coherent detection. To improve the detection sensitivity, it is necessary to make the local-oscillator light and signal light parallel and coincidental on the photosensitive surface as much as possible because the signal light is affected by atmospheric turbulence when transmitted through the atmosphere, resulting in a distortion of its wavefront when it reaches the receiving end. Currently, scientists use various methods to reduce the impact of wavefront distortion on the detection sensitivity. The adaptive-optics system is the most popular wavefront-correction method.

3.4.3 Polarization Mismatch Between the Signal Light and Local-Oscillator Light

In a general coherent-optical communication system, the signal light emitted by the transmitter is generally a circularly polarized Gaussian beam. After the signal light is transmitted through the atmospheric channel, the amplitude of the optical signal received by the receiver does not have a Gaussian distribution, and its polarization state is also transformed. Thus, the polarization states of the signal light and the LO light are no longer consistent, leading to a reduction in the mixing efficiency and detection sensitivity. In coherent optical communication systems, the optical transceiver is an important component, and its function is to receive the signal light. Figure 3.7 shows the structural diagram of a typical Cassegrain-type optical antenna transceiver.

If both the intrinsic light and signal light are circularly polarized and coherently mixed with the signal light transmitted through the atmospheric channel, the mixing efficiency can be expressed as [72]:

$$\eta = \frac{||\cos\alpha| \int\int |E_s'||E_l| \sim \mathrm{d}s|^2}{\iint |E_l|^2 \sim \mathrm{d}s \iint |E_l'|^2 \sim \mathrm{d}s} \tag{3.7}$$

where α is the polarization-mismatch angle between the signal light and the local-oscillator light, E_s' is the outgoing light field, E_l' is the local-oscillator light field, s is the photosensitive area of the photoelectric receiver, and η is the mixing efficiency.

According to Eq. (3.7), the mixing efficiency η is proportional to $|\cos\alpha|^2$. When the α value is 0°–180°, cosα is inversely proportional to α, and the mixing efficiency η is inversely proportional to $|\cos\alpha|^2$.

In summary, it can be seen that the polarization mismatch between the local-oscillator and signal lights has a significant impact on the mixing efficiency. In practical engineering, polarization-control compensation devices should be used to maintain the polarization state of the signal light and local-oscillator light as stable as possible, which is very important for improving the detection sensitivity.

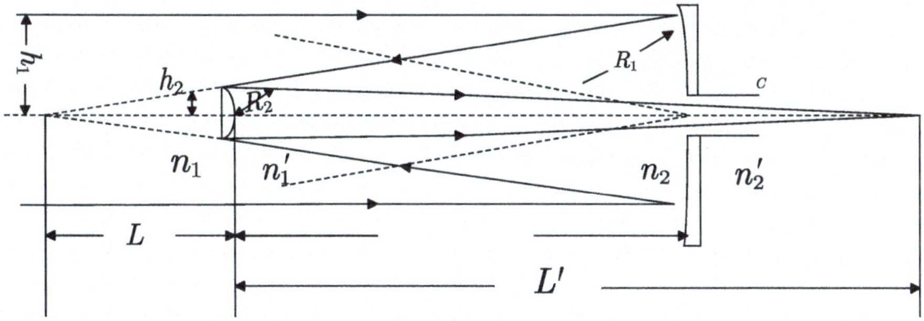

Fig. 3.7 Structural diagram of a Cassegrain-type optical antenna [72]

3.4.4 Spot-Size Deviation and Optical-Axis Deflection

The deviation in spot size is caused by errors in the optical-component manufacturing process, leading to the divergence or contraction of the incident beam. Thus, it does not match the size of the spot generated by the local-oscillator light. Therefore, the size of the spot receiving the target signal light is different from that of the local-oscillator light. To obtain high-quality beams, spatial light mixers often place the receiving target surface at the waist of the incident beam; therefore, the spot size is often described by the waist radius of the intrinsic and signal beams, and the mixing-efficiency formula is obtained [73]:

$$\eta_{\text{coherent}} = \frac{\left[\int_0^{n} \exp\left(-\frac{r^2}{r_x}\right)\exp\left(-\frac{r^2}{r_{lo}^2}\right) r\mathrm{d}r\right]^2}{\int_0^{r_0}\left[\exp\left(-\frac{r^2}{r_{lo}}\right)\right]^2 r\mathrm{d}r \int_0^{\infty}\left[\exp\left(-\frac{r^2}{r_x}\right)\right]^2 r\mathrm{d}r}, \tag{3.8}$$

where r is the radius of the receiving target surface, r_x and r_{lo} are the waist radii of the signal light and local-oscillator light, and exp (·) is an exponential function with e as the base.

Processing errors in the optical components may cause the incident beam to diverge and compress, resulting in different spot sizes for the signal light and local light at the receiving target surface. Thus, there is a deviation in the spot size. Using the spot size of the local oscillator as a reference, the deviation in the spot size can be calculated using $\Delta r = r_x - r_{lo}$. Substituting the deviation into the optical mixing-efficiency Eq. (3.8) yields a new optical mixing-efficiency formula considering the spot-size deviation [73]:

$$\eta_{\text{coherent}} = \frac{\left[\int_0^{n} \exp\left(-\frac{r^2}{(r_{lo}+\Delta r)^2}\right)\exp\left(-\frac{r^2}{r_{lo}^2}\right) r\mathrm{d}r\right]^2}{\int_0^{r_0}\left[\exp\left(-\frac{r^2}{r_{lo}}\right)\right]^2 r\mathrm{d}r \int_0^{\infty}\left[\exp\left(-\frac{r^2}{(r_{lo}+\Delta r)^2}\right)\right]^2 r\mathrm{d}r}, \tag{3.9}$$

where Δr is the spot-size deviation.

In practical applications, both displacements and size mismatches of the light spot exist on the receiving target surface because the processing and adjustment errors of the optical components inside the spatial light mixer cannot be eliminated. Therefore, both of these factors must be considered simultaneously. The mixing-efficiency expression of the mixer when both factors exist simultaneously is obtained as follows [73]:

$$\eta_{ls} = \frac{\left[\int_0^{n} \exp(-r^2/r_{lo}^2)\cdot\exp(-r^2/(r_{lo}+\Delta r)^2)\cdot J_0(C_p k r\delta) r\,\mathrm{d}r\right]}{\left[\int_0^{r_0}\left[\exp(-r^2/r_{lo}^2)\right]^2 r\,\mathrm{d}r\right]\left[\int_0^{\infty}\left[\exp(-r^2/(r_{lo}+\Delta\omega)^2)\right]^2 r\,\mathrm{d}r\right]} \tag{3.10}$$

where J_0 (·) is the zero-order Bessel function, r_x is the waist radius of the signal light, r_{lo} is the waist radius of the local oscillator, Δr is the spot-size deviation, r is the radius of the receiving target surface, δ is the deflection angle of the optical axis, exp (·) is the exponential function with e as the base, k is the Boltzmann constant, and C_p is a constant.

From the above equation, it can be seen that the deviation in the spot size and optical-axis deflection can affect the mixing efficiency of the mixer to some extent. In practical coherent-light detection, methods have often been adopted to suppress this error. Commonly used methods include improving the level of hardware processing or determining the relationship between the spot-size deviation, optical-axis deflection, and mixing efficiency to achieve the maximum mixing efficiency.

3.4.5 Coupling Efficiency

The ability of optical fibers to couple signal light affects, to a certain extent, the sensitivity of coherent-detection systems. This ability to couple optical fibers is called the coupling efficiency of fiber couplers. Figure 3.8 shows a schematic of a fiber-optic coupler. When all conditions are ideal, the light rays converge at the focal point of the lens after passing through the lens, and then advance along the direction of the axis in the fiber until they reach the next module.

The coupling efficiency is expressed as follows:

$$\eta_{Coupling} = P/P_0, \tag{3.11}$$

where P represents the optical power entering the fiber, and P_0 represents the total incident power.

As shown in Fig. 3.9, the coupling efficiency of the fiber coupler $\eta_{Coupling}$ can be expressed as in [74] when analyzed using the mode-field matching method:

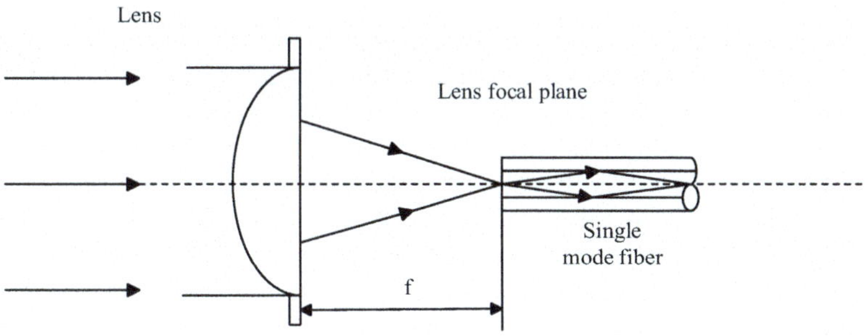

Fig. 3.8 Schematic diagram of a fiber-optic coupling

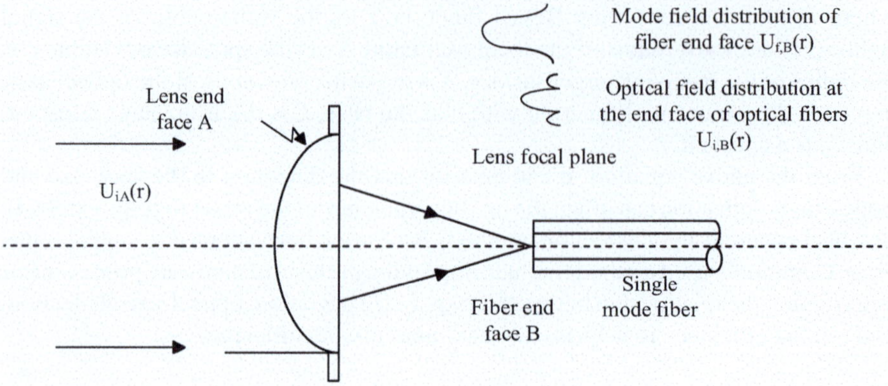

Fig. 3.9 Schematic diagram of a fiber-coupling mode-field analysis

$$\eta_{Coupling} = \frac{\left| \iint U_i^*(r) U_f(r) r\mathrm{d}r\mathrm{d}\gamma \right|^2}{\iint U_i(r) U_i^*(r) r\mathrm{d}r\mathrm{d}\gamma \times \iint U_f(r) U_f^*(r) r\mathrm{d}r\mathrm{d}\gamma} \tag{3.12}$$

where $U_i\,(r)$ is the amplitude distribution of the signal light; $U_i^*(r)$ is the conjugate function of $U_i\,(r)$; $U_f\,(r)$ is the distribution of the electromagnetic field of a single-mode fiber on the lens end face; $U_f^*(r)$ is the conjugate function of $U_f\,(r)$; r is the radius of the lens end face; γ is the receiving angle; and $\eta_{Coupling}$ is the coupling efficiency.

In summary, it can be seen that the coupling efficiency of fiber-optic couplers is influenced by a series of factors, such as lens parameters, reception angle, signal light amplitude, etc. In future research, these factors should be improved to improve the coupling efficiency of the couplers and increase the coherent-detection sensitivity.

3.4.6 Detector Performance

A photodetector converts optical signals into electrical signals. In the study of coherent-detection limits, the detection sensitivity largely depends on the performance of the detector. The detectors commonly used at present include photomultiplier tubes, single-photon detectors, and superconducting detectors. Figure 3.10 shows a schematic diagram of the photoelectric effect.

Figure 3.10a shows the change in the P–N junction depletion region when the externally applied electric field is in the same direction as the internal electric field, and Fig. 3.10b shows the energy-band diagram near the P–N junction depletion region. It can be seen that the electrons move in the energy band, the potential of the location where the electrons gather is low, and the energy band is high.

With the development of science and technology, new detectors with higher sensitivities and faster responses, such as nanowire and graphene detectors, have emerged

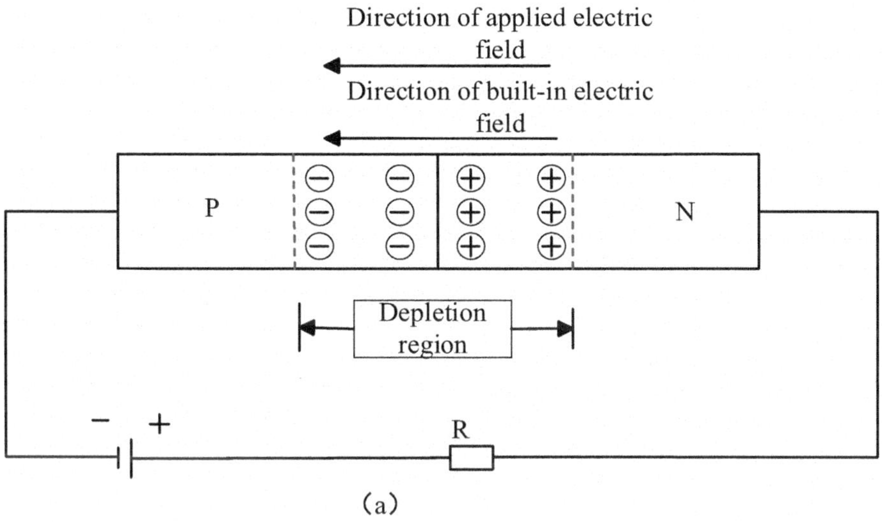

(a) Depletion-region structure diagram of a P-N junction

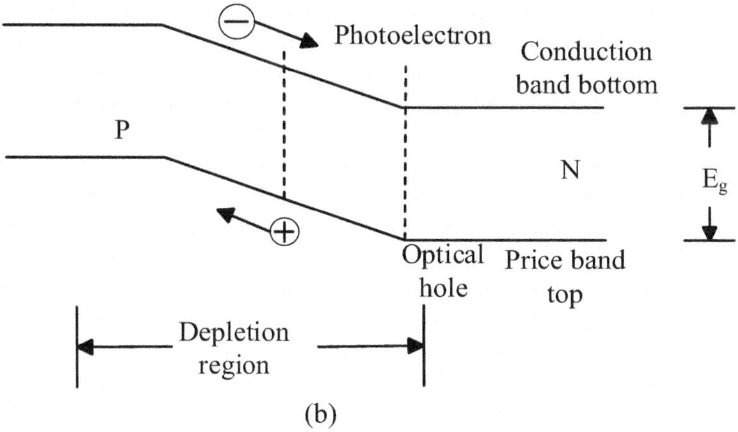

(b) Energy-band diagram around the P-N depletion region

Fig. 3.10 Schematic diagram of the photoelectric effect

in recent years. Nanowire detectors use semiconductor nanowire materials for detection and sensing. A charge transfer can occur when the target material reacts with the surface of the nanowire, thereby changing its electrical properties and facilitating detection.

Graphene photodetectors utilize the electronic and optical properties of graphene for photoelectric conversion. The electronic properties of graphene enable high-speed electron-transmission capability, high sensitivity, high transmittance, and low reflectivity. Improvements in various aspects of photodetector performance are of great significance for improving the sensitivity of coherent detection.

3.4.7 Pre-optical Amplifier

A preamplifier is commonly used in optical communications and sensor systems. It amplifies weak optical signals at the input end of optical fibers or detectors. Its primary function is to amplify the optical signal at the input end to a sufficiently high signal-to-noise ratio. The device does not undergo any internal photoelectric or electro-optical conversions. In coherent-detection systems, optical amplifiers play a significant role in improving the detection sensitivity and increasing detection limits. There are two types of commonly used optical amplifiers: semiconductor and fiber. This section introduces an erbium-doped fiber amplifier (EDFA) with excellent performance.

As shown in Fig. 3.11, the structure diagram of an EDFA consists of two parts: an optical path and an auxiliary circuit. The optical path is composed of an optical isolator, pump light source, optical coupler, erbium-doped fiber, and optical filter. The auxiliary circuit is composed of a power supply, warning and protection circuit, and automatic microprocessor control.

The function of the auxiliary circuit is to ensure the normal operation of the optical amplifier. The function of the pump source is to provide sufficient energy for the erbium-doped fiber to invert the particle beam. The core device of the optical amplifier is an erbium-doped fiber, and its main material is quartz fiber. In practice, to improve the performance of an optical amplifier, the quartz is often doped with the rare-earth element Er^{+3}.

In digital optical communication systems, regardless of the circuit thermal noise and extinction ratio, the detection sensitivity can be expressed as follows:

$$P_s = 2h\nu n_{sp}B(Q^2 + Q \cdot \sqrt{0.5\mu B_0/B_e - 0.5})P_s = 2h\nu n_{sp}B(Q) \tag{3.13}$$

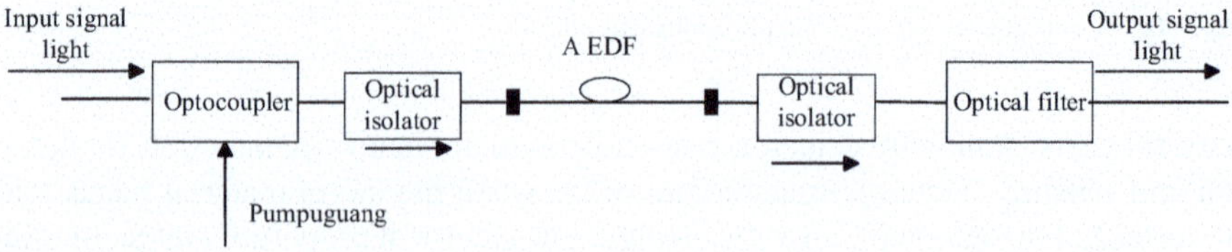

Fig. 3.11 Structural block diagram of an EDFA

where h is Planck's constant, η_{sp} represents the spontaneous emission factor, B_0 represents the bandwidth of the optical filter, B_e is the bandwidth of the electric filter, the value of μ (1 or 2) depends on whether a polarization filter is used, and ν is the carrier frequency. When the bit-error rate is 10^{-9}, $Q = 6$; when the bit-error rate is 10, Q = 5.2. If the receiving end adopts error-correction coding and the preamplifier uses a silver-doped fiber amplifier, the sensitivity of coherent detection can reach -50 dBm.

3.5 Development Trends

Coherent-detection technology is used to measure coherent properties and is one of the most important technologies in the field of optics. With continuous progress in science and technology, coherent-detection technology is constantly developing. This section provides a detailed explanation of the development trends in coherent-detection technology.

3.5.1 High Detection Sensitivity and Resolution

The highest sensitivity of coherent detection is the single-photon counting of photon counters or single-photon detectors. However, the resolution of this method is extremely low. Therefore, high-sensitivity methods are required to improve the resolution of coherent detection. The emergence of new technologies has improved the sensitivity and resolution of coherent detection. The main approach is to use higher detector sensitivity, more efficient photon splitters, and better coherent interferometers.

3.5.2 Wider Applications

Coherent-detection technology has wide applications in fields such as optical communication, military, optical imaging, and microscopy. In the future, as the technology develops, it will gradually be applied to fields such as quantum computing, high-speed communication, and biomedicine. The demand for efficient and accurate coherent-detection technology in these fields is increasing; therefore, the application of coherent-detection technology will become increasingly widespread.

3.5.3 More Efficient and Simpler Detection Methods

With the continuous development of science and technology, coherent detection methods are constantly being developed. In the future, equipment will become more efficient and precise. The emergence of new coherent-detection technologies has made coherent detection more convenient. For example, automation-control technology for optical interferometers will greatly improve the efficiency of coherent detection. Thus, coherent detection will no longer be a difficult task.

3.5.4 Further Research

Coherent-detection technology is an active research field, and further research may be conducted in the future to expand its application fields and improve the performance of detection technology. With the rapid development of communication technologies, research on coherent-detection technology has received increasing attention. Therefore, the development of coherent-detection technology will continue to receive attention from scientists and researchers.

In summary, coherent-detection technology will continue to develop and play an important role in many fields. The development of future coherent-detection technology will include higher detection sensitivity, wider application fields, more efficient and simplified detection methods, and deeper research. With the advancement of these technologies, coherent-detection technology will become indispensable in modern optical fields.

3.6 Future Prospects

This chapter summarized the domestic and foreign research progress in coherent detection. Coherent detection was divided into heterodyne detection and homodyne detection, according to the detection method, and the methods were discussed. Then, the factors affecting the detection sensitivity were analyzed and suppression schemes were described to address these effects and achieve a high detection sensitivity. Although the current coherent-detection technology is relatively mature, the following issues are still worth studying and discussing.

3.6.1 Using Better Laser Sources

By using a laser source with a higher power, narrower linewidth, and higher spot quality factor, the laser-beam coherence can be improved, thereby improving the sensitivity of the detector.

3.6.2 Using Higher Detection Efficiency

Maximizing the response efficiency of the detector can improve its sensitivity and accuracy. Common methods include increasing the detector area and gain, and improving the detector response time.

3.6.3 Reducing Noise

Using low-noise preamplifiers and signal processors to reduce noise can improve the sensitivity and accuracy of detectors.

3.6.4 Optimizing the Detector Structure and Materials

Changing the process, structure, and materials of a detector can improve its sensitivity and accuracy. For example, new semiconductor materials can be used to increase the doping concentration and change the detector structure.

3.6.5 Using Adaptive-Optics Technology

Using adaptive-optics, the phase perturbation of coherent beams can be eliminated, thereby improving the sensitivity and accuracy of coherent detectors.

References

1. Gudmundsen RA (1955) Photoelectric mixing of incoherent light waves. Phys Rev 99(6):1691–1700
2. Forrester AT, Gudmundsen RA, Johson PO (1955) Photoelectric mixing of inherent light. Phys Rev 99(6):1691
3. Xizheng K, Jiali W (2019) Principles and applications of wireless optical coherent communication. Science Press, Beijing
4. Ling Z, Ying F (2007) Research progress and trends in satellite coherent optical communication. Progress in Laser and Optoelectronics 44(6):49–53
5. Yao Zhoushi HuYu (2005) Development history and trends of intersatellite coherent optical communication technology. Opt Commun Technol 29(08):44–46
6. Sodnik ZFBLH (2010) Optical intersatellite communication. IEEE J Sel Top Quantum Electron 16(5):1051–1057
7. Zhenkun T (2019) Factors affecting the performance of heterodyne detection in wireless optical communication and experimental research. Xi'an: Xi'an University of Technology
8. Meng H (2016) High sensitivity homodyne coherent detection of laser pulse echoes. University of Electronic Science and Technology, Chengdu
9. Xizheng K, Lijun D (2016) Wireless optical communication. Science Press, Beijing, pp 7–17
10. Liu C, Chen SQ, Li XY et al (2014) Performance evaluation of adaptive optics for atmospheric coherent laser communications. Opt Express 22(13):15554–15563
11. Barry JR, Lee EA (1990) Performance of coherent optical receivers. Proc IEEE 78(8):1369–1394
12. Teich MC (1966) Optimum heterodyne detection at 10.6 μm in photoconductive Ge:Cu. Appl Phys Lett 9(10):357–360
13. Popoola WO, Ghassemlooy Z (208) Performance of subcarrier modulated freespace optlce omunicaition link in negative exponential atmospheric turbulenceenvironment. Int J Autonomous Adapt Commun Syst 1(3):342–355
14. Cohen SC (1975) Heterodyne detection: phase front alignment, beam spot size, and detector uniformity. Appl Opt 14(8):1953–1959
15. Abbas GL, Chan V, Yee TK (1983) Local-oscillator excess-noise suppression for homodyne and heterodyne detection. Opt Lett 8(8):419–421
16. Tanaka K, Ohta N (1987) Effects of tilt and offset of signal field on heterodyne eficiency. ApplidOptics 26(4):627–632
17. Calvani R, Caponi R (1988) Polarisation phase-shift keying: a coherent transmission technique with differential heterodyne detection. Electron Lett 24(10):642–643
18. Tanaka T (1992) Heterodyne efficiency of a partially coherent optical signal. Appl Opt 46(56):789–799
19. Bar-David I, Salz J (2002) On dual optical detection: homodyne and transmitted-reference heterodyne reception. IEEE Trans Commun 36(12):1309–1315
20. Wang L, Choa FS (1999) Counterreceiving heterodyne detection with an integrated coherent transceiver and its applications in bandwidth-on-demand access networks. J f Lightwave Technol 17(10): 1724–1731
21. Sato M, Onodera K, Akiba M et al (2000) Phase-drift suppression in heterodyne detection and its application to optical coherent tomography. Filtr Ind Anal 41(15):388–391
22. Gol'tsman GN, Okunev et al (2001) Picosecond superconducting single-photon optical detector. Appl Phys Lett 25(8):115:125
23. Ricklin JC (2002) Free-space laser communication using a partially coherent source beam. The Johns Hopkins University

24. Sobolewski R, Verevkin et al (2003) Ultrafast superconducting single-photon optical detectors and their applications. IEEE Trans Appl Superconductivity 13(2):1151–1157
25. Fernandez MM, Vilnrotter VA (2004) Coherent optical receiver for PPM signals received through atmospheric turbulence: performance analysis and preliminary experimental results. Opt Eng 56(62):5338–5345
26. Yang Y, Lu Y, Chen Y et al (2005) A radial-shearing interference system of testing laser-pulse wavefront distortion and the original wavefront reconstructing. Opt Eng 5638(1):200–204
27. Dauler EA, Robinson BS, Kerman AJ et al (2007) Multi-element superconducting nanowire single-photon detector. IEEE Trans Appl Supercond 17(2):279–284
28. Hyundai P, Alexander et al (2007) A hybrid AlGaInAs-silicon evanescent waveguide photodetector. Opt Express 15(10):6044–6052
29. Xu Q, Arvizu Mondragn A, Gallion P et al (2009) Homodyne in-phase and quadrature detection of weak coherent states with carrier phase tracking. IEEE J Select Top Quantum Electron 15(6):1581–1590
30. Gregory M, Hemmati H, Heine F et al (2010) Inter-satellite and satellite-ground laser communication links based on homodyne BPSK. Proc Spie 65(68):7587–75870
31. Salem M, Rolland JP (2010) Heterodyne eficiency of a detection system for partially coherent beams. J Opt Soc Am A 27(5):111–119
32. Gan X, Shiue RJ, Gao Y et al (2013) Chip-integrated ultrafast graphene photodetector with high responsivity. Nat Photonics 7(11):883–887
33. Boroson MD, Robinson SB (2014) The lunar laser communication demonstration: NASA's first step toward very high data rate support of science and exploration missions. Space Sci Rev 185(4):115–128
34. Chengqiang L, Tingfeng W, Heyong Z et al (2015) The performance of heterodyne detection system for partially coherent beams in turbulent atmosphere. Opt Commun 356(56):620–627
35. Yan W, Chengqiang L, Tingfeng W et al (2016) The effects of polarization changes of stochastic electromagnetic beams on heterodyne detection in turbulence. Laser Phys Lett 13(11):116–126
36. Yufei Y, Changxiang Y, Chunhui H et al (2017) Modified heterodyne efficiency for coherent laser communication in the presence of polarization aberrations. Opt Express 25(7):7567–7591
37. Chen M, Liu C, Rui D et al (2019) Highly sensitive fiber coupling for free-space optical communications based on an adaptive coherent fiber coupler. Optics Communications 430(87):223–226
38. Maiti R, Patil C, Saadi M et al (2020) Strain-engineered high-responsivity MoTe2 photodetector for si-licon photonic integrated circuits. Nat Photonics 14(9):578–586
39. Ji H (2022) Complementary polarization-diversity coherent receiver for self-coherent homodyne detection with rapid polarization tracking. Lightwave Technol,64(8):2773–2779
40. Da N, Mingjia P, Liming F (1990) CO_ Study on the sensitivity of 2 laser heterodyne detection systems. J Opt 10(08):714–720
41. Hui H, Qi W, Lei L (2002) Long wavelength and high sensitivity InP/InGaAs resonant cavity photodetectors. Optoelectron Laser 03:221–224
42. Qi W, Chunhui W, Tieliang S (2003) Heterodyne efficiency of coherent detection of Gaussian local oscillator light and Airy spot signal light. China Laser 30(89):183–186
43. Tao J, Lei G (2007) Research on detection error of coherent detection and orientation system. Acta Photonica Sinica, 36(02):340–343
44. Changzheng Z (2007) Preliminary scheme and heterodyne efficiency study of spatial coherent optical communication. Yanshan University, Qinhuangdao
45. Huan L, Hongtao Z, Fuchang Y (2008) Adaptive compensation method for atmospheric turbulence in space laser communication systems. J Changchun Univ Technol (Natural Science Edition) 98(02):1–3
46. Shu C, Hang J, Yong M (2008) Design and simulation study of optical microwave subcarrier multiplexing coherent detection system. Ship Electron Eng 56(02):61–165

47. Chunying Z (2010) Research on heterodyne asynchronous demodulation technology for coherent optical communication. Changchun University of Technology, Changchun
48. Hongzhan L, Yuefeng J, Nan X (2011) The influence of amplitude distribution of signal light and local oscillator light on the mixing efficiency of inter satellite coherent optical communication systems. J Opt 31(10):63–68
49. Liqiang H, Qi Y, etc (2010) Study on the performance of free space optical communication system based on adaptive optics compensation. Appl Opt 31(2):301–303
50. Longchao C, Wenhui F (2012) Research on high sensitivity and low noise terahertz electrooptic detectors. J Electron 40(9):1705–1709
51. Tong Z, Shoufeng T (2013) The effect of collimation mismatch on the mixing efficiency of spatial coherent laser communication. J Changchun Univ Technol (Nat Sci Edition) 36(Z1):13–15
52. Tingting M, Shoufeng T, Hang N et al (2017) The influence of signal polarization characteristics on the mixing efficiency of spatial coherent detection. Progress in Laser and Optoelectron 54(2):20–64
53. Tie L, Xizheng K, Juan C (2012) Polarization control in coherent light detection systems. Infrared and Laser Eng 41(11):3069–3074
54. Xiang X (2013) Space light receiving technology based on pre amplifier. Changchun University of Technology
55. Pengcheng H, Hongxing Y, Jianting M (2014) Research on high sensitivity detectors for long range high speed laser heterodyne signals. Optoelectron. Laser 25(05):925–931
56. Binbin L (2014) Space wireless digital coherent optical communication. Beijing University of Posts and Telecommunications, Beijing
57. Xiangyang L, Zongfeng M, Dele S (2015) Influence of Gaussian beam field distribution on coherent detection mixing efficiency. Infrared Laser Eng 44(02):539–543
58. Yingxiu K, Xizheng K, Yuan Y (2016) The influence of local oscillator power on signal-to-noise ratio in spatial coherent optical communication. Infrared and Laser Eng 45(02):242–247
59. Airlines S, Peng Z, Shoufeng T (2017) Analysis of the influence of spot size deviation and optical axis deflection on the mixing efficiency of space optical mixers. Infrared Laser Eng 46(04):205–212
60. Yingxiu K, Xizheng K, Yuan Y (2017) The effect of laser linewidth on the performance of spatial coherent optical communication systems. J Instrum 38(07):1668–1674
61. Xu T (2020) High efficiency and high-speed characteristics of superconducting nanowire single photon detectors. Nanjing University
62. Dajian C, Lang Z, Shuiqing X, Li R, Feng L, Jingzhong G (2020) Research on spaceborne high sensitivity balanced photodetectors. Semiconductor Optoelectron 41(04):480–484
63. Shuang C (2021) Research on correction technology for large amplitude distorted Wavefront in free space coherent optical communication. Xi'an: Xi'an University of Technology
64. Jiali W, Xizheng K, Shangjun Y (2022) Calibration of alignment errors in spatial light single mode fiber coupling using multidimensional couplers. J Opt 42(07):40–50
65. Anqi F (2022) Research on graphene composite structure photodetectors. University of Electronic Science and Technology, Chengdu
66. Zhenli L, Feng H, Yiji S, Ding D, Guofang Z (2022) High performance graphene/silicon nanowire array heterojunction photodetector. Laser and Infrared 52(11):1671–1677
67. Lingling X, Xiaoyan Y, Tianzhu Z (2023) High comprehensive performance superconducting nanowire single photon detector. J Phys 1–18 [2023–05–09]
68. Yanxu Z, Jianwei L, Qixuan L (2023) ZnO nanowire photosensitive gate photodetector based on AlGaN/GaN HEMT structure. J Beij Univ Technol 49(02):188–196
69. Liu Y, Tong SF, Chang SA et al (2018) Design of a phase sensor applied in the optical phase-loceked loop based on a high speed coherent laser communication system. IEEE Access 6(10):22131–22139

70. Barry JR, Lee EA (1990) Performance of coherent optical receivers. Proc IEEE 78(8):1369–1394
71. Yonghong D, Yanfei L, Haotian Z (2015) Sensitivity testing experiment of balanced detectors in space coherent optical communication. Infrared Laser Eng 44(10):3110–3116
72. Weixia Z, Jiayou W (2018) The effect of polarization mismatch on the performance of spatial coherent optical communication systems. Semiconductor Optoelectron 39(03):435–439
73. Hang N, Peng Z, Shoufeng T (2017) Analysis of the influence of spot size deviation and optical axis deflection on the mixing efficiency of space optical mixers. Infrared Laser Eng 46(04):205–212
74. Leeb WR, Winzer PJ, Kudielka KH (1998) Aperture dependence of the mixing efficiency, the signal-to-noise ratio, and the speckle number in coherent Lidar receiver. Appl Opt 37(15):3143–3148

Chapter 4
Analysis of Factors Affecting the Detection Sensitivity of Coherent-Detection Systems

In practical applications, the antenna in a coherent-detection system receives the signal beam, couples it into an optical fiber, and then mixes the frequency with that of the local-oscillator beam. This process is affected by the performance of the detector itself, as well as the influence of external factors on the detection performance of the system. Consequently, the optical power of the signal reaching the surface of the detector is smaller than that of the received signal. Therefore, it is necessary to study the factors affecting the detection sensitivity of a coherent-detection system and their influence on the system performance.

4.1 Detector-Performance Factors Affecting the System Detection Sensitivity

4.1.1 Influence of the Local-Oscillator Optical Power on the System Sensitivity

When considering thermal noise, shot noise, relative-intensity noise, etc. in a balanced-detection system, the thermal noise can be expressed as [1]

$$\langle i_t^2 \rangle = \frac{4KT_bB}{R_L}, \tag{4.1}$$

where R_L is the detector resistance, K is the Boltzmann constant, B is the detector bandwidth, T_b is the temperature in Kelvin, and <·> is the ensemble average.

Shot noise is primarily caused by random variations in the number of photons; this can be caused by medium scattering and absorption, beam-source instability, and other factors in the optical path. Owing to the random variation in the number

X. Ke and C. Ke, *Performance-Improvement Limits in Optical Wireless Coherent Communication*, Optical Wireless Communication Theory and Technology,
https://doi.org/10.1007/978-981-96-2000-5_4

of photons, different numbers of photons strike the detector each time, creating shot noise. The mean-square value of the shot-noise current of the detector can be expressed as [1]

$$\langle i_s^2 \rangle = 2e\frac{e\eta}{hv}P_L B. \tag{4.2}$$

The relative-intensity noise (RIN) is the fluctuation of the relative intensity of the output optical power of the laser, which is measured using the average output power of the laser as the normalized benchmark. Its unit is decibel/hertz (dB/Hz). The RIN expression is [1]

$$\langle i_{RIN}^2 \rangle = R_{in} \times B \times (\Re P_L)^2, \tag{4.3}$$

where R_{in} is the RIN figure and its range depends on the performance and working conditions of the laser. The RIN level of a laser is typically between −155 and −130 dB/Hz. It can be seen from Eqs. (4.2) and (4.3) that the optical power of the local oscillator has a linear and square ratio relationship with the shot noise and the excess-intensity noise of the local oscillator, respectively. Therefore, when the optical power of the local oscillator is gradually increased, the main noise of the detector is gradually converted from thermal noise to shot noise and excess noise from the local oscillator.

When the average laser power is increased further, the excess-intensity noise of the local oscillator becomes the main noise source for the detector, resulting in a decrease in the signal-to-noise ratio of the coherent communication link. However, in a dual-tube balanced detector, when the responsivities of the two photodetector tubes are perfectly matched, the excess-intensity noise of the local oscillator is completely eliminated, and the main noises of the detector system are shot noise and thermal noise. The signal-to-noise ratio of the system heterodyne detection is expressed as

$$SNR = \frac{2(e\eta/hv)^2 P_S P_L R_L}{2e(e\eta/hv)BP_L R_L + 4KT_b B}. \tag{4.4}$$

In an actual balanced-detection system, because the responsivity of the two photodetectors is difficult to match completely, the responsivities of the two photodiodes are set to $\Re_1$ and $\Re_2$, respectively, and the balance-consistency coefficient of the balanced detector is introduced to reduce the impact of inconsistencies on the responsivity. The equilibrium-consistency coefficient can be defined as [1].

$$\Re' = \left[1 - \frac{2|\Re_1 - \Re_2|}{\Re_1 + \Re_2}\right] \times 100\%. \tag{4.5}$$

When the two photodiodes do not match, the excess-intensity noise of the local oscillator introduced by the system can be expressed as [1]

$$P_{in} = \left(1 - \Re'\right)R_{in}\left(\frac{e\eta}{hv}\right)^2 P_L^2 BR_L, \tag{4.6}$$

where R_{in} represents the RIN. Typically, the RIN level of a laser is between -155 dB/Hz and -130 dB/Hz. The signal-to-noise ratio of the balanced detector can be expressed as

$$SNR = \frac{2(e\eta/hv)^2 P_S P_L R_L}{2e(e\eta/hv)BP_L R_L + 4KT_b B + (1 - \Re')R_{in}(e\eta/hv)^2 P_L^2 BR_L}. \tag{4.7}$$

When the optical power of the local oscillator continuously increases, the response characteristic of the detector becomes nonlinear, and the output current of the detector does not change with the optical power of the local oscillator. In this case, the output current of the detector will no longer be linear, but become a nonlinear function, which can be expressed as [2]

$$i = \frac{e\eta}{hv}P_L(1 - \upsilon P_L) + \frac{e\eta}{hv}(1 - 2\upsilon P_L)\sqrt{2P_S P_L}|_{P_L \gg P_S}, \tag{4.8}$$

where υ is the detector parameter.

The signal-to-noise ratio of a detector working in the nonlinear region is

$$SNR = \frac{2(e\eta/hv)^2(1 - 2\upsilon P_L)^2 P_S P_L R_L}{2e(e\eta/hv)B(1 - \upsilon P_L)P_L R_L + 4KT_b B + (1 - \Re')R_{in}(e\eta/hv)^2 P_L^2(1 - \upsilon P_L)^2 BR_L}, \tag{4.9}$$

where D is the balance-consistency coefficient, and R_{in} is the RIN figure.

Adjusting the optical power of the local oscillator can affect the performance of a balanced-detection system. Increasing the optical power of the local oscillator can increase the conversion gain of the detector, thereby enhancing the detector's ability to detect optical signals and improving the signal-to-noise ratio of the system. However, with an increase in the optical power of the local oscillator, the noise generated by the optical signal of the local oscillator will also increase accordingly. This limits the detector's ability to detect the signal, resulting in a decrease in detection sensitivity.

Therefore, the optimal optical power of the local oscillator can be determined through numerical simulations and experimentally measuring system parameters. In practical applications, it is necessary to improve the signal-to-noise ratio and detection sensitivity of the system as much as possible to ensure that the bit-error rate of the system satisfies these requirements.

Figure 4.1 shows that the detection sensitivity of the available system varies with the local-oscillator optical power when the bit-error rate is 10^{-9}. When the ideal state of the detector-saturation effect is not considered, the electronic charge $e = 1.6 \times 10^{-19}$ C, the quantum efficiency $\eta = 0.8$, the detector bandwidth $B = 200$ MHz, Planck's constant $h = 6.626 \times 10^{-34}$ J·s, carrier frequency $\upsilon = 1.9 \times 10^{14}$ m/s, R_L

$= 50\ \Omega$, $K = 1.38 \times 10^{-23}$ J/K, temperature $T_b = 300$ K, signal optical power $P_S =$ 10 nW, and local-oscillator optical power $P_L = 5$ mW.

Figure 4.1 shows the variation in the detection sensitivity with the local-oscillator optical power under different conditions. It can be seen from Fig. 4.1a that under a certain local-oscillator power, the greater the quantum efficiency η, the greater the detection sensitivity of the system. When the quantum efficiency increases, the detector can convert more photons into electrical signals every time it receives more photons, which increases the signal-intensity output by the detector, thereby improving the detection sensitivity.

As can be seen in Fig. 4.1b, at a certain local-oscillator optical power, the larger the detector bandwidth B is, the smaller the detection sensitivity is. This is because

Fig. 4.1 Detection sensitivity as a function of the local-oscillator optical power

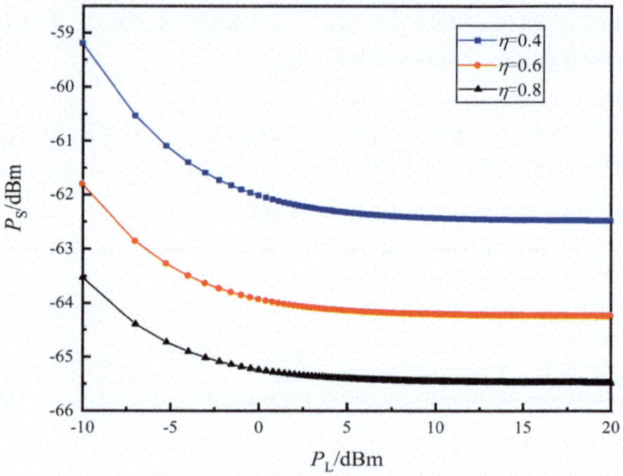

(a) Different quantum efficiencies η

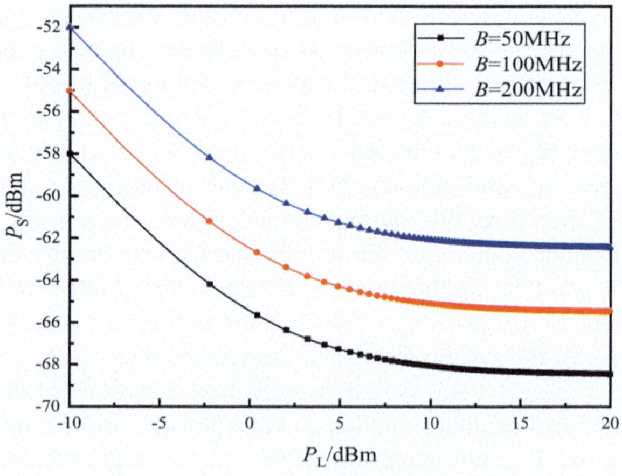

(b) Different detector bandwidths B

the larger the detector bandwidth, the better the detector can detect higher-frequency optical signals. It will also increase the sensitivity of the detector to noise because the higher-frequency signal will also increase the impact of thermal and other noise, thus decreasing the sensitivity of the system.

In addition, it can be seen that the sensitivity of the system increases with an increase in the optical power of the local oscillator, because when the two photodiodes of the balanced detector are completely matched, the excess-intensity noise of the local oscillator can be completely eliminated. Therefore, under the same quantum efficiency or bandwidth condition, increasing the local-oscillator power can improve the detection sensitivity. However, when the shot noise of the system is significantly larger than the thermal noise, increasing the local-oscillator power does not significantly improve the detection sensitivity of the system.

When the responsivities of the two photodiodes of a balanced detector are not completely equal, when the bit-error rate is 10^{-9}, according to Eq. (4.3) and using the same parameters as in Fig. 4.1, the calculated system detection sensitivity increases with the local-oscillator power change, as shown in Fig. 4.2. It can be seen from Fig. 4.2a that with the increase in the local-oscillator power, the detection sensitivity first increases and then decreases because, with the increase in the local-oscillator power, the main noise of the system gradually changes from thermal noise to local-oscillator excess-intensity noise.

When the local-oscillator power is low, the thermal noise of the system dominates. With an increase in the local-oscillator power, the excess-intensity noise of the local oscillator begins to dominate, resulting in a decrease in the detection sensitivity. In addition, when the optical power of the local oscillator is constant, the greater the balance-consistency coefficient and the greater the detection sensitivity of the system.

Figure 4.2b shows the variation in the detection sensitivity of the system with respect to the optical power of the local oscillator when the relative-intensity noise coefficient R_{in} changes. With an increase in the optical power of the local oscillator, the detection sensitivity of the system first increases and then decreases. In addition, when the local-oscillator power is constant, the larger the value of $\mathrm{R_{in}}$, the smaller the detection sensitivity.

With a further increase in the local-oscillator power, the detector operates in the nonlinear region. When the bit-error rate is 10^{-9}, using the same parameters as in Fig. 4.1, the change curve of the detection sensitivity with the local-oscillator power can be obtained from Eq. (4.5), as shown in Fig. 4.3.

As shown in Fig. 4.3, when the nonlinear parameter υ of the detector remains unchanged, with an increase in the local-oscillator power, the detection sensitivity first increases and then decreases. In addition, as υ increases, the detection-sensitivity curve decreases faster because a larger nonlinear parameter υ causes a stronger nonlinear effect on the detector, thereby affecting the detection sensitivity of the system. Therefore, in practical applications, it is necessary to select an appropriate local-oscillator power and nonlinear parameter υ to maintain the linearity of the detector and obtain the best detection sensitivity.

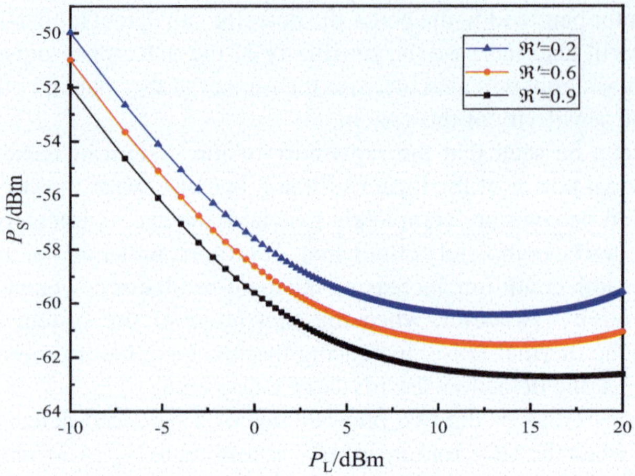

(a) Different equilibrium-consistency coefficient values $\Re'$

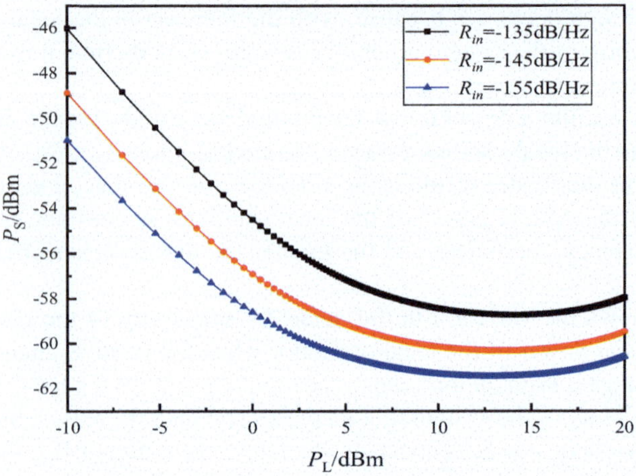

(b) Different relative intensity noises R_{in}

Fig. 4.2 Detection sensitivity as a function of local-oscillator optical power

4.1.2 Influence of Thermal Noise on the System Detection Sensitivity

According to Eq. (4.4), the thermal-noise power is related to factors such as the detector bandwidth, temperature, and load resistance. As the temperature increases, the thermal-noise power of the detector increases. Therefore, lowering the temperature is an effective method to reduce thermal noise. The detector bandwidth is also an important factor affecting the thermal noise. The smaller the bandwidth, the smaller

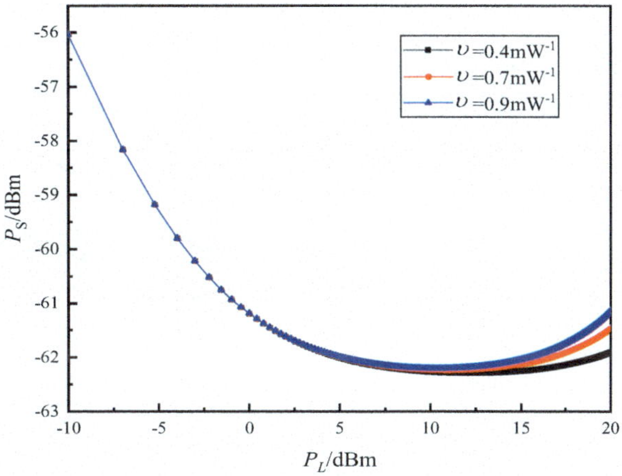

Fig. 4.3 Curve plot of the detection sensitivity with the local-oscillator power under different nonlinear parameters

the thermal-noise power. Therefore, reducing the bandwidth of the detector can also effectively reduce the thermal noise.

In Eq. (4.4), we set the following values: electronic charge $e = 1.6 \times 10^{-19}$ C, Planck's constant $h = 6.626 \times 10^{-34}$ J·s, carrier frequency $\upsilon = 1.9 \times 10^{14}$ m/s, $R_L = 50\ \Omega$, effective noise bandwidth $B = 200$ MHz, and $k = 1.38 \times 10^{-23}$ J/K. Figure 4.4 shows the variation of the detection sensitivity with the working background temperature under different detector quantum efficiencies η when the bit-error rate is 10^{-9}.

It can be seen from the simulation results in Fig. 4.4 that the system detection sensitivity decreases as the background temperature increases. This is because an increase in the background temperature increases the thermal-noise power. Moreover, the dark current of the optical mixer increases, thus introducing more shot

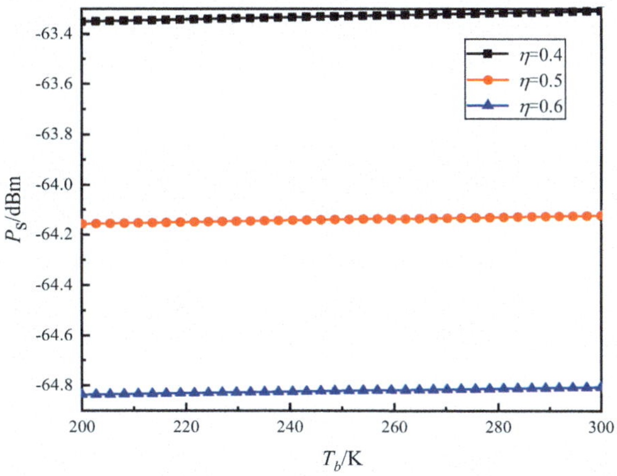

Fig. 4.4 Changes in detection sensitivity with the background temperature under different η values

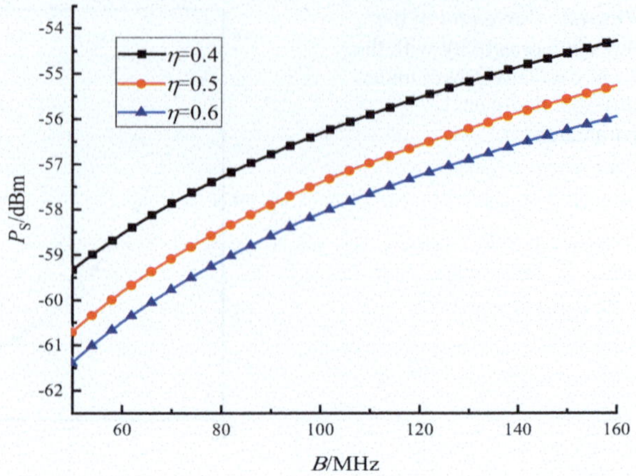

Fig. 4.5 Detection sensitivity varies with the detector bandwidth under different η values

noise. These noises affect the performance of the detector and reduce the detection sensitivity.

In addition, the quantum efficiency of the detector can be increased to improve the detection sensitivity. At the same temperature, the higher the quantum efficiency, the more beam signals the detector can receive; hence, the detection sensitivity increases accordingly.

In addition, limiting the bandwidth of the circuit has a significant inhibitory effect on the thermal noise. Using Eq. (4.4), we set $T_b = 300$ K and select the same parameters as those shown in Fig. 4.4. Figure 4.5 shows the system detection sensitivity varying with the detector bandwidth when the bit-error rate is 10^{-9} under different detector quantum efficiencies η.

The simulation results in Fig. 4.5 show that the detection sensitivity decreases with an increase in the detector bandwidth *B*. This is because the larger the detector bandwidth, the greater the corresponding thermal noise, which has a greater impact on the detection performance of the system, resulting in a gradual decrease in its detection sensitivity. In addition, if the quantum efficiency of the detector increases, the detection sensitivity increases at the same temperature.

4.1.3 Influence of the Proportional Coefficient of Photodiode Quantum Efficiency on the Detection Sensitivity of the System

The quantum efficiency of a photodiode refers to its ability to convert photoelectricity into an electrical signal. It is also an important factor affecting the detection sensitivity of a balanced-detection system. When the quantum efficiencies of the two detectors differ, the detection sensitivity of the balanced-detection system will be seriously affected.

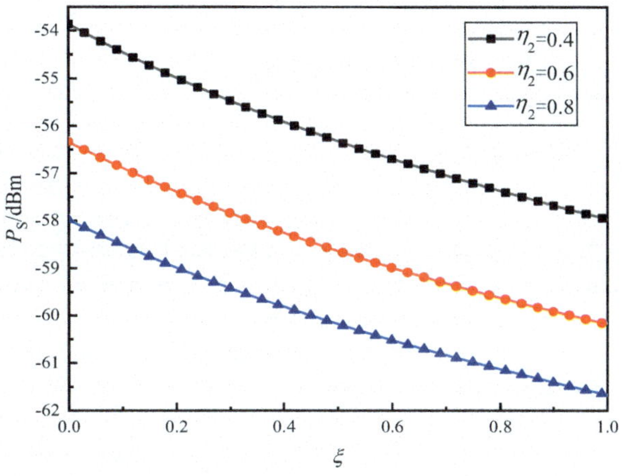

Fig. 4.6 Change curve of the detection sensitivity with k for different η_2 values

When studying the influencing factors of a balanced-detection system, it is necessary to consider the quantum efficiencies of the two detectors. We assume that they are η_1 and η_2, respectively, and there is a proportional relationship between them; that is, $\eta_1 = \xi\eta_2$ $(0 < \xi \leq 1)$. Then, the signal-to-noise ratio can be written as

$$SNR = \frac{(\xi\eta_2 + \eta_2)^2(e/hv)^2 P_S P_L R_L}{2e[e(\xi\eta_2 + \eta_2)/hv]BP_L R_L + 8KT_b B}. \tag{4.10}$$

When the bit-error rate is 10^{-9}, we select the same parameters as in Fig. 4.1. By Eq. (4.10), it can be obtained that under different detector quantum efficiencies η_2, the detection sensitivity of the system increases with the proportional coefficient of the photodiode quantum efficiency. The change curve is shown in Fig. 4.6.

According to the simulation results in Fig. 4.6, when the quantum efficiency of detector 2 is constant, the detection sensitivity of the system increases with the increase of the photodiode quantum-efficiency proportional coefficient ξ, and reaches the maximum value when $\xi = 1$. Therefore, it is best to select two detectors with equal quantum efficiencies to maximize the detection sensitivity of the system. Using detectors with different quantum efficiencies requires a more complex system design and optimization to achieve better detection sensitivity and performance. Other detector parameters also affect the detection sensitivity of the system. Therefore, in practical applications, various factors must be considered when selecting the optimal detector and system design scheme.

4.2 Influence of the Beam-Splitting Ratio on the System Detection Sensitivity

A balanced-detection system uses two photodiodes to detect the intensity of the optical signal. The intensity is determined by comparing the difference in the optical power received by the two photodiodes. A beam splitter is an important part of a balanced-detection system, and the beam-splitting ratio directly affects the detection sensitivity of the system. Figure 4.7 shows the equivalent diagram of one of the I and Q channels in the balanced detector.

After passing through the beam splitter, as shown in Fig. 4.7, the four beams formed are [2]

$$E_{S1} = \sqrt{(1-\varepsilon)}A_S \cos(\omega_S t) \tag{4.11}$$

$$E_{S2} = \sqrt{\varepsilon}A_S \cos(\omega_S t) \tag{4.12}$$

$$E_{L2} = \sqrt{(1-\varepsilon)}A_L \cos(\omega_L t) \tag{4.13}$$

$$E_{L1} = \sqrt{\varepsilon}A_L \cos(\omega_L t), \tag{4.14}$$

where ε is the beam-splitting ratio. The intermediate-frequency currents generated on the two photodetectors are, respectively,

$$i_1 = \sqrt{\varepsilon(1-\varepsilon)}A_S A_L \cos(\Delta\omega t) \tag{4.15}$$

$$i_2 = -\sqrt{\varepsilon(1-\varepsilon)}A_S A_L \cos(\Delta\omega t). \tag{4.16}$$

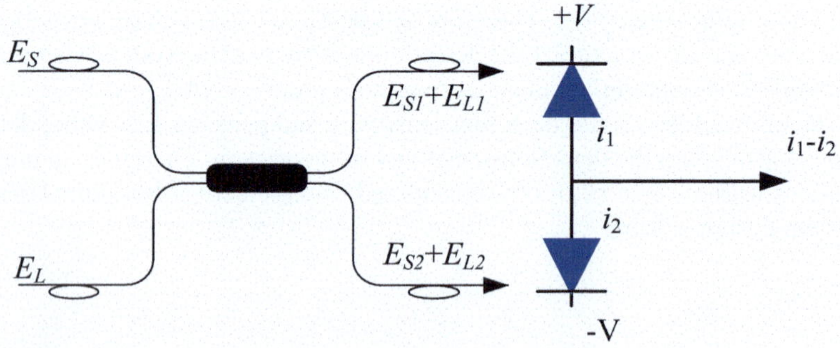

Fig. 4.7 Balanced heterodyne detection schematic

The power of the intermediate-frequency signal output by the balanced detector is

$$S = 8(e\eta/hv)^2 P_S P_L R_L \varepsilon(1-\varepsilon). \tag{4.17}$$

When the parameters of the two photodiodes are equal and the excess-intensity noise is not considered, the noise power can be expressed as

$$\begin{aligned} N &= N_1 - N_2 \\ &= 2e(e\eta/hv)BP_L R_L(1-\varepsilon) - 2e(e\eta/hv)BP_L R_L \varepsilon \\ &= 2e(e\eta/hv)BP_L R_L(1-2\varepsilon). \end{aligned} \tag{4.18}$$

N_1 and N_2 respectively represent the shot noise caused by the local-oscillator beam in the two photodetectors. Considering the thermal noise, the signal-to-noise ratio can be expressed as

$$SNR = \frac{8(e\eta/hv)^2 P_S P_L R_L \varepsilon(1-\varepsilon)}{2e(e\eta/hv)BP_L R_L(1-2\varepsilon) + 4KT_b B}. \tag{4.19}$$

When the bit-error rate is 10^{-9}, according to Eq. (4.19), using the same parameters as in the problem and Fig. 4.1, Fig. 4.8 shows the change curve of the detection sensitivity with the beam-splitting ratio.

As shown in Fig. 4.8, when the beam-splitting ratio is 0.5, that is, close to 1:1, the optical power received by the two detectors is equal and the system sensitivity is higher. If the beam-splitting ratio is too low or high, the detection sensitivity is reduced. Therefore, the beam-splitting ratio must be selected according to the specific

Fig. 4.8 Change curve of the detection sensitivity with beam-splitting ratio ε

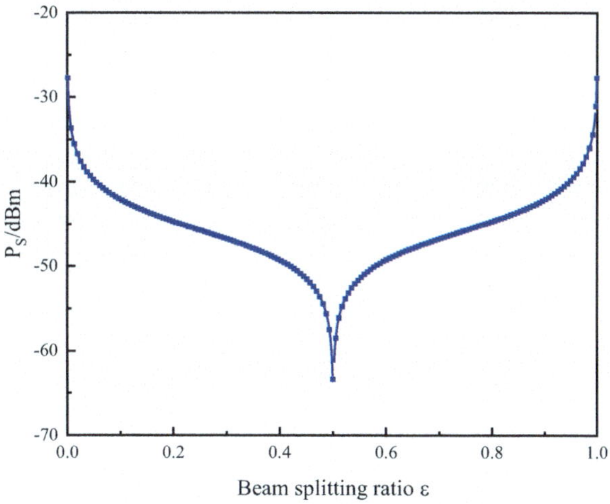

experimental conditions and required sensitivity. If a higher sensitivity is required, a beam splitter with a beam-splitting ratio close to 1:1 should be selected.

4.3 Influence of Beam-Field Mode Matching on the System Detection Sensitivity

In optical wireless communication systems, matching the optical-field modes is crucial for system performance. If the beam-field mode does not match, it leads to an uneven energy distribution of the received beam spot, which affects the detection sensitivity of the system. Only by ensuring that the beam-field modes match can the stable operation and efficient transmission of a spatial coherent laser-communication system be guaranteed.

In addition, if the coincidence degree of the beam spots is low, that is, their central positions do not coincide, it will cause the energy distribution of the beam spots received by the detector to be uneven, thereby reducing the detection sensitivity. Therefore, when designing a coherent laser-communication system, it is necessary to pay special attention to the problems of optical-field mode matching and spot coincidence to fully exploit the performance advantages of the system.

4.3.1 Local-Oscillator Beam Is Uniformly Distributed

An incident signal beam can be regarded as a plane wave, when transmitted over a long distance. When the signal is focused by a lens, an Airy disk can be formed at the focal point; the local-oscillator beam typically uses a Gaussian beam. Therefore, the Airy–Gaussian superposition field is the most common superimposition method. In addition, changing the spot diameter and spatial distribution of the local-oscillator beam results in different beam-field superposition effects, such as Airy–uniform plane waves or Airy–Airy-type superposition fields.

In the following analysis, the phases of the beams of the default signal beam and local-oscillator beam are matched; the center positions of the two beams are coincident when they simultaneously irradiate the photosensitive surface of the detector, and there is no angular deviation in the direction. The mixing-efficiency formula can be written as

$$\eta_{mixing} = \frac{\left|\int_0^{r_0} U_S(r)U_L(r)rdr\right|^2}{\int_0^{r_0} |U_L(r)|^2 rdr \int_0^{\infty} |U_S(r)|^2 rdr}, \tag{4.20}$$

where $U_S(r)$ and $U_L(r)$ are the amplitudes of the signal beam and local-oscillator beam, respectively. When the amplitude distributions of the signal beam and local-oscillator beam are an Airy disk and plane wave, respectively, the amplitudes can be written as [3]

$$U_S(r_0) = \frac{2J_1\left(\frac{\pi r_0}{F\lambda}\right)}{\frac{\pi r_0}{F\lambda}} \tag{4.21}$$

$$U_L(r_0) = 1, \tag{4.22}$$

where $J_1(\cdot)$ is the first-order Bessel function, r_0 is the radius of the photosensitive area of the detector, λ is the wavelength, and F is the F-number of the lens, which usually determines the size of the Airy spot. $F = f / d$, where f is the focal length and d is the effective aperture.

By substituting the beam-field amplitude distribution functions of the signal beam and local-oscillator beam (4.21) and (4.22) into Eq. (4.20), the mixing efficiency of the superposition of the two beam fields can be obtained as

$$\eta_{AU}(r_0) = 4\frac{\left[1 - J_0\left(\frac{\pi r_0}{F\lambda}\right)\right]^2}{\left(\frac{\pi r_0}{F\lambda}\right)^2}, \tag{4.23}$$

where $J_0(\cdot)$ is the 0th-order Bessel function, the electron charge $e = 1.6 \times 10^{-19}$ C, quantum efficiency $\eta = 0.8$, Planck's constant $h = 6.626 \times 10^{-34}$ J·s, carrier frequency $\upsilon = 1.9 \times 10^{14}$ m/s, detector bandwidth $B = 200$ MHz, and wavelength $\lambda =$ 1550 nm. According to Eqs. (2.48) and (4.23), under the condition of sensitivity 10^{-9}, the detection sensitivity of the system varies with the diameter of the photosensitive area. The change curve is shown in Fig. 4.9.

As shown in Fig. 4.9, with an increase in the radius of the photosensitive area of the detector, the detection sensitivity first increases to an extreme value and then decreases, indicating that the size of the detector influences the detection sensitivity.

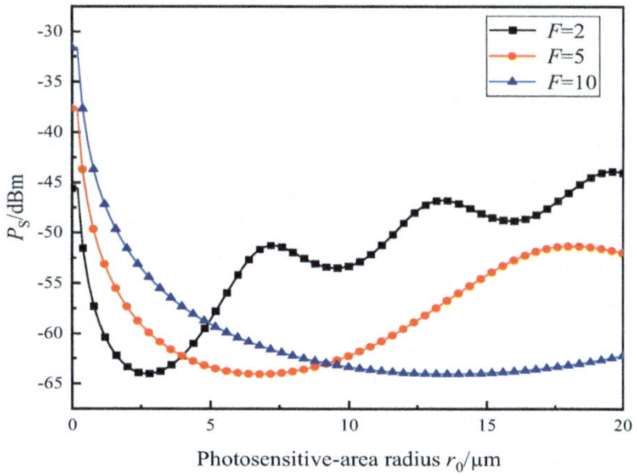

Fig. 4.9 Curve of the detection sensitivity as a function of the radius r_0 of the photosensitive area

In addition, when the radius of the photosensitive area of the detector remains unchanged, the larger the F-number of the lens, the larger the radius of the photosensitive area that is required to achieve maximum sensitivity. This is because when the F-number of the lens is large, the corresponding focusing distance is long, and the detector must have a large photosensitive area to fully receive the signal. In this simulation experiment, the best achievable sensitivity is -65 dBm.

4.3.2 Local-Oscillator Beam Has an Airy Distribution

When the signal beam and local-oscillator beam both have Airy distributions, their amplitude distributions are

$$U_S(r_0) = \frac{2J_1\left(\frac{\pi r_0}{F\lambda}\right)}{\frac{\pi r_0}{F\lambda}} \tag{4.24}$$

$$U_L(r_0) = \frac{2J_1\left(\frac{\pi r_0}{F\lambda}\right)}{\frac{\pi r_0}{F\lambda}}. \tag{4.25}$$

This allows us to obtain a formula in which the mixing efficiency varies with the radius r_0 of the photosensitive area of the detector:

$$\eta_{AA}(r_0) = 1 - J_0^2\left(\frac{\pi r_0}{F\lambda}\right) - J_1^2\left(\frac{\pi r_0}{F\lambda}\right). \tag{4.26}$$

Selecting the same parameters as in Fig. 4.9, from Eqs. (2.48) and (4.26), under the condition that the sensitivity is 10^{-9}, Fig. 4.10 shows the change curve of the detection sensitivity with the radius of the photosensitive area of the detector.

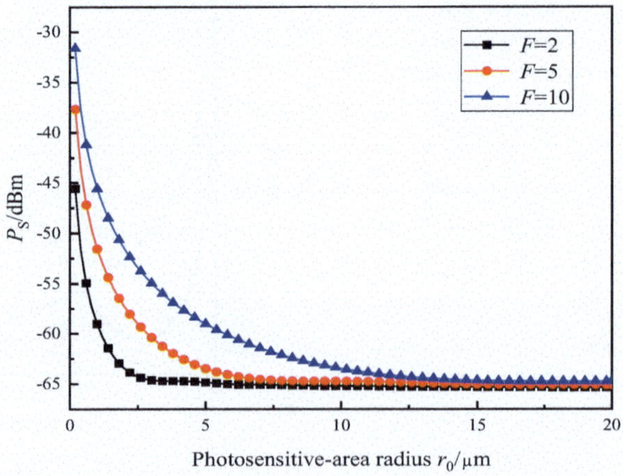

Fig. 4.10 Curve of the detection sensitivity as a function of the radius r_0 of the photosensitive area

As shown in Fig. 4.10, the detection sensitivity of the system increases rapidly with an increase in the radius of the photosensitive area of the detector, after which the change in sensitivity tends to stabilize. In addition, when the radius of the photosensitive area of the detector is constant, as the F-number of the lens increases, the radius of the photosensitive area of the detector that is needed to achieve the maximum sensitivity also increases. This is because the larger the F-number of the lens, the longer its focal length. The detector needs a larger size to be able to receive sufficient energy to achieve the best detection sensitivity.

4.3.3 Local-Oscillator Beam Has a Gaussian Distribution

When the signal beam and local-oscillator beam amplitude distributions are Gaussian and Airy, respectively, the amplitude distributions are [3]

$$U_S(r_0) = \exp\left(\frac{-r_0^2}{w_0^2}\right) \tag{4.27}$$

$$U_L(r_0) = \frac{2J_1\left(\frac{\pi r_0}{F\lambda}\right)}{\frac{\pi r_0}{F\lambda}}, \tag{4.28}$$

where w_0 is the beam waist radius of the Gaussian beam. The beam-field amplitude-distribution functions of the two beams are introduced into Eq. (4.20). The formula for the change in the mixing efficiency with the radius r_0 of the photosensitive area of the detector is obtained as follows:

$$\eta_{AG}(r_0) = \frac{8\left|\int_0^{r_0} J_1\left(\frac{\pi r_0}{F\lambda}\right) \cdot \exp\left(-\frac{r_0^2}{w_0^2}\right) dr\right|^2}{w_0^2\left[1 - \exp\left(-2\frac{r_0^2}{w_0^2}\right)\right]}. \tag{4.29}$$

Using the same parameters as in Fig. 4.9, the beam waist radius $w_0 = 1.6\lambda$ of the Gaussian beam, according to Eqs. (2.48) and (4.29), under the condition that the sensitivity is 10^{-9}, the detection sensitivity of the system varies with the radius of the photosensitive area of the detector. The change curve is shown in Fig. 4.11.

As shown in Fig. 4.11, the detection sensitivity of the system increases with an increase in the radius of the photosensitive area of the detector, as the Airy–Gaussian superposition field increases and then gradually tends to stabilize. In addition, the different F-numbers of the focusing lens affect the detection sensitivity. The changing law has an influence; with an increase in the F-number, the detection sensitivity gradually decreases. Based on the above analysis, the Airy–Gaussian pattern-matching mode is an effective choice for achieving stable and high sensitivity.

As shown in Figs. 4.9, 4.10and4.11, in coherent detection, the different mode-matching methods have different effects on the sensitivity and stability of the detector.

Fig. 4.11 Curve of the detection sensitivity as a function of the radius r_0 of the photosensitive area

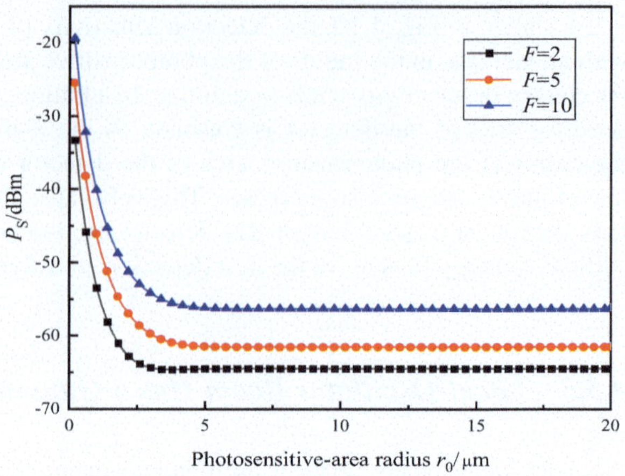

Airy–uniform mode matching has high requirements for the detector and its sensitivity is unstable; therefore, it is not suitable for high-sensitivity detection. Airy–Airy mode matching requires the photosensitive area of the detector to have a larger radius to achieve stable sensitivity. In contrast, Airy–Gaussian mode-matched systems have relatively high detection sensitivity and stability and are thus a better choice for stable, high-sensitivity spatial beam mixing.

4.4 Effect of the Spot Size on the Detection Sensitivity of the System

In addition to considering the beam-field mode matching of the signal and local-oscillator beams, their size ratios on the photosensitive surface of the detector are also an important factor that affects the detection sensitivity. Therefore, these factors must be comprehensively considered during system design to achieve the best detection effect.

Taking the common Airy–Gaussian superimposed field as an example, the influence of the size of the superimposed field of the local-oscillator and signal beams on the detection sensitivity is discussed. Assuming that the center of the signal beam is aligned exactly with the direction of the incident beam, the detector has a sufficiently large photosensitive area to receive all the energy of the incident beam. Usually, the spot size of an Airy beam is determined by the focal length and effective aperture of the lens, that is, the F-number. The spot size of a Gaussian beam is determined by its waist radius. Therefore, it is not directly related to the F-number.

To facilitate the analysis of practical applications, a parameter can be defined to compare the spot sizes of the Gaussian and Airy beams. This parameter is usually the ratio of the beam waist radius of the Gaussian beam to the F-number of the Airy beam, and can be written as [4]

$$\kappa = \frac{\pi r}{F\lambda} / \frac{r}{w_0} = \frac{\pi w_0}{F\lambda}. \tag{4.30}$$

Using X and Z to denote the variables in the signal beam and local-oscillator beam amplitude-distribution functions, respectively, where $X = \frac{\pi r}{F\lambda}$ and $Z = \frac{r}{w_0}$, according to Eq. (4.20), the following can be obtained:

$$\eta_{mixing} = \frac{8}{\kappa^2} \cdot \frac{\left[\int_0^{X_0} J_1(X) \exp\left(-\frac{X^2}{\kappa}\right)^2 dX\right]^2}{1 - \exp\left(-2\frac{X_0^2}{\kappa^2}\right)}. \tag{4.31}$$

Using Eqs. (2.48) and (4.31), we set the following values: electron charge $e = 1.6 \times 10^{-19}$ C, quantum efficiency $\eta = 0.8$, Planck's constant $h = 6.626 \times 10^{-34}$ J·s, carrier frequency $\upsilon = 1.9 \times 10^{14}$ m/s, detector bandwidth is $B = 200$ MHz, and the wavelength of the signal beam and the local-oscillator beam is $\lambda = 1550$ nm. Figure 4.12 shows the change curve of the detection sensitivity of the system with parameter κ when the bit-error rate is 10^{-9}.

It can be seen from Fig. 4.12 that as parameter κ increases, the detection sensitivity of the system first increases and then decreases. When the κ value is approximately 2.23, an optimal matching size exists that can maximize the signal on the photosensitive surface of the detector and improve the detection sensitivity of the system. Therefore, the relative size of the beam can be appropriately adjusted to further improve the detection sensitivity, thereby improving the communication performance of the coherent optical communication system.

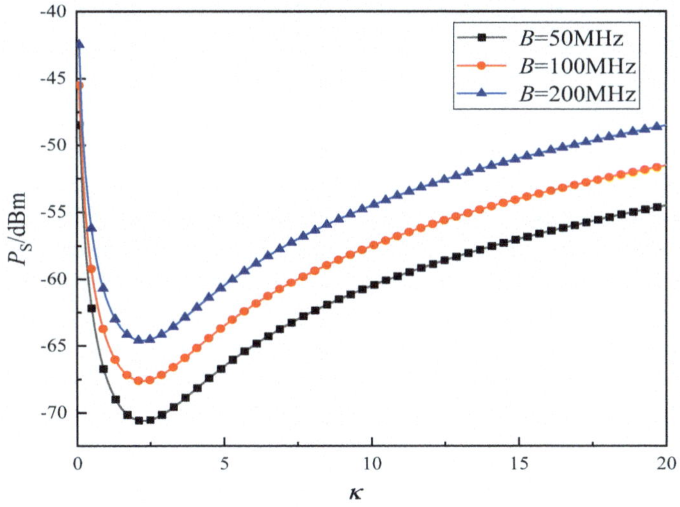

Fig. 4.12 Variation curve of the detection sensitivity with parameter κ

4.5 Influence of Alignment Errors on the Detection Sensitivity of the System

In an actual spatial optical coupling system, the diameter of the single-mode fiber is only at the micron level; owing to the existence of alignment errors, axial, radial, and deflection errors may occur during the process of assembling the optical-fiber coupling system. These errors may cause the beam position to deviate from the focal plane, thereby affecting the detection sensitivity of the system. In this section, the influence of these alignment errors on the detection sensitivity is analyzed.

In real situations, because the signal beam passes through the coupling of the antenna system and mixes with the local-oscillator beam before entering the detector, the power of the signal beam reaching the surface of the detector is less than the actual received power. Therefore, the expression for the quantum-noise limit can be modified by considering the influence of the coupling efficiency and mixing efficiency on the signal optical power, and a new expression for the signal-to-noise ratio is obtained:

$$SNR = \frac{2\eta P_S}{h\nu B} \cdot \eta_{mixing} \cdot \eta_o, \tag{4.32}$$

where η_{mixing} is the heterodyne mixing efficiency of the signal beam and the local-oscillator beam and η_o is the coupling efficiency of the receiving antenna to the optical fiber.

4.5.1 Influence of Axial Errors on the System Detection Sensitivity

In a wireless laser communication system, the signal beam is distorted by factors such as atmospheric turbulence during space transmissions; therefore, it is necessary to shape the beam using a shaping system to ensure the transmission quality. In general, after the shaping system, the beam is Gaussian because it has better transmission and focusing characteristics. When the signal beam is focused by a lens, it is coupled into the single-mode fiber at the focal plane of the lens, and an Airy-disk diffraction pattern is formed on the rear focal plane of the lens; that is, the spot of the signal beam coincides with the core diameter of the fiber, as shown in Fig. 4.13.

The plane wave at the aperture in Fig. 4.13 can be expressed as [5]

$$E_A(r) = circ(r)\exp[j\varphi(r)], \tag{4.33}$$

where $\varphi(r)$ is the distortion-phase difference and $circ(r)$ is the aperture function of the lens, which can be expressed as [5]

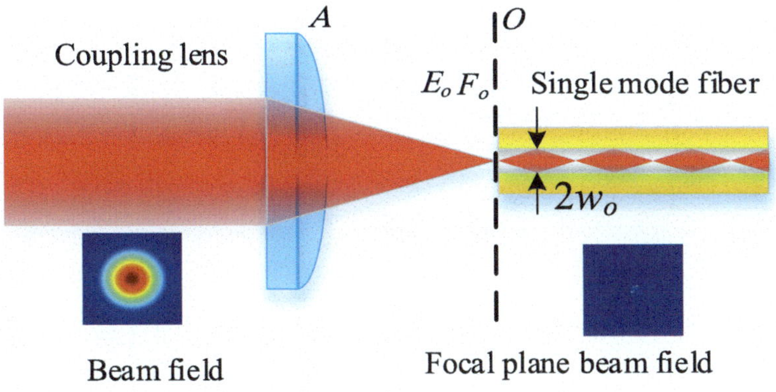

Fig. 4.13 Schematic diagram of optical fiber coupling

$$circ(r) = \begin{cases} 1 & r^2 \le D^2 \\ 0 & r^2 > D^2 \end{cases}, \quad (4.34)$$

where D is the diameter of the coupling lens and r is the radial distance from any point on the fiber cross section to the center.

Assuming that the fiber cross section is perpendicular to the incident-beam field and located at the center of the focal plane of the lens, because the Fourier change of a Gaussian function is still a Gaussian function, the mode-field distribution, converted from the normalized-fiber mode-field distribution to the surface of the receiving lens, can be expressed as [6]

$$F_o(r) = \sqrt{\frac{2}{\pi}} \frac{\pi w_0}{\lambda f} \exp\left[-r^2 \left(\frac{\pi w_0}{\lambda f}\right)^2\right], \quad (4.35)$$

where w_0 is the radius of the fiber mode field and λ is the wavelength. In fiber coupling, the beam radius and the angle of incidence of the incident-beam field should match the fiber mode. The polar coordinates of the coupling efficiency of a spatial beam coupled into a single-mode fiber can be expressed as

$$\eta_o = \frac{\left|\int\int E_o^*(r) F_o(r) r dr d\varphi\right|^2}{\int\int E_o^*(r) E_o(r) r dr d\varphi \cdot \int\int F_o^*(r) F_o(r) r dr d\varphi}, \quad (4.36)$$

where $E_o(r)$ is the beam-field distribution at the focal plane.

In optics, if the optical field in the entrance pupil plane is expressed as a time-domain signal, the optical field in the focal plane is expressed as a frequency-domain signal. According to Parseval's theorem, the energies in the time and frequency domains are equal. Therefore, the coupling efficiency calculated in the entrance pupil plane can be regarded as an expression of the coupling efficiency in the corresponding frequency domain, where the two are equivalent. Substituting Eqs. (4.33) and (4.35) into Eq. (4.36) and setting $\varphi(r) = 0$, the coupling efficiency can be expressed as

$$\eta_o = \left| \iint circ(r) \exp[j\varphi(r)] \frac{w_0\sqrt{2\pi}}{\lambda f} \exp\left[-r^2 \left(\frac{\pi w_0}{\lambda f} \right)^2 \right] r dr d\varphi \right|^2 = 2\frac{\left(1 - \exp\left(-\frac{\pi D w_0}{2\lambda f}\right)\right)^2}{\left(\frac{\pi D w_0}{2\lambda f}\right)^2}. \tag{4.37}$$

Let $\beta = \pi D w_0 / 2\lambda f$ be the coupling parameter; then, Eq. (4.37) can be written as

$$\eta_o = 2\frac{\left[1 - \exp(-\beta^2)\right]^2}{\beta^2}, \tag{4.38}$$

where f is the focal length and λ is the wavelength. When there is an error in the coupling-system assembly, the fiber may not be located on the focal plane of the incident beam, but may deviate from the focal plane in the axial direction, resulting in a Δz axial error. This is shown in Fig. 4.14.

When the influence of a lens aberration is not considered, the electromagnetic-field distribution at the fiber end face remains unchanged when there is an axial error Δz; however, the optical-field distribution of the incident beam changes at the fiber end face. This is because an axial misalignment changes the angle and position of the incident beam. When the aperture limitation is considered, the Fresnel diffraction theory can be used to describe the beam-field distribution at the axial error Δz of the focal point after the beam wave passes through a single lens.

According to the Fresnel diffraction formula, the following expression can be obtained:

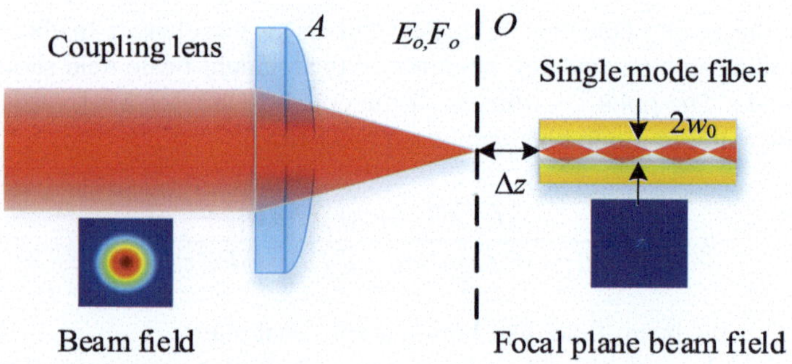

Fig. 4.14 Schematic diagram of an axial error

$$E_o(r) = \frac{1}{j\lambda(f+\Delta z)} \exp[jk(f+\Delta z)] \exp\left[\frac{jkr^2}{2(f+\Delta z)}\right]$$
$$\cdot F\left\{circ\left(\frac{r}{D/2}\right) \cdot \exp\left[-\frac{jkr^2}{2f}\right] \cdot \exp\left[\frac{jkr^2}{2(f+\Delta z)}\right]\right\}, \tag{4.39}$$

where F is a Fourier transform. When there is an axial error, the mode field distribution remains unchanged. After performing the Fourier transform on the optical-field distribution, the definitive expression of the coupling efficiency can be obtained by inserting it into Eq. (4.36):

$$\eta_o = 8\beta^2 \left| \frac{1}{R^2} \int_0^R \exp\left[-\frac{\beta^2 r^2}{R^2}\left(1 - j\frac{\lambda \Delta z}{\pi w_0^2}\right) rdr\right] \right|^2, \tag{4.40}$$

where R is the radius of the coupling lens. When the coupling parameter $\beta = 1.1209$, the coupling efficiency has the maximum extreme value [8]. Owing to the Fresnel diffraction, the coupling efficiency can be approximated as

$$\eta_o = 0.8145 \exp\left[-\left(\frac{\lambda \Delta z}{2\sqrt{2}\pi w_0^2}\right)^2\right]. \tag{4.41}$$

Using Eqs. (2.48), (4.32), and (4.41), we set the bit-error rate as 10^{-9}, electron charge $e = 1.6 \times 10^{-19}$ C, quantum efficiency $\eta = 0.8$, Planck's constant $h = 6.626 \times 10^{-34}$ J·s, carrier frequency $\upsilon = 1.9 \times 10^{14}$ m/s, detector bandwidth $B = 200$ MHz, and wavelength $\lambda = 1550$ nm. Figure 4.15 shows the change curve of the system detection sensitivity with axial error Δz.

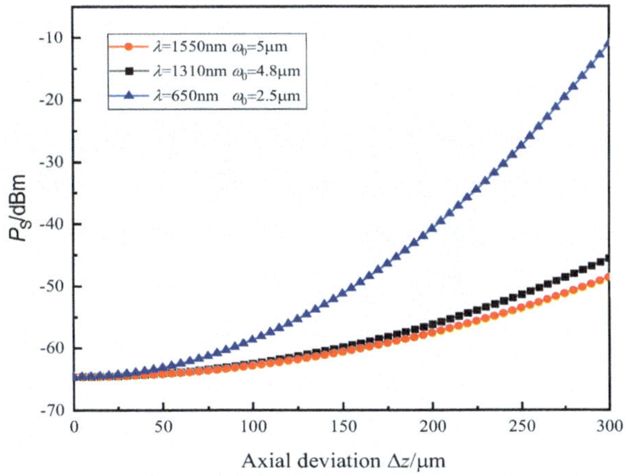

Fig. 4.15 Curve of the detection-sensitivity variation with an axial error

As shown in Fig. 4.15, the detection sensitivity of the system decreases continuously with an increase in the axial error. When the axial error exceeds 50 μm, the detection sensitivity decreases gradually. This is due to the increase in the axial error and the increase in the distance between the fiber end face and the strongest point of the Airy-disc beam intensity, resulting in the weakening of the average optical power in the fiber. In addition, when $\beta = 1.1209$, the greater the wavelength λ, the greater the change of the system-detection sensitivity with the axial error. When the axial error is constant, the greater the wavelength λ, the greater the detection sensitivity of the system.

4.5.2 *Influence of Radial Errors on the System Detection Sensitivity*

Figure 4.16 shows an example of when the optical fiber at the receiving end of the coupling system has a radial error.

The electromagnetic-field distribution at the fiber end face is

$$F_o(r) = \sqrt{\frac{2}{\pi w_0^2}} \exp\left[-\left(\frac{r + \Delta r}{w_0}\right)^2\right]. \tag{4.42}$$

When there is a radial error, the beam-field distribution at the focal plane of the parallel beam remains unchanged after passing through the lens. By substituting the electromagnetic-field distribution at the end face of the fiber and the optical-field distribution at the focal plane into Eq. (4.32), the coupling efficiency is obtained as follows:

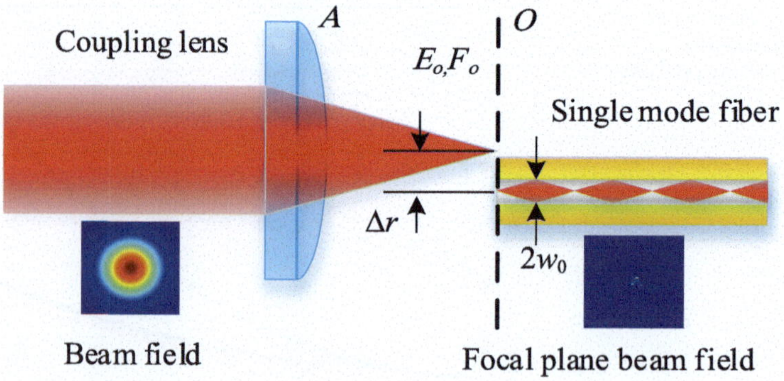

Fig. 4.16 Schematic diagram of a radial error

$$\eta_o = 8\beta^2 \left| \frac{1}{R^2} \int_0^R \exp(-\beta^2) J_0\left(\frac{2r\Delta r}{w_0 R}\beta\right) r dr \right|^2, \tag{4.43}$$

where $J_0(\cdot)$ is the zero-order modified Bessel function of the first kind. Setting $\beta = 1.1209$ and considering the Fresnel reflection, the coupling efficiency can be approximately expressed as

$$\eta_o = 0.8145 \exp\left[-\left(\frac{\Delta r}{w_0}\right)^2\right]. \tag{4.44}$$

Using Eqs. (2.48), (4.32), and (4.44), under the condition of sensitivity of 10^{-9}, Fig. 4.17 shows the change curve of the detection sensitivity of the system with radial error Δr.

As shown in Fig. 4.17, the detection sensitivity of the system decreases continuously with an increase in the radial error Δr. When the axial error exceeds 1 μm, the detection sensitivity decreases gradually. In addition, when $\beta = 1.1209$, the greater the wavelength λ, the greater the change of system detection sensitivity with a radial error. When the radial error is constant, the greater the wavelength λ, the greater the detection sensitivity.

Comparing Figs. 4.15 and 4.17, it can be seen that the system sensitivity is more easily affected by radial errors; therefore, it is necessary to be more precise when adjusting the coupling device to minimize the radial error and ensure the best detection sensitivity within the allowable error range.

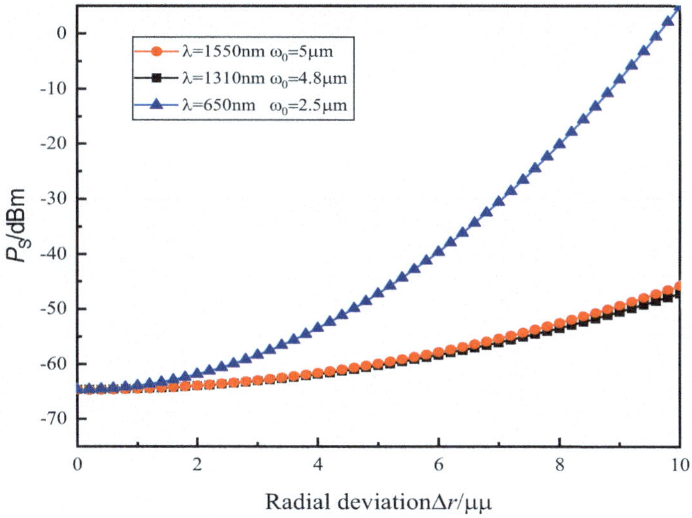

Fig. 4.17 Curve of the detection-sensitivity variation with a radial error

4.5.3 Influence of Deflection Errors on the System Detection Sensitivity

When the cross section of an optical fiber is located at the focal plane of the lens, the beam spot converges on the focal point after passing through the lens; an angle error will appear between the straight line where the focal point is located and the end face of the optical fiber. This error is the angular error; that is, the angle between the end face of the fiber and the focal point of the spot. This is shown in Fig. 4.18.

At this time, the optical-field distribution of the incident beam at the fiber end face remains unchanged. However, under the ideal condition of the fiber end face, the lens single-mode fiber coupling electromagnetic-field distribution will be deflected, and the optical-field distribution of the fiber-output mode field reversely transmitted to the receiving aperture; it can be expressed as follows [5]:

$$E_o = \sqrt{\frac{2}{\pi}}\frac{1}{w_L}\exp\left(-\frac{(r-\Delta\theta f)^2}{w_L^2}\right), \tag{4.45}$$

where $w_L = \lambda f / \pi w_0$ denotes the beam radius of the receiving aperture. When there is an angular error, the coupling-efficiency expression is

$$\eta_o = 8\beta^2\exp\left[-2\left(\frac{\pi\Delta\theta w_0}{\lambda}\right)^2\right]\left|\int_0^1 \exp(-\beta^2\gamma^2)J_0\left[\frac{D\Delta\theta}{f}\left(\frac{\pi w_0}{\lambda}\right)^2 r\right]rdr\right|^2, \tag{4.46}$$

where $J_0(\cdot)$ is a zero-order modified Bessel function of the first kind, and $\gamma = \pi\Delta\theta w_0 / \lambda$ is the normalized angular error. When $\beta = 1.1209$, and considering the Fresnel reflection phenomenon, the fiber-coupling efficiency can be approximately expressed as

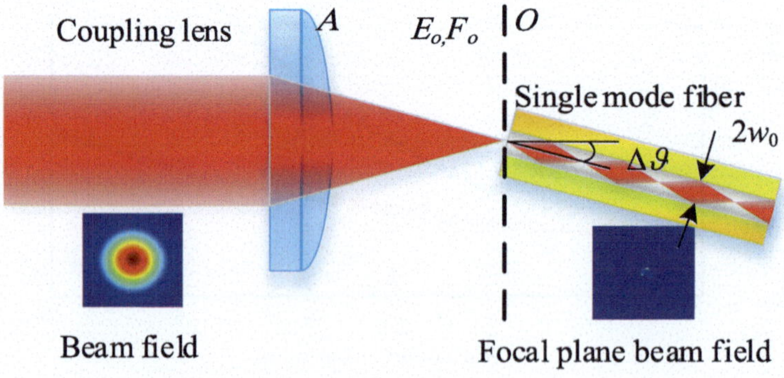

Fig. 4.18 Schematic diagram of an angular error

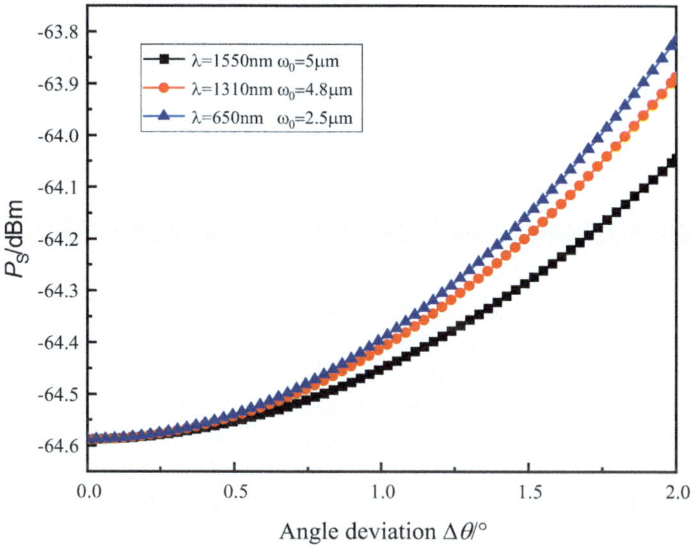

Fig. 4.19 Curve of the detection-sensitivity variation with an angular error

$$\eta_o = 0.8145 \exp\left[-\left(\frac{\pi w_0 \Delta\theta}{\lambda}\right)^2\right]. \tag{4.47}$$

Using Eqs. (2.48), (4.28), and (4.47), when the bit-error rate is 10^{-9}, Fig. 4.19 shows the change curve of the detection sensitivity of the system with angular error $\Delta\theta$.

As shown in Fig. 4.19, the detection sensitivity of the system decreases continuously with an increase in the angular error $\Delta\theta$. When the axial error exceeds 0.5°, the detection sensitivity of the system decreases gradually. This is because the larger the angle of the error, the more the beam is shifted from the plane of focus. In addition, when $\beta = 1.1209$, the greater the wavelength λ, the greater the change of the system detection sensitivity with the angle error. When the angular error is constant, the greater the wavelength λ, the greater the detection sensitivity of the system.

4.6 Influence of the Partially Coherent EGSM Beam Polarization State on the Coherent-Detection Sensitivity

In addition to the amplitude and phase properties of coherent beams, the polarization properties of partially coherent beams are equally important. In this section, the expression for the sensitivity of a heterodyne detection system is derived when a partially coherent EGSM beam travels obliquely through atmospheric turbulence. The influences of the power-spectrum power law, turbulence scale, transmission distance, receiving height, zenith angle, and detector photosensitive-area diameter on the detection sensitivity of electromagnetic Gaussian–Schell-model beams transmitted through uplinks and downlinks under different polarization states were analyzed.

4.6.1 Basic Theory of Partially Coherent EGSM Beams

A Gaussian beam is a scalar beam that involves only the amplitude and phase of an electromagnetic-wave field, and does not include the vibration directions of the electric and magnetic fields. Therefore, the Gaussian beam itself has no polarization properties. However, when the beam passes through some media, such as turbulent media, the turbulence will increase the randomness of the wavefront, which will cause the original coherent beam to become partially coherent. Among partially coherent beam sources, Gaussian–Schell sources produce beams with the same beam-field intensity distribution in the far field as fully coherent Gaussian beams.

In addition, the laser itself may have multimode oscillations; therefore, the laser-beam output may exhibit partially coherent characteristics. The Gaussian–Schell model (GSM) can be used to describe the shape of this partially coherent beam. Typically, partially coherent GSM beams are used to describe scalar-beam fields, ignoring their vector properties; thus, they can be regarded as completely polarized. However, the beam field is often not absolutely polarized, and partial polarization of the electromagnetic field occurs; therefore, the beam field usually exists in the form of a partially polarized beam [7].

The electromagnetic Gaussian–Schell model (EGSM) is a vector-beam field that represents the electromagnetic-wave field as a complex vector field, where each vector represents the vibration direction and strength of the electric and magnetic fields. Thus, partially coherent EGSM beams can directly describe the polarization properties of a beam. This section introduces the beam coherence–polarization matrix, unified theory of polarization coherence, and cross-plane density matrix for partially coherent vector beams.

4.6.1.1 Coherence–Polarization Matrix of an EGSM Beam

Gori et al. proposed using the cross-coherence function of the Cartesian components of a beam field to describe its polarization state, and thereby constructed a beam coherence–polarization (BCP) matrix [8]. This method can accurately describe the polarization characteristics of beams and can be applied to various complex polarization problems.

The Cartesian component of the beam field includes electric-field components in the horizontal and vertical directions. The mutual-coherence function between these two components can be expressed as a 2×2 matrix, and its expression is

$$\Gamma(\boldsymbol{r}_1, \boldsymbol{r}_2, z; \tau) = \begin{bmatrix} \Gamma_{xx}(\boldsymbol{r}_1, \boldsymbol{r}_2, z; \tau) & \Gamma_{xy}(\boldsymbol{r}_1, \boldsymbol{r}_2, z; \tau) \\ \Gamma_{yx}(\boldsymbol{r}_1, \boldsymbol{r}_2, z; \tau) & \Gamma_{yy}(\boldsymbol{r}_1, \boldsymbol{r}_2, z; \tau) \end{bmatrix}, \tag{4.48}$$

where

$$\Gamma_{\alpha\beta}(r_1, r_2, z; \tau) = \left\langle u_\alpha^*(r_1, r_2, z; t) u_\beta(r_1, r_2, z; t + \tau)\right\rangle, (\alpha, \beta = x, y). \tag{4.49}$$

The Cartesian components are expressed as

$$u_\alpha(r_1, r_2, z; t) = E_\alpha(r_1, r_2, z; t) \exp\{i[\phi_\alpha(t) - 2\pi \overline{v} t]\}, \tag{4.50}$$

where E_α is the amplitude of the beam field, ϕ_α is the random phase of the beam field, $\overline{v}$ is the average frequency of the beam field, and Δv is the effective bandwidth. It can be seen from Eq. (4.49) that the time statistics of the beam field are only related to the random phase, and are not related to the amplitude; therefore, according to Eq. (4.50), Eq. (4.49) can be further expressed as [9]

$$\begin{aligned} \Gamma_{\alpha\beta}(\boldsymbol{r}_1, \boldsymbol{r}_2, z; \tau) &= \left\langle E_\alpha^*(\boldsymbol{r}_1, \boldsymbol{r}_2, z; t) E_\beta(\boldsymbol{r}_1, \boldsymbol{r}_2, z; t)\right\rangle \exp(-i2\pi \overline{v} \tau) \\ &\quad \times \langle \exp\{i[\phi_\alpha(t + \tau) - \phi_\alpha(t)]\}\rangle \end{aligned} \tag{4.51}$$

$$\begin{aligned} \Gamma_{\alpha\beta}(\boldsymbol{r}_1, \boldsymbol{r}_2, z; \tau) &= \left\langle E_\alpha^*(\boldsymbol{r}_1, \boldsymbol{r}_2, z; t) E_\beta(\boldsymbol{r}_1, \boldsymbol{r}_2, z; t)\right\rangle \\ &\quad \times \exp(-i2\pi \overline{v} \tau) exp(-\pi \Delta v |\tau|). \end{aligned} \tag{4.52}$$

When the time delay τ of the beam field is very small relative to the coherence time 1/Δv, Eq. (4.5) can be written as

$$\Gamma_{\alpha\beta}(\boldsymbol{r}_1, \boldsymbol{r}_2, z; \tau) = J_{\alpha\beta}(\boldsymbol{r}_1, \boldsymbol{r}_2, z) \exp(-i2\pi \overline{v} \tau), \tag{4.53}$$

where

$$J_{\alpha\beta}(\boldsymbol{r}_1, \boldsymbol{r}_2, z) = \left\langle E_\alpha^*(\boldsymbol{r}_1, \boldsymbol{r}_2, z; t) E_\beta(\boldsymbol{r}_1, \boldsymbol{r}_2, z; t)\right\rangle. \tag{4.54}$$

Equation (4.54) is also part of the mutual coherence function:

$$J(\boldsymbol{r}_1, \boldsymbol{r}_2, z) = \begin{bmatrix} J_{xx}(\boldsymbol{r}_1, \boldsymbol{r}_2, z) & J_{xy}(\boldsymbol{r}_1, \boldsymbol{r}_2, z) \\ J_{yx}(\boldsymbol{r}_1, \boldsymbol{r}_2, z) & J_{yy}(\boldsymbol{r}_1, \boldsymbol{r}_2, z) \end{bmatrix}. \tag{4.55}$$

Matrix formula (4.55) can be used to describe the polarization characteristics and spatial-coherence properties of the beam field. The polarization characteristics describe the intensity distribution of the beam field in different directions, whereas the spatial coherence describes the intensity distribution of the beam field at different positions. These two characteristics can be analyzed through a coherence–polarization matrix, namely Eq. (4.55), whose normalized expression is [8]

$$j_{\alpha\beta}(\boldsymbol{r}_1, \boldsymbol{r}_2, z) = \frac{J_{\alpha\beta}(\boldsymbol{r}_1, \boldsymbol{r}_2, z)}{\sqrt{J_{\alpha\alpha}(\boldsymbol{r}_1, \boldsymbol{r}_2, z) J_{\beta\beta}(\boldsymbol{r}_1, \boldsymbol{r}_2, z)}}. \tag{4.56}$$

It can be seen that when $\left|j_{\alpha\beta}\right| \leq 1$, the normalized matrix elements j_{xx} and j_{yy} respectively represent the complex coherence of the beam field in the x- and y-directions. The mutual-coherence functions corresponding to these matrix elements describe the relationship between the two polarization components at different positions in the beam field. Using these matrix elements, the polarization characteristics of a beam at any position in space can be derived and represented by the BCP matrix [8]:

$$P(\boldsymbol{r}, z) = \sqrt{1 - \frac{4\mathrm{Det}J(\boldsymbol{r}, \boldsymbol{r}, z; \omega)}{\mathrm{Tr}J(\boldsymbol{r}, \boldsymbol{r}, z; \omega)}}. \tag{4.57}$$

As this chapter studies a partially coherent electromagnetic Gaussian–Schell model beam, the BCP matrix of an EGSM beam can be expressed as

$$J_{ij}(r_1, r_2) = I_{ij} \exp\left[-\frac{r_1^2 + r_2^2}{4w_i^2} - \frac{(r_1 - r_2)^2}{2\delta_{ij}^2}\right], \tag{4.58}$$

where $\boldsymbol{r}_1$ and $\boldsymbol{r}_2$ are two position vectors on the source plane, I_{ij} is the beam intensity, w_i is the beam waist radius in the x-direction or y-direction of the beam field, and δ_{ij} is the spatial-coherence length, $i, j = x, y$.

4.6.1.2 Unified Theory of Polarization Coherence and Cross-spectral Density Matrix of EGSM Beams

The coherence–polarization matrix of the beam is used to describe the polarization state of the beam, mainly the polarization distribution of the beam in space; however, for the temporal characteristics of the beam field, the coherence–polarization matrix cannot directly describe the temporal coherence characteristics of the beam. To solve this problem, Wolf proposed a unified theory of polarization coherence based on the coherence–polarization matrix of the beam combined with the cross-spectral density function of the mutual-coherence function, which can describe the temporal coherence and polarization characteristics of the beam [10].

This theory introduces the concept of the polarization state, and the degree of coherence unifies the polarization and coherence properties of a beam. The cross-spectral density matrix of the two components of the beam field is obtained according to the Fourier transform of the mutual-coherence matrix in Eq. (4.49), which can be written as

$$W(\boldsymbol{r}_1, \boldsymbol{r}_2, z; \omega) = \frac{1}{2\pi} \int_{-\infty}^{\infty} \Gamma(\boldsymbol{r}_1, \boldsymbol{r}_2, z; \tau) \exp(i\omega\tau) d\tau. \tag{4.59}$$

The spectral density matrix $W(\boldsymbol{r}_1, \boldsymbol{r}_2, \omega)$ is specifically expressed as [10]

$$W(\boldsymbol{r}_1, \boldsymbol{r}_2, z; \omega) = \begin{bmatrix} W_{xx}(\boldsymbol{r}_1, \boldsymbol{r}_2, z; \omega) & W_{xy}(\boldsymbol{r}_1, \boldsymbol{r}_2, z; \omega) \\ W_{yx}(\boldsymbol{r}_1, \boldsymbol{r}_2, z; \omega) & W_{yy}(\boldsymbol{r}_1, \boldsymbol{r}_2, z; \omega) \end{bmatrix}, \tag{4.60}$$

where

$$W_{ij}(\boldsymbol{r}_1, \boldsymbol{r}_2, z; \omega) = \langle E_i^*(\boldsymbol{r}_1, z; \omega) E_j(\boldsymbol{r}_2, z; \omega) \rangle, \ (i, j = x, y). \tag{4.61}$$

For random electromagnetic beams, the cross-spectral density function can be used to describe the interactions between two random beam fields. By analyzing the cross-spectral density function, information such as the power-spectral density, spectral-coherence degree, and spectral-polarization degree of each frequency component of a random electromagnetic beam can be obtained.

The power-spectral density of a random electromagnetic beam can be expressed as

$$S(\boldsymbol{r}, z; \omega) = \mathrm{Tr} W(\boldsymbol{r}, \boldsymbol{r}, z; \omega), \tag{4.62}$$

where Tr is the trace of matrix $W(\boldsymbol{r}, \boldsymbol{r}, z; \omega)$ and the spectral coherence is [10]

$$\mu(\boldsymbol{r}_1, \boldsymbol{r}_2, z; \omega) = \frac{\mathrm{Tr} W(\boldsymbol{r}_1, \boldsymbol{r}_2, z; \omega)}{\sqrt{\mathrm{Tr} W(\boldsymbol{r}_1, \boldsymbol{r}_1, z; \omega)} \sqrt{\mathrm{Tr} W(\boldsymbol{r}_2, \boldsymbol{r}_2, z; \omega)}}, \tag{4.63}$$

where *Det* represents the determinant of the matrix. In addition,

$$P(\boldsymbol{r}, z; \omega) = \sqrt{1 - \frac{4 Det W(\boldsymbol{r}, \boldsymbol{r}, z; \omega)}{\mathrm{Tr} W(\boldsymbol{r}, \boldsymbol{r}, z; \omega)}}. \tag{4.64}$$

For an electromagnetic Gaussian–Schell-model beam, the form of the cross-spectral density matrix element can be expressed as [11, 12]

$$\begin{aligned} W_{ij}(\boldsymbol{r}_1, \boldsymbol{r}_2, z = 0; \omega) &= \sqrt{S_i(\boldsymbol{r}, z = 0; \omega)} \sqrt{S_j(\boldsymbol{r}, z = 0; \omega)} \\ &\times \mu_{ij}(\boldsymbol{r}_1 - \boldsymbol{r}_2, z = 0; \omega), \end{aligned} \tag{4.65}$$

where $S_i(\boldsymbol{r}_1, \boldsymbol{r}_2, z = 0; \omega)$ and $S_j(\boldsymbol{r}_1, \boldsymbol{r}_2, z = 0; \omega)$ are the power-spectral densities of the two Cartesian components of the electromagnetic beam, and $\mu_{ij}(\boldsymbol{r}_1 - \boldsymbol{r}_2, z = 0; \omega)$ is the spectral correlation of the two Cartesian components of the electromagnetic beam. The EGSM beams are all Gaussian and can be written as [13]

$$S_i(\boldsymbol{r}, z = 0; \omega) = A_i^2 \exp\left(-\frac{2\boldsymbol{r}^2}{w_i^2}\right), (i = x, y) \tag{4.66}$$

$$\mu(\boldsymbol{r}_1 - \boldsymbol{r}_2, z = 0; \omega) = B_{ij} \exp\left[-\frac{(\boldsymbol{r}_1 - \boldsymbol{r}_2)^2}{2\delta_{ij}^2}\right], \tag{4.67}$$

where A_i is the amplitude in the x-direction or y-direction of the beam field, w_i is the beam waist radius in the x-direction or y-direction of the beam field, δ is the spatial-coherence length, describes the degree of correlation between two points in the beam field, and B_{ij} is the Cartesian beam-field component correlation coefficient.

4.6.2 Influence of the Polarization State of Partially Coherent EGSM Beams on the Coherent-Detection Sensitivity Under Atmospheric Turbulence

When the beam is transmitted in an atmospheric channel, it is affected by the random fluctuation of the atmospheric refractive index and experiences atmospheric scintillation, angle-of-arrival fluctuation, and beam drift. These effects significantly degrade the performance of optical wireless communication links. Our study found that compared with coherent beams, partially coherent beams are less affected by turbulence and have stronger stability.

In this section, an expression for the sensitivity of a heterodyne-detection system is derived when a partially coherent EGSM beam is transmitted obliquely under atmospheric turbulence, and the influence of different polarization states on the detection sensitivity of the system is analyzed.

4.6.2.1 Mathematical Model of the Influence of the Polarization State of a Partially Coherent EGSM Beam on the Detection Sensitivity Under Atmospheric Turbulence

This section considers the electromagnetic Gaussian–Schell-model (EGSM) beam as the research object and discusses the system performance when the EGSM beam is used for heterodyne detection under a non-Kolmogorov turbulent flow. As shown in Fig. 4.20, it is assumed that the signal and local-oscillator beams are both partially coherent EGSM beams that are incident perpendicular to the detector surface and mixed at z = 0. k_o and k_s are the waves of the local-oscillator beam and signal beam, respectively.

Then, the instantaneous beam fields of the local-oscillator beam $U_o(\rho, t)$ and signal beam $U_s(\rho, t)$ on the detector surface are expressed as [14]

$$U_o(\boldsymbol{\rho}, t) = \left[U_{ox}(\boldsymbol{\rho})\boldsymbol{e}_x + U_{oy}(\boldsymbol{\rho})\boldsymbol{e}_y\right] \exp(j\omega_o t) \tag{4.68}$$

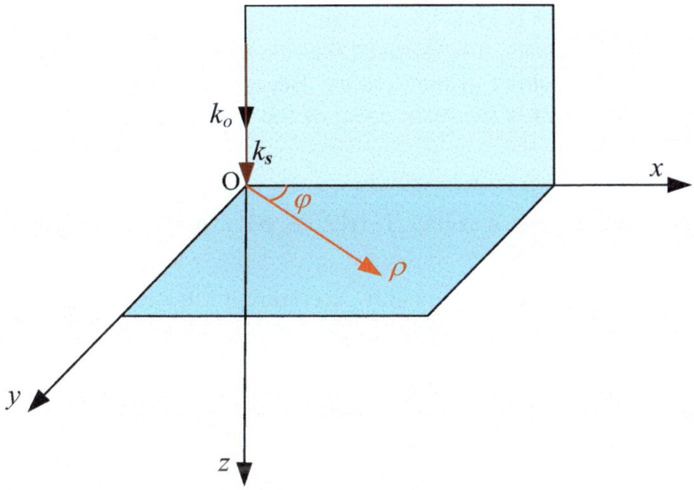

Fig. 4.20 Heterodyne beam-detection model on the z = 0 plane

$$U_s(\boldsymbol{\rho}, t) = \left[U_{sx}(\boldsymbol{\rho})\boldsymbol{e}_x + U_{sy}(\boldsymbol{\rho})\boldsymbol{e}_y\right]\exp(j\omega_s t), \tag{4.69}$$

where e_x and e_y are unit vectors, ω_o and ω_s respectively represent the central angular frequencies of the local-oscillator beam and signal beam, and ρ represents a two-dimensional transverse vector perpendicular to the beam-propagation direction.

The polarization characteristics and spatial-coherence properties of partially coherent beams can be represented by the beam coherence–polarization (BCP) matrix J. BCP matrices $J_s(r_1, r_2)$ and $J_o(r_1, r_2)$ of the signal and local-oscillator beams, respectively, propagating in the z-direction can be expressed as follows [14]:

$$J_s(\boldsymbol{\rho}_1, \boldsymbol{\rho}_2) = \begin{bmatrix} \left\langle U_{sx}^*(\boldsymbol{\rho}_1)U_{sx}(\boldsymbol{\rho}_2)\right\rangle & \left\langle U_{sx}^*(\boldsymbol{\rho}_1)U_{sy}(\boldsymbol{\rho}_2)\right\rangle \\ \left\langle U_{sy}^*(\boldsymbol{\rho}_1)U_{sx}(\boldsymbol{\rho}_2)\right\rangle & \left\langle U_{sy}^*(\boldsymbol{\rho}_1)U_{sy}(\boldsymbol{\rho}_2)\right\rangle \end{bmatrix} \tag{4.70}$$

$$J_o(\boldsymbol{\rho}_1, \boldsymbol{\rho}_2) = \begin{bmatrix} \left\langle U_{ox}^*(\boldsymbol{\rho}_1)U_{ox}(\boldsymbol{\rho}_2)\right\rangle & \left\langle U_{ox}^*(\boldsymbol{\rho}_1)U_{oy}(\boldsymbol{\rho}_2)\right\rangle \\ \left\langle U_{oy}^*(\boldsymbol{\rho}_1)U_{ox}(\boldsymbol{\rho}_2)\right\rangle & \left\langle U_{oy}^*(\boldsymbol{\rho}_1)U_{oy}(\boldsymbol{\rho}_2)\right\rangle \end{bmatrix}, \tag{4.71}$$

where * indicates the complex conjugate and $< \cdot >$ indicates the statistical mean.

According to the principles of heterodyne and balanced detection, the power of the intermediate-frequency signal is obtained as [15]

$$P_{IF} = 2\Re^2 \iint \iint \mathrm{Re}\{\mathrm{Tr}[J_o(\boldsymbol{\rho}_1, \boldsymbol{\rho}_2)J_s(\boldsymbol{\rho}_1, \boldsymbol{\rho}_2)]\}d^2\boldsymbol{\rho}_1 d^2\boldsymbol{\rho}_2, \tag{4.72}$$

where Tr represents the trace of the matrix, Re represents the real part of the complex number, $\Re = e\eta / h\nu$ is the response of the detector, e is the electron charge, η is the quantum efficiency, h is Planck's constant, and v is the optical frequency. Here, for convenience of analysis, it is assumed that the response $\Re$ of the detector does not change with the position of the detector. Because the shot noise generated by the local-oscillator beam is the main noise of the system, the noise power can be expressed as [14]

$$P_n = eB\Re \iint \mathrm{Tr}\left[J_o(\boldsymbol{\rho}, \boldsymbol{\rho})\right] d^2\boldsymbol{\rho}, \tag{4.73}$$

where B denotes the detector bandwidth. According to its definition, the SNR of a coherent-detection system for a partially coherent EGSM beam is [14]

$$SNR_{EGSM} = \frac{2\Re \iint \iint \mathrm{Re}\left\{\mathrm{Tr}[J_o(\boldsymbol{\rho}_1, \boldsymbol{\rho}_2) J_s(\boldsymbol{\rho}_1, \boldsymbol{\rho}_2)]\right\} d^2\boldsymbol{\rho}_1 d^2\boldsymbol{\rho}_2}{eB \iint \mathrm{Tr}\left[J_o(\boldsymbol{\rho}, \boldsymbol{\rho})\right] d^2\boldsymbol{\rho}}. \tag{4.74}$$

According to the definition of heterodyne efficiency, the expression for the heterodyne efficiency of a partially coherent EGSM beam for heterodyne detection is [13]

$$\eta_{EGSM} = \frac{\iint \iint \mathrm{Re}\left\{\mathrm{Tr}[J_o(\boldsymbol{\rho}_1, \boldsymbol{\rho}_2) J_s(\boldsymbol{\rho}_1, \boldsymbol{\rho}_2)]\right\} d^2\boldsymbol{\rho}_1 d^2\boldsymbol{\rho}_2}{\iint \mathrm{Tr}[J_o(\boldsymbol{\rho}_1, \boldsymbol{\rho}_2)] d^2\boldsymbol{\rho}_1 \iint \mathrm{Tr}[J_s(\boldsymbol{\rho}_1, \boldsymbol{\rho}_2)] d^2\boldsymbol{\rho}_2}. \tag{4.75}$$

The expression for each element of the BCP matrix of a partially coherent EGSM beam is [15]

$$J_{\beta ij}(\boldsymbol{r}_1, \boldsymbol{r}_2) = I_{\beta ij} \exp\left[-\frac{\boldsymbol{r}_1^2 + \boldsymbol{r}_2^2}{4w_\beta^2} - \frac{(\boldsymbol{r}_1 - \boldsymbol{r}_2)^2}{2\delta_{\beta ij}^2}\right], \tag{4.76}$$

where r_1 and r_2 are two position vectors on the source plane, $I_{\beta ij}$ is the beam intensity, w_α is the beam waist radius, $\delta_{\beta ij}$ is the spatial-coherence length, $i, j = x, y$. When α is s, it represents the signal beam; when α is o, it represents the local-oscillator beam.

Assuming that the EGSM propagates forward along the z axis in atmospheric turbulence, according to the generalized Huygens–Fresnel principle, the expression for each element of the BCP matrix of the beam at the receiving end is [16]

$$\begin{aligned} J_{\beta ij}(\boldsymbol{\rho}_1, \boldsymbol{\rho}_2, z_\beta) = {} & I_{\beta ij}\left(\frac{k}{2\pi z_\beta}\right)^2 \iint d^2\boldsymbol{r}_1 \iint d^2\boldsymbol{r}_2 \\ & \times \exp\left[-ik\frac{(\boldsymbol{\rho}_1 - \boldsymbol{r}_1)^2 - (\boldsymbol{\rho}_2 - \boldsymbol{r}_2)^2}{2z_\beta}\right] \\ & \times \exp\left[-\frac{\boldsymbol{r}_1^2 + \boldsymbol{r}_2^2}{4w_\beta^2} - \frac{(\boldsymbol{r}_1 - \boldsymbol{r}_2)^2}{2\delta_{\beta ij}^2}\right] \\ & \times < \exp\left[\psi^*(\boldsymbol{\rho}_1, \boldsymbol{r}_1, z_\beta) + \psi(\boldsymbol{\rho}_2, \boldsymbol{r}_2, z_\beta)\right] >, \end{aligned} \tag{4.77}$$

where z_β is the transmission distance; $k = 2\pi/\lambda$ is the wavenumber, and λ is the wavelength; $\psi(\boldsymbol{\rho}, \boldsymbol{r}, z_\beta)$ is the complex random disturbance of the turbulent medium.

$< \exp[\psi * (\boldsymbol{\rho}_1, \boldsymbol{r}_1, z_\beta) + \psi(\boldsymbol{\rho}_2, \boldsymbol{r}_2, z_\beta)] >$, describing the effect of turbulence in Eq. (4.77), can be expressed as [17]

$$< \exp\left[\psi^*(\boldsymbol{\rho}_1, \boldsymbol{r}_1, z_\beta) + \psi(\boldsymbol{\rho}_2, \boldsymbol{r}_2, z_\beta)\right] >= \exp\left[-\frac{1}{2}D_\phi(\boldsymbol{\rho}_1 - \boldsymbol{\rho}_2, \boldsymbol{r}_1 - \boldsymbol{r}_2)\right], \tag{4.78}$$

where $D_\phi(\boldsymbol{\rho}_1 - \boldsymbol{\rho}_2, \boldsymbol{r}_1 - \boldsymbol{r}_2)$ is the wave-structure function. When the beam is transmitted to the receiving end with an oblique path, the wave-structure function can be expressed as [17]

$$< \exp\left[\psi^*(\boldsymbol{\rho}_1, \boldsymbol{r}_1, z_\beta) + \psi(\boldsymbol{\rho}_2, \boldsymbol{r}_2, z_\beta)\right] >_m \cong \exp[-\frac{1}{2}(M_1\boldsymbol{r}_d^2 + M_2\boldsymbol{r}_d\boldsymbol{\rho}_d + M_3\boldsymbol{\rho}_d^2)], \tag{4.79}$$

where $\boldsymbol{\rho}_d = \boldsymbol{\rho}_1 - \boldsymbol{\rho}_2$; $\boldsymbol{r}_d = \boldsymbol{r}_1 - \boldsymbol{r}_2$; and M_1, M_2, and M_3 are, respectively,

$$\begin{cases} M_1 = 2\pi^2k^2\sec\theta \int\limits_{h_0}^{H}\int\limits_{0}^{\infty} \Phi_n(\kappa, h)(1-\xi)^2\kappa^3 d\kappa dh \\ M_2 = 4\pi^2k^2\sec\theta \int\limits_{h_0}^{H}\int\limits_{0}^{\infty} \Phi_n(\kappa, h)(1-\xi)\xi\kappa^3 d\kappa dh, \\ M_3 = 2\pi^2k^2\sec\theta \int\limits_{h_0}^{H}\int\limits_{0}^{\infty} \Phi_n(\kappa, h)\xi^2\kappa^3 d\kappa dh \end{cases} \tag{4.80}$$

where, in the uplink, $\xi = 1-(h-h_0)/(H-h_0)$, H is the height of the receiving end, and h_0 is the height of the beam source. In the downlink, $\xi = (h-h_0)/(H-h_0)$, H is the height of the beam source, and h_0 is the height of the receiving end. θ is the zenith angle. When the atmospheric-turbulent refractive-index spectral-density function $\Phi_n(\kappa, h)$ is a non-Kolmogorov spectrum, its expression is [18]

$$\Phi_n(\kappa, h, \alpha) = A(\alpha)\tilde{C}_n^2\frac{\exp\left[-\kappa^2/\kappa_m^2\right]}{(\kappa^2+\kappa_0^2)^{\alpha/2}}, 0 \le \kappa < \infty, 3 < \alpha < 4, \tag{4.81}$$

where κ is the spatial wavenumber, $\kappa_m = c(\alpha)/l_0$, $\kappa_0 = 2\pi/L_0$, l_0 and L_0 are the inner and outer atmospheric-turbulence scales, respectively; $\tilde{C}_n^2$ is the structural constant of the non-Kolmogorov turbulent-atmospheric refractive-index; α is the power-spectrum power law, and the range is $3 < \alpha < 4$. $c(\alpha)$ is a function of the power-spectrum power law and $A(\alpha)$ is the generalized amplitude. The expressions for $A(\alpha)$ and $c(\alpha)$ are as follows [18]:

$$A(\alpha) = \frac{\Gamma(\alpha - 1)\cos(\alpha\pi/2)}{4\pi^2} \tag{4.82}$$

$$c(\alpha) = \left[\Gamma\left(5 - \frac{\alpha}{2}\right)A(\alpha)\frac{2}{3}\pi\right]^{1/(\alpha-5)}, \tag{4.83}$$

where $\Gamma(x)$ is the gamma function.

Under non-Kolmogorov turbulence, $\int_0^\infty \Phi_n(\kappa, h, \alpha)\kappa^3 d\kappa$ can be obtained by scoring:

$$\int_0^\infty \Phi_n(\kappa, h, \alpha)\kappa^3 d\kappa = \frac{A(\alpha)}{2(\alpha - 2)}\tilde{C}_n^2\left[\kappa_m^{2-\alpha}\beta \exp\left(\frac{\kappa_0^2}{\kappa_m^2}\right)\Gamma\left(2 - \frac{\alpha}{2}, \frac{\kappa_0^2}{\kappa_m^2}\right) - 2\kappa_0^{4-\alpha}\right], \tag{4.84}$$

where $\chi = 2\kappa 2\ 0 - 2\kappa 2\ m + \alpha\kappa 2\ m$.

When the beam is transmitted obliquely, the atmospheric refractive-index structure constant model proposed by the ITU-R is used, and the expression is [19]

$$C_n^2(h) = 8.148 \times 10^{-56} v_{RMS}^2 h^{10} e^{-h/1000} + 2.7 \times 10^{-16} e^{-h/1500} + C_0 e^{-h/100}, \tag{4.85}$$

$$v_{RMS} = \sqrt{v_g^2 + 30.69 v_g + 348.91}, \text{ and } \tilde{C}_n^2 = \gamma C_n^2,$$

where γ is a constant 1, v_{RMS} is the vertical-path wind speed, v_g is the wind speed near the surface, and C_0 represents the atmospheric structure constant near the surface.

Substituting Eq. (4.78) into Eq. (4.77) yields

$$\begin{aligned} J_{\beta ij}(\rho_1, \rho_2, z_\beta) = I_{\beta ij}\left(\frac{k}{2\pi z_\beta}\right)^2 \iint d^2 r_1 \iint d^2 r_2 \times \exp\left[-ik\frac{(\rho_1 - r_1)^2 - (\rho_2 - r_2)^2}{2z_\beta}\right] \\ \times \exp\left[-\frac{r_1^2 + r_2^2}{4w_\beta^2} - \frac{(r_1 - r_2)^2}{2\delta_{\beta ij}^2}\right] \\ \times \exp[-\frac{1}{2}(M_1 r_d^2 + M_2 r_d \rho_d + M_3 \rho_d^2)] \end{aligned} \tag{4.86}$$

Using integral expressions,

$$\int_{-\infty}^{+\infty}\left[e^{-ax^2}\right]e^{-i2\pi w_x x}dx \cdot \int_{-\infty}^{+\infty}\left[e^{-ay^2}\right]e^{-i2\pi w_y y}dy = \frac{\pi}{a} \cdot e^{-\frac{\pi^2 w^2}{a}}. \tag{4.87}$$

Equation (4.86) can be obtained by an integral operation:

$$J_{\beta ij}(\rho_1, \rho_2, z_\beta) = \frac{I_{\beta ij}}{Q_{\beta ij}} \exp\left[-H_{\beta ij}(\rho_1 - \rho_2)^2\right] \times \\ \exp\left[-\frac{1}{8w_\beta^2 Q_{\beta ij}}(\rho_1 + \rho_2)^2\right] \times \exp\left[-ikT_{\beta ij}(\rho_1^2 - \rho_2^2)\right] \tag{4.88}$$

where

$$\begin{aligned} Q_{\beta ij} &= 1 + \left(\frac{z_\beta}{kw_\beta \theta_{\beta ij}}\right)^2 + \frac{z_\beta^2 M_3}{k^2 w_\beta^2} \\ H_{\beta ij} &= \frac{1}{2\theta_{\beta ij}^2 Q_{\beta ij}} + \frac{1}{2Q_{\beta ij}}(M_3 + M_2) + \frac{M_1}{2} - \frac{M_2^2 z_\beta^2}{8w_\beta^2 k^2 Q_{\beta ij}}. \\ \frac{1}{\theta_{\beta ij}^2} &= \frac{1}{4w_\beta^2} + \frac{1}{\delta_{\beta ij}^2} \\ T_{\beta ij} &= \frac{2k^2 w_\beta^2 Q_{\beta ij} + M_2 z_\beta^2 - 2k^2 w_\beta^2}{4k^2 z_\beta w_\beta^2 Q_{\beta ij}} \end{aligned} \tag{4.89}$$

For a random electromagnetic beam, the degree of polarization is defined as [16]

$$\mathrm{P}(\boldsymbol{\rho}_1, \boldsymbol{\rho}_2, z_\beta) = \sqrt{1 - \frac{4\mathrm{Det}J(\boldsymbol{\rho}_1, \boldsymbol{\rho}_2, z_\beta)}{[\mathrm{Tr}J(\boldsymbol{\rho}_1, \boldsymbol{\rho}_2, z_\beta)]^2}}, \tag{4.90}$$

where Det represents the determinant of the matrix.

Substituting Eq. (4.88) into Eq. (4.90) and setting $\boldsymbol{\rho}_1 = \boldsymbol{\rho}_2 = 0$, the polarization-degree expression of a point on the beam axis on the receiving plane is as follows:

$$\mathrm{P}(\boldsymbol{\rho} = 0, z_\beta) = \sqrt{1 - \frac{4\left(\frac{I_{\beta xx} I_{\beta yy}}{Q_{\beta xx} Q_{\beta yy}} - \frac{I_{\beta xy} I_{\beta yx}}{Q_{\beta xy} Q_{\beta yx}}\right)}{\left(\frac{I_{\beta xx}}{Q_{\beta xx}} + \frac{I_{\beta yy}}{Q_{\beta yy}}\right)^2}}. \tag{4.91}$$

It can be seen that the degree of polarization satisfies $0 \le P \le 1$. When $P = 1$, the beam is fully polarized; when $0 < P < 1$, the beam is partially polarized; and when $P = 0$, the beam is unpolarized.

Substituting Eq. (4.88) into Eq. (4.72) and using the polar-coordinate operation, the power of the intermediate-frequency signal is obtained as

$$P_{IFij} = 2\Re^2 \int_0^{2\pi}\int_0^{2\pi}\int_0^{D/2}\int_0^{D/2} \mathrm{Re}[J_{oij}(\boldsymbol{\rho}_1, \boldsymbol{\rho}_2, z_\beta) \times J_{sij}^*(\boldsymbol{\rho}_1, \boldsymbol{\rho}_2, z_\beta)]\rho_1\rho_2 d\rho_1 d\rho_2 d\varphi_1 d\varphi_2, \tag{4.92}$$

where D is the effective diameter of the detector and φ is the angle between the position vector $\boldsymbol{\rho}$ and the x-axis. For calculation simplicity, the soft-aperture approximation is used, and the hard aperture D is approximated as a Gaussian aperture or a soft aperture with a radius of W. The relationship between the two is [14] $W^2 = D^2 / 8$, and Eq. (4.92) can be organized as follows:

$$\begin{aligned} P_{IFij} = {} & 2\Re^2 \frac{I_{oij}I_{sij}}{Q_{oij}Q_{sij}} \int_0^{2\pi}\int_0^{2\pi}\int_0^{\infty}\int_0^{\infty} \mathrm{Re}\{\frac{I_{oij}}{Q_{oij}} \exp[-H_{oij}(\boldsymbol{\rho}_1 - \boldsymbol{\rho}_2)^2]\exp[-\frac{1}{8w_o^2 Q_{oij}}(\boldsymbol{\rho}_1+\boldsymbol{\rho}_2)^2] \\ & \times \exp[-ikT_{oij}(\boldsymbol{\rho}_1^2 - \boldsymbol{\rho}_2^2)] \times \frac{I_{sij}}{Q_{sij}} \exp[-H_{sij}(\boldsymbol{\rho}_1 - \boldsymbol{\rho}_2)^2]\exp[-\frac{1}{8w_s^2 Q_{sij}}(\boldsymbol{\rho}_1+\boldsymbol{\rho}_2)^2] \\ & \times \exp[ikT_{sij}(\boldsymbol{\rho}_1^2 - \boldsymbol{\rho}_2^2)]\exp[-\frac{\boldsymbol{\rho}_1^2+\boldsymbol{\rho}_2^2}{W^2}]\}\rho_1\rho_2 d\rho_1 d\rho_2 d\varphi_1 d\varphi_2. \end{aligned} \tag{4.93}$$

Then, from the relational expression $\boldsymbol{\rho}_1\boldsymbol{\rho}_2 = \rho_1\rho_2 \cos(\varphi_1 - \varphi_2)$, Eq. (4.93) can be organized as

$$\begin{aligned} P_{IFij} = {} & 2\Re^2 \frac{I_{oij}I_{sij}}{Q_{oij}Q_{sij}} \int_0^{2\pi}\int_0^{2\pi}\int_0^{\infty}\int_0^{\infty} \mathrm{Re}\{\exp[-g_{ij}(\rho_1^2 + \rho_2^2)] \times \exp[ip_{ij}(\rho_1^2 - \rho_2^2)] \\ & \times \exp[2q_{ij}\rho_1\rho_2 \cos(\varphi_1 - \varphi_2)]\}\rho_1\rho_2 d\rho_1 d\rho_2 d\varphi_1 d\varphi_2, \end{aligned} \tag{4.94}$$

where

$$\begin{aligned} g_{ij} &= H_{oij} + \frac{1}{8w_o^2 Q_{oij}} + H_{sij} + \frac{1}{8w_s^2 Q_{sij}} + \frac{1}{W^2} \\ q_{ij} &= H_{oij} - \frac{1}{8w_o^2 Q_{oij}} + H_{sij} - \frac{1}{8w_s^2 Q_{sij}} \\ p_{ij} &= k(T_{sij} - T_{oij}). \end{aligned} \tag{4.95}$$

Simplifying the calculation yields

$$P_{IFij} = \frac{2\pi^2\Re^2 I_{oij}I_{sij}}{Q_{oij}Q_{sij}(g_{ij}^2 + p_{ij}^2 - q_{ij}^2)}, \quad (i, j = x, y). \tag{4.96}$$

The optical power of the local oscillator is

$$\begin{aligned} P_{oii} &= \Re \iint J_{oii}(\boldsymbol{\rho}, \boldsymbol{\rho}) d^2\boldsymbol{\rho} = \Re \int_0^{2\pi} \int_0^{\infty} J_{oii}(\boldsymbol{\rho}, \boldsymbol{\rho}) e^{-(\rho^2)/W^2} \rho d\rho d\varphi \\ &= \frac{2\Re\pi I_{oii} W^2 w_o^2}{W^2 + 2w_o^2 Q_{oii}}, (i = x, y). \end{aligned} \tag{4.97}$$

According to the above results, the following can be obtained:

$$SNR_{EGSM} = \frac{\sum_{i,j=x,y} P_{IFij}}{eB \sum_{i=x,y} P_{oii}}. \tag{4.98}$$

Similarly, the signal optical power is

$$\begin{aligned} P_{sii} &= \Re \iint J_{sii}(\boldsymbol{\rho}, \boldsymbol{\rho}) d^2\boldsymbol{\rho} = \Re \int_0^{2\pi} \int_0^{\infty} J_{sii}(\boldsymbol{\rho}, \boldsymbol{\rho}) e^{-(\rho^2)/W^2} \rho d\rho d\varphi \\ &= \frac{2\Re\pi I_{sii} W^2 w_s^2}{W^2 + 2w_s^2 Q_{sii}}, (i = x, y). \end{aligned} \tag{4.99}$$

The heterodyne-efficiency expression of heterodyne detection is

$$\eta_{EGSM} = \frac{\sum_{i,j=x,y} P_{IFij}}{2\sum_{i=x,y} P_{oii} \sum_{i=x,y} P_{sii}}. \tag{4.100}$$

From Eqs. (4.99) and (4.100), $\mathrm{SNR}_{\mathrm{EGSM}}$ can be written as

$$\begin{aligned} SNR_{EGSM} &= \frac{\sum_{i=x,y} P_{sii} \sum_{i,j=x,y} P_{IFij}}{eB \sum_{i=x,y} P_{oii} \sum_{i=x,y} P_{sii}} \\ &= \frac{2\sum_{i=x,y} P_{sii}}{eB} \eta_{EGSM}. \end{aligned} \tag{4.101}$$

The sensitivity of the detection system refers to the minimum input signal power that can be detected under a certain bit-error rate, which can be obtained from Eq. (4.101):

$$\sum_{i=x,y} P_{sii} = \frac{eB}{2\eta_{EGSM}} SNR_{EGSM}. \tag{4.102}$$

The bit-error rate (BER) of the receiver is an important parameter for measuring the performance of coherent optical communication systems. In this study, the binary phase-shift keying modulation method is used, and the expression for the bit-error rate is [20]

$$BER = \frac{1}{2} erfc\left(\frac{SNR_{EGSM}}{2}\right)^{1/2}. \tag{4.103}$$

4.6.2.2 Effect of the Polarization State of a Partially Coherent EGSM Beam on the Detection Sensitivity of the System

Based on the above theoretical analysis, the Monte Carlo and theoretical-calculation methods were used to simulate the EGSM-beam heterodyne-detection system based on binary phase-shift keying modulation and analyze its bit-error performance. The parameter values are shown in Table 4.1. To maintain constant the polarization characteristics of the beam in the atmospheric channel, the beam-space lengths of the signal beam and local-oscillator beam are, respectively, $\delta_{oxy} = \delta_{oyx} = 0.2$ mm, $\delta_{oxx} = \delta_{oyy} = 0.2$ mm, $\delta_{sxx} = \delta_{syy} = 0.4$ mm, and $\delta_{sxy} = \delta_{syx} = 0.4$ mm.

Figure 4.21 shows the curve of the polarization degree and system bit-error rate during the theoretical derivation and Monte Carlo simulation. As shown in the figure, as the degree of polarization P increases, the bit-error rate of the system gradually decreases. The Monte Carlo simulation curve is in good agreement with the theoretical bit-error rate curve, which verifies the correctness of the numerical simulation results.

Next, a numerical-analysis method was used to study the influence of the EGSM beam on the heterodyne-detection sensitivity under different polarization states, when the EGSM beam was transmitted obliquely under non-Kolmogorov atmospheric

Table 4.1 Simulation parameters

Parameter name	Parameter value	Parameter name	Parameter value
Wavelength λ	1550 nm	Receiver or beam-source height H	2000 m
Inner turbulent scale l_0	0.001 m	Receiver or beam-source height h_0	0 m
Outer turbulent scale L_0	1 m	Atmospheric structure constant C_0	1.7×10^{-14} $m^{-2/3}$
Power-spectrum power law α	3.6	Signal beam waist radius w_s	3 mm
Near-surface wind speed v_g	2.8 m/s	Local-oscillator beam waist radius w_o	3 mm
Detector photosensitive-area diameter D	2 mm	Carrier optical frequency v	1.9×10^{14} Hz
Planck's constant h	6.623×10^{-34} J·s	Detector bandwidth B	200 MHz
Quantum efficiency η	0.8	Electron charge e	1.6×10^{-19} C
Transfer bit points	10^7	Transmission distance z_s	5000 m

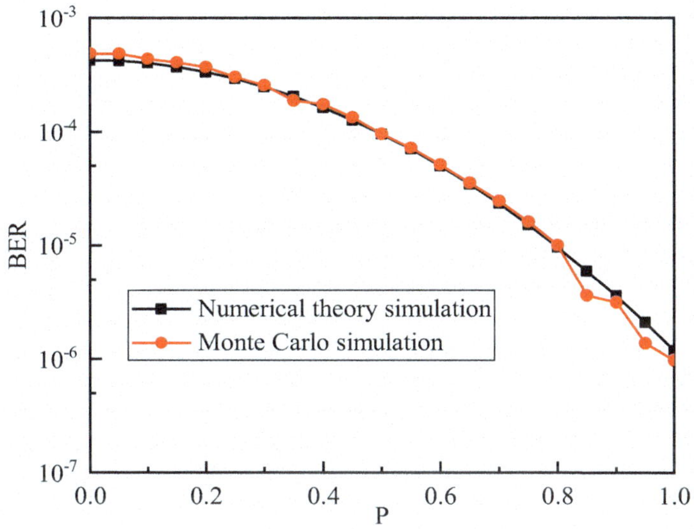

Fig. 4.21 Variation of the uplink system error rate of the EGSM beam used for heterodyne detection with polarization degree P

turbulence. It can be seen from Eq. (4.102) that the system sensitivity is determined by the turbulence size, turbulence scale, zenith angle, detector parameters, transmission distance, receiver height or beam-source height H, etc. The method analyzed how each factor affects the sensitivity of the heterodyne detection system when the bit-error rate is 10^{-9}. The numerical values of the simulation parameters are listed in Table 4.2.

As shown in Fig. 4.22, when the EGSM beam was used for heterodyne detection, the sensitivity of the uplink and downlink systems varied with the degree of polarization. Since it is assumed in this study that the polarization characteristics of the beam remain unchanged in the atmospheric channel, according to Eq. (4.91), when $I_{oxy} = I_{sxy} = I_{oyx} = I_{syx}$ are 0, 0.2, 0.4, 0.6, 0.8, and 1, the degree of polarization P is 0, 0.2, 0.4, 0.6, 0.8, and 1, respectively.

From these five different values, it can be seen that the system sensitivity of heterodyne detection increases with the increase of polarization degree P. Because it can be seen from Eq. (4.91), when the beam intensity increases, the degree of polarization increases and the detection sensitivity increases.

In addition, when the degree of polarization remains constant, the sensitivity of the uplink is greater than that of the downlink, which is caused by the difference in the turbulence terms M_1, M_2, and M_3 in Eq. (4.76). In Eq. (4.102), if the value of the channel turbulence term is smaller than the value of the downlink turbulence term, then the beam is less affected by turbulence during the uplink transmission than the downlink transmission; therefore, the sensitivity of the beam during the uplink transmission is greater.

Table 4.2 Simulation parameters

Parameter name	Parameter value	Parameter name	Parameter value
Wavelength λ	1550 nm	Atmospheric structure constant C_0	1.7×10^{-14} m$^{-2/3}$
Inner turbulent scale l_0	0.005 m	Signal beam waist radius w_s	3 mm
Outer turbulent scale L_0	1 m	Local-oscillator beam waist radius w_o	3 mm
Power-spectrum power law α	3.7	Detector photosensitive-area diameter D	3 mm
Near-surface wind speed v_g	2.8 m/s	Planck's constant h	6.623×10^{-34} J·s
Receiver or beam-source height h_0	0 m	Carrier optical frequency v	1.9×10^{14} Hz
Quantum efficiency η	0.8	Detector bandwidth B	200 MHz
Receiver or beam-source height H	800 m		

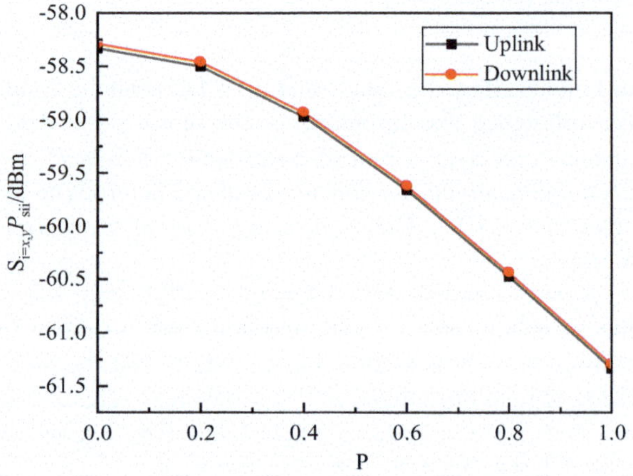

Fig. 4.22 Variation of the uplink and downlink system sensitivity with polarization degree P when an EGSM beam is used for heterodyne detection

To more intuitively analyze the influence of various factors on the sensitivity of heterodyne detection under different polarization states during the oblique transmission of EGSM beams, the sensitivities of EGSM beams in fully polarized, non-polarized, and partially polarized states for heterodyne detection were analyzed. A detailed analysis is shown in Figs. 4.23, 4.24 and 4.25.

From Eq. (4.91), we set $I_{oxy} = I_{sxy} = 1$, $I_{oyx} = I_{syx} = 1$, $I_{oxx} = I_{oyy} = 1$, $I_{sxx} = I_{syy} = 1$, $\mathrm{I}_{sxx} = \mathrm{I}_{syy} = 1$, and degree of polarization $P = 1$, at which time the signal beam and local-oscillator beam are fully polarized. Figure 4.23a–f respectively show the system-sensitivity variation curves with the transmission distance of a fully polarized EGSM beam used for heterodyne detection, for different power-spectrum power laws,

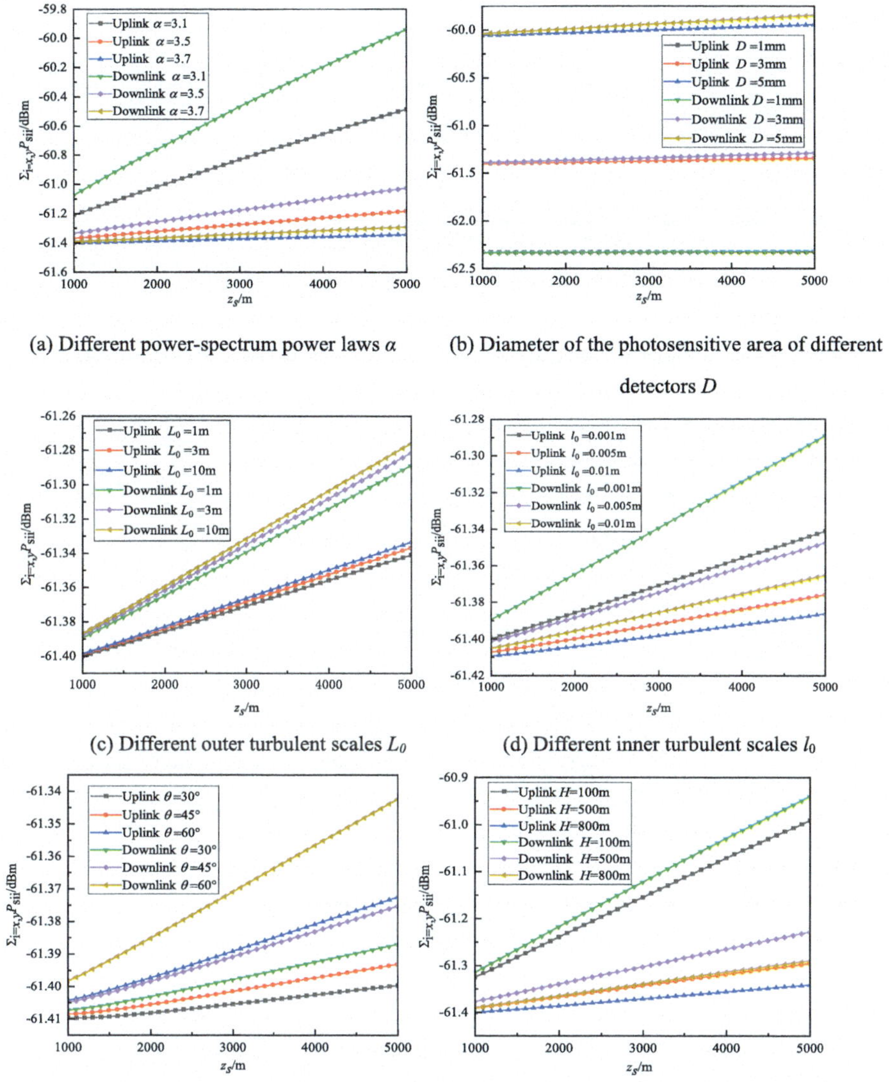

(a) Different power-spectrum power laws α

(b) Diameter of the photosensitive area of different detectors D

(c) Different outer turbulent scales L_0

(d) Different inner turbulent scales l_0

(e) Different zenith angles θ

(f) Different receiver heights or beam-source heights H

Fig. 4.23 Variation of the uplink and downlink system sensitivity with transmission distance when a fully polarized EGSM beam is used for heterodyne detection

detector photosensitive-area diameters, outer turbulent scales, inner turbulent scales, zenith angles, and receiving-end heights or beam-source heights H.

As shown in Fig. 4.23, the detection sensitivity of the system decreases with an increase in the transmission distance. This is because the beam is affected by the refractive index of the atmosphere when transmitted during atmospheric turbulence.

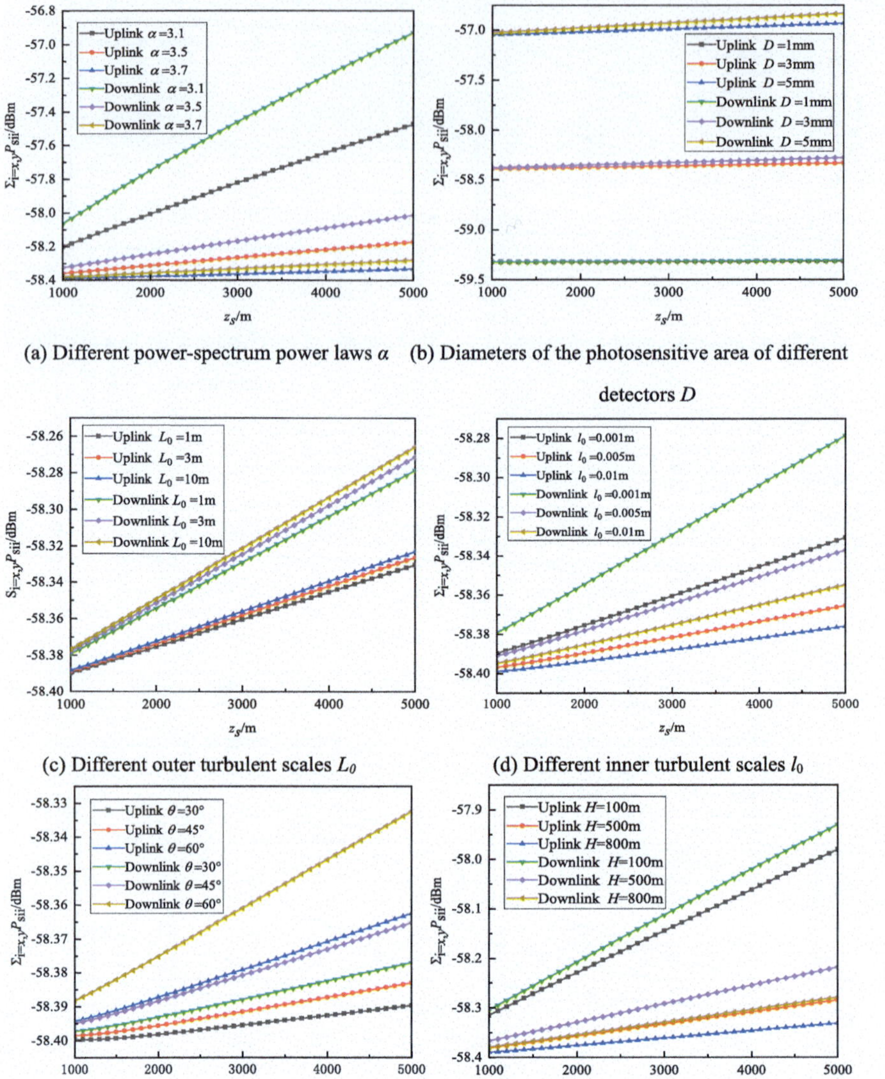

(a) Different power-spectrum power laws α (b) Diameters of the photosensitive area of different detectors D

(c) Different outer turbulent scales L_0 (d) Different inner turbulent scales l_0

(e) Different zenith angles θ (f) Different receiver heights or beam-source heights H

Fig. 4.24 Variation of uplink and downlink system sensitivity with the transmission distance when a non-polarized EGSM beam is used for heterodyne detection

When the transmission distance is greater, the generated wavefront distortion will be more severe and the sensitivity will decrease.

As shown in Fig. 4.23a, at a certain transmission distance, the detection sensitivity of the system increases with an increase in the power-spectrum power law. When the power-spectrum power law α is 3.1, the turbulence has a greater impact on the beam,

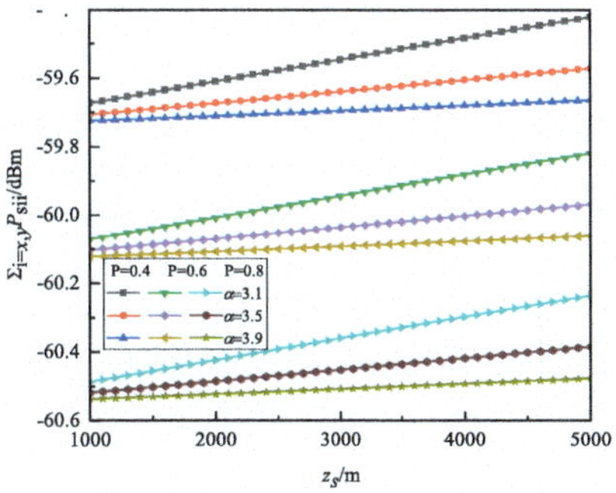

(a) Different power-spectrum power laws α

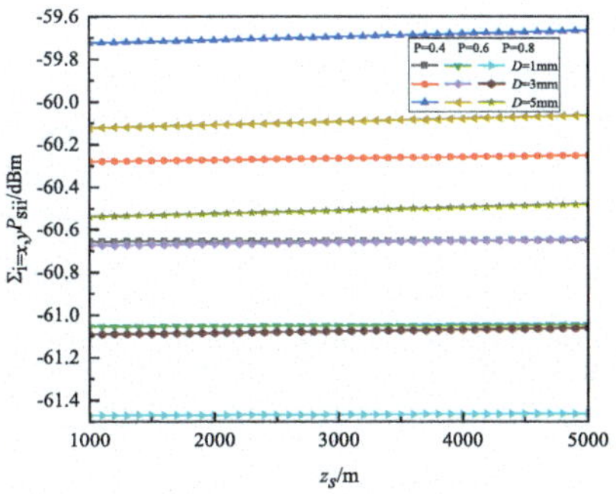

(b) Diameters of the photosensitive area of different detectors D

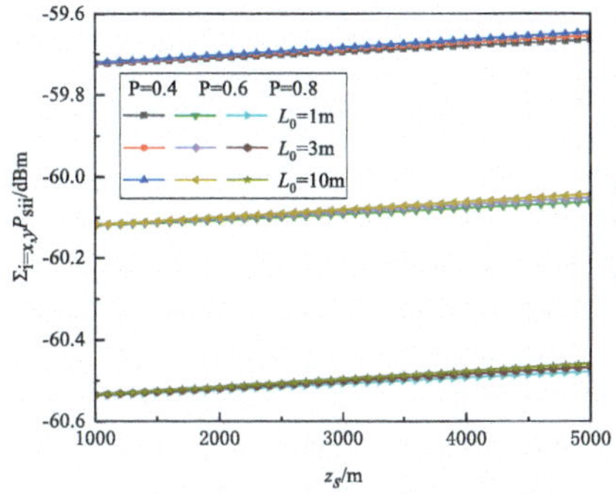

(c) Different outer turbulent scales L_0

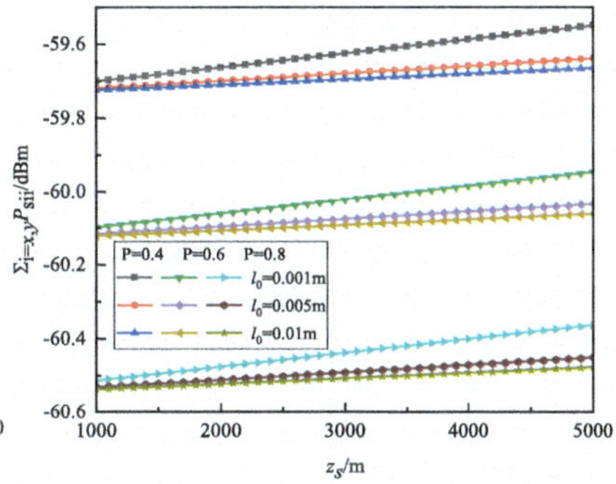

(d) Different inner turbulent scales l_0

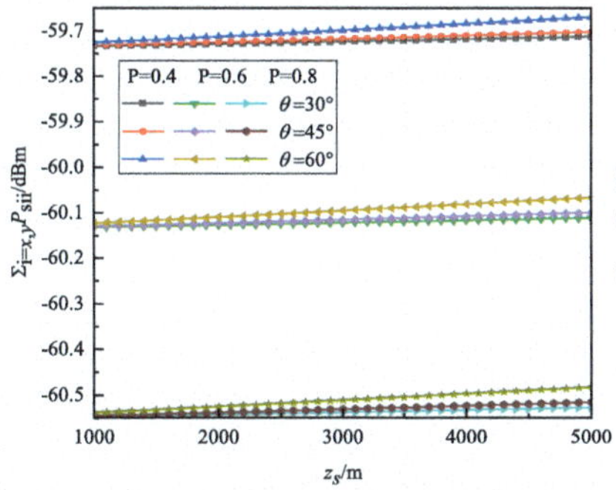

(e) Different zenith angles θ

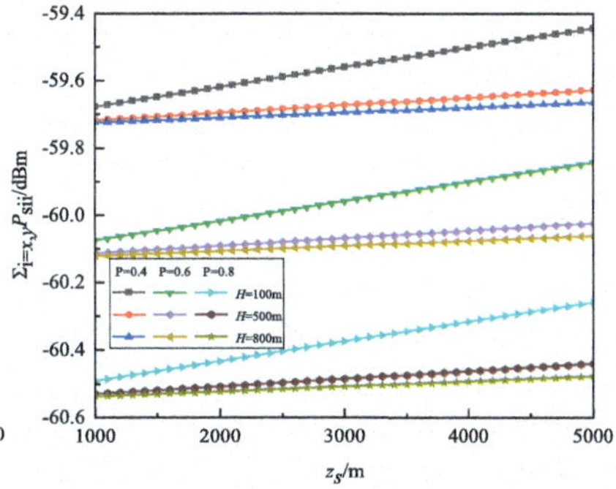

(f) Different receiver heights or beam-source heights H

Fig. 4.25 Variation of the uplink system sensitivity with transmission distance when a partially polarized EGSM beam is used for heterodyne detection resulting in minimal sensitivity. In addition, the sensitivity change in the downlink transmission is larger than that in the uplink transmission.

Figure 4.23b shows that for a certain transmission distance, the detection sensitivity of the system decreases with an increase in the diameter of the photosensitive area. Because of the limitation of the spatial coherence of the signal beam, when the diameter of the photosensitive area is smaller, the signal beam and the local-oscillator beam match more closely; thus, the smaller the diameter of the photosensitive area of the detector, the greater the detection sensitivity. In addition, at a certain diameter of the photosensitive area, when the transmission distance gradually increases, the system sensitivity changes slowly with the transmission distance.

From Fig. 4.23c, d, it can be seen that at a certain transmission distance, the system detection sensitivity decreases with the increase in the outer turbulence scale and the decrease in the inner scale, because the outer scale is smaller; the smaller the inner turbulent scale, the greater the intensity of the atmospheric turbulence, resulting in a decrease in sensitivity.

From Fig. 4.23e, f, it can be seen that under a certain transmission distance, the detection sensitivity of the system decreases with an increase in the zenith angle and a decrease in the height of the receiving end or the height of the beam source H because, during the transmission process, when the transmission height is lower than 10 km, the atmospheric-turbulence refractive-index structure constant increases with a decrease in height. For the same transmission distance, the larger the zenith angle and the lower the height H, the closer it is to lower than the level, the beam is more affected by the atmospheric turbulence, resulting in reduced system sensitivity.

In addition, as shown in Fig. 4.23, the sensitivity of the uplink is greater than that of the downlink, which is consistent with the results shown in Fig. 4.22. Therefore, the detection sensitivity of the system can be optimized under different power-spectrum power laws, detector photosensitive-area diameters, outer turbulent scales, inner turbulent scales, zenith angles, receiver heights, and beam-source heights H, thereby improving the performance of the optical communication system.

According to Eq. (4.91), we set $I_{oxy} = I_{sxy} = 0$, $I_{oyx} = I_{syx} = 0$, $I_{oxx} = I_{oyy} = 1$, $I_{sxx} = I_{syy} = 1$, and degree of polarization $P = 0$. At this time, the signal beam and local-oscillator beam are both unpolarized beams. Figure 4.24a–f show the variation curves of the system sensitivity with transmission distance when an unpolarized EGSM beam is used for heterodyne detection, with different power-spectrum power laws, diameters of the photosensitive area of the detector, outer turbulence scales, inner turbulence scales, the zenith angles, and receiving ends of receiver heights or beam-source heights H.

As shown in Fig. 4.24, when the transmission distance increases, the smaller the power-spectrum power law is, the larger the diameter of the photosensitive area of the detector is, the larger the outer turbulence scale is, the smaller the inner turbulence scale is, the larger the zenith angle is, and the larger the height of the receiving end or the lower the beam-source height H is, and the lower the detection sensitivity of the system is.

By comparing Figs. 4.23 and 4.24, it can be observed through numerical analysis that under the same link conditions, the detection sensitivity of the system when the fully polarized EGSM beam is used for heterodyne detection is 3 dBm greater than that of the non-polarized beam.

According to Eq. (4.91), when $I_{oxx} = I_{oyy} = 1$, $I_{sxx} = I_{syy} = 1$, and $I_{oxy} = I_{sxy} = I_{oyx} = I_{syx}$ are taken as 0.4, 0.6, and 0.8, respectively, the degrees of polarization P are 0.4, 0.6, and 0.8, respectively. At this time, the signal beam and local-oscillator beam are partially polarized beams. Figure 4.25a–f show the variation curves of the system sensitivity with the transmission distance when partially polarized EGSM beams with different degrees of polarization are used for heterodyne detection, with different power laws of different power spectra, diameters of the photosensitive area of the detector, outer turbulence scales, inner turbulence scales, zenith angles, receiver heights, or beam-source heights H.

To more intuitively analyze the influence of various factors on the system sensitivity when EGSM beams with different polarization degrees are used for heterodyne detection, the uplink that was less affected by turbulence during the oblique transmission of the beam is selected for detailed analysis. It can be seen from Fig. 4.25 that the greater the degree of polarization, the greater the sensitivity of the system, which is consistent with the conclusion in Fig. 4.22.

It can be seen from Figs. 4.23, 4.24, and 4.25 that under the same transmission conditions, the changing trends of the different polarization states are basically similar when the various beams are used for heterodyne detection. The difference is that the system detection sensitivity is the largest when the fully polarized EGSM beam is used, followed by the partially polarized EGSM beam, and then the unpolarized EGSM beam. The partially polarized beam is a combination of fully polarized and unpolarized beams, so when a partially polarized EGSM beam is used for heterodyne detection, its detection sensitivity is between unpolarized and fully polarized beam.

The numerical analysis revealed that, for example, when the bit-error rate is 10^{-9}, the other conditions are the same, the diameter D of the photosensitive area is 1 mm, and the partially polarized EGSM beam in Fig. 4.24b is used for a heterodyne-detection uplink transmission, the detection sensitivity of the system is −61.5 dBm, which is between the -62.3 dBm for the fully polarized beam in Fig. 4.23b and the -59.3 dBm for the unpolarized beam in Fig. 4.25b.

4.7 Experimental Study on the Effect of the Partially Coherent EGSM Beam Polarization State on the System Performance

Based on the theoretical analysis, experiments were conducted on the influence of the polarization state of the partially coherent beam on the system performance. This experiment consists of a 1550-nm laser, homogenizer, convex lens, transmitting/receiving antenna, 90° mixer, and balanced detector to determine the influence of different polarization states on the system performance.

4.7.1 Generating the Unpolarized State of a Partially Coherent EGSM Beam

Figure 4.26 shows a schematic diagram of the experimental system for generating the non-polarized state of the electromagnetic Gaussian–Schell model beam. The phase, amplitude, and other parameters of the two laser beams are independent random variables. The collimated laser beams respectively pass through polarizers 1 and 2, whose polarization directions are orthogonal to each other, to form a BCP matrix satisfying $J_{xy}(\rho_1, \rho_2) = J_{yx}(\rho_1, \rho_2) = 0$, and then combine them into a beam of partially coherent vector beams through a beam splitter [21].

Because the trace of the coherence matrix is equal to the average intensity of the corresponding beam wave, using Eq. (4.57), the intensities of the beam wave and polarization part can be expressed as

$$I_0 = TrJ(\boldsymbol{\rho}, \boldsymbol{\rho}, z_\beta; \omega) = J_{xx}(\boldsymbol{\rho}_1, \boldsymbol{\rho}_2, z_\beta) + J_{yy}(\boldsymbol{\rho}_1, \boldsymbol{\rho}_2, z_\beta) \tag{4.104}$$

$$\begin{aligned} I_{polarization} &= Tr(J(\rho, \rho, z_\alpha; \omega)) \\ &= \sqrt{\left(J_{xx}(\rho_1, \rho_2, z_\beta) + J_{yy}(\rho_1, \rho_2, z_\beta)\right)^2 - 4\left|J(\rho, \rho, z_\beta; \omega)\right|} \end{aligned} \tag{4.105}$$

According to the definition of the spectral-polarization degree, the spectral-polarization degree expressed by the coherence matrix is obtained:

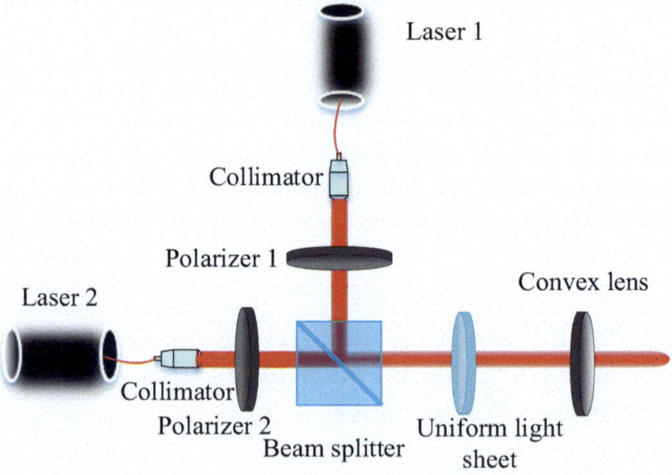

Fig. 4.26 Diagram of the experimental setup for generating a non-polarized EGSM beam

$$P = \frac{I_{polarization}}{I_0} = \sqrt{1 - \frac{4\left|J(\rho, \rho, z_\beta; \omega)\right|}{\left(J_{xx}(\rho_1, \rho_2, z_\beta; \omega) + J_{yy}(\rho_1, \rho_2, z_\beta; \omega)\right)^2}}. \tag{4.106}$$

When $J_{xy}(\rho_1, \rho_2) = J_{yx}(\rho_1, \rho_2) = 0$, Eq. (4.106) can be written as

$$P = \left|\frac{J_{xx}(\boldsymbol{\rho}_1, \boldsymbol{\rho}_2, z_\beta) - J_{yy}(\boldsymbol{\rho}_1, \boldsymbol{\rho}_2, z_\beta)}{J_{xx}(\boldsymbol{\rho}_1, \boldsymbol{\rho}_2, z_\beta) + J_{yy}(\boldsymbol{\rho}_1, \boldsymbol{\rho}_2, z_\beta)}\right|. \tag{4.107}$$

Because the output power of the laser is adjustable, two beams with the same output power are selected for combination through the beam splitter. After the two beams are transmitted and reflected, the optical power on the same optical axis is equal, that is, the beam intensities $I_{xx} = I_{yy}$ are equal. $J_{xx}(\rho_1, \rho_2) = J_{yy}(\rho_1, \rho_2)$ can be obtained from Eq. (4.91), which can be substituted into Eq. (4.107) to obtain a beam with a polarization degree $P = 0$. The uniform light sheet can evenly scatter the incident beam in all directions, so that the beam source is more uniform and stable. Finally, the beam-intensity distribution of the uniform spot is transformed into an approximate Gaussian distribution using a convex lens.

In this experiment, a convex lens was used to transform the beam-intensity distribution of a uniform spot into an approximate Gaussian distribution. Convex lenses refract and defocus the beam such that it changes on either side of the lens. If a uniform spot of the beam passes through a convex lens, the lens focuses the beam to a point, thereby reducing the spot. At this time, the beam-intensity distribution of the spot is similar to a Gaussian distribution; therefore, the unpolarized EGSM beam can be approximated after the combined beam passes through a uniform light sheet and convex lens.

4.7.2 Influence of the Polarization State of a Partially Coherent EGSM Beam on the System Performance

To verify the influence of partially coherent EGSM beams on the system performance in different polarization states, the experimental schematic shown in Fig. 4.27 was built indoors. The system is composed of a laser, laser collimator, polarizer, homogenizer, convex lens, beam splitter, transmitting/receiving antenna, 90° mixer, balanced detector, and oscilloscope.

After the EGSM beam is split using a beam splitter, two beam paths are generated. One beam path is sent out as the signal beam through the transmitting antenna, and the other beam path is coupled into the optical fiber as a local-oscillator beam through the coupling lens and then mixed with the signal beam received by the receiving antenna through 90°. The mixed-frequency beam is processed by the balanced detector and then input to the oscilloscope.

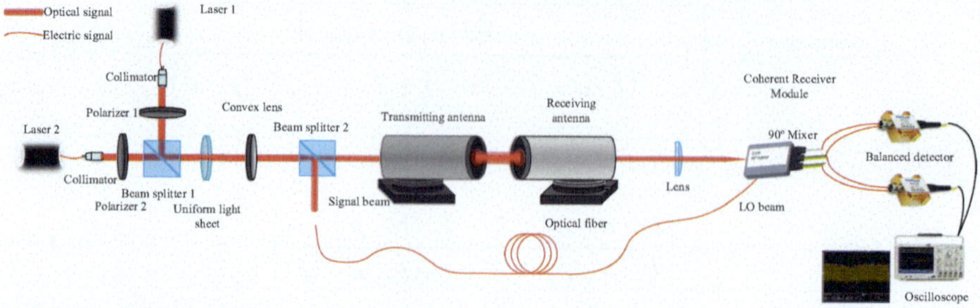

Fig. 4.27 Experimental schematic diagram

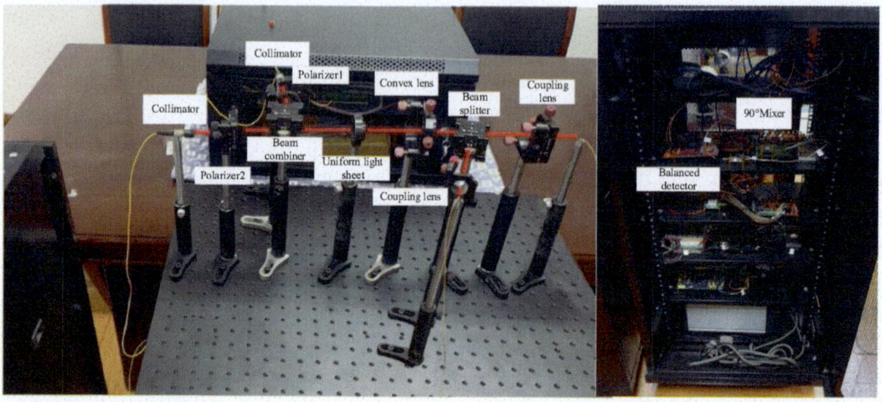

Fig. 4.28 Experimental physical diagram

When polarizer 3 is added between the convex lens and the beam splitter, a fully polarized EGSM beam can be obtained. When the unpolarized beam passes through the polarizer, only the beam waves whose vibration direction is the same as that allowed by the polarizer can pass through. The beam waves in the vertical direction are completely blocked. The beam waves in non-polarized beams have random vibration directions; therefore, after passing through the polarizer, only a portion of the beam waves can pass through. The vibration direction of these beam waves is the same as that allowed by the polarizer, thus forming a linearly polarized beam, that is, a fully polarized beam. This is shown in Fig. 4.29.

When a reflector is added between the convex lens and beam splitter, a partially polarized EGSM beam can be obtained. When an unpolarized beam is incident on the surface of the reflector, it is divided into two perpendicularly polarized beams in an equal manner. The amplitudes and phases of these two polarized beams are equal and are called *s*-polarized and *p*-polarized beams, respectively. Because their amplitudes and phases are equal, their strengths are also equal.

However, when these polarized beams are reflected, their relative amplitudes and phases change such that their intensities are no longer equal, and the vibration

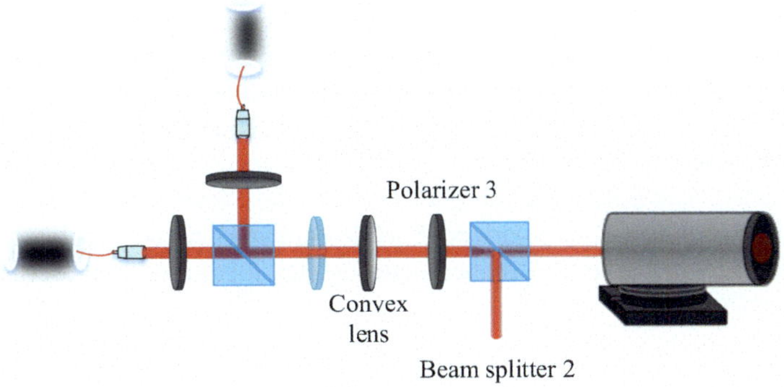

Fig. 4.29 Experimental-setup diagram for generating a completely polarized EGSM beam

direction of the beam is not restricted to a specific direction; thus, the reflected beam becomes a partially polarized beam. This is shown in Fig. 4.30.

The output wavelength of the laser used in this experiment was 1550 nm. Because a beam of this wavelength is invisible to the human eye, when building the experimental beam path, a laser with a visible 650-nm wavelength beam was first used for construction.

In this experiment, the 1550-nm lasers emitted by laser 1 and laser 2 are connected to the laser collimator through optical fibers. The collimator turns the output beam into a parallel beam, and then it passes through polarizers 1 and 2 to form a BCP matrix that satisfies a $J_{xy}(\rho_1, \rho_2) = J_{yx}(\rho_1, \rho_2) = 0$ beam. Two beams with the same output power are combined using a beam splitter.

After the two beams pass through the transmission and reflection, the optical powers on the same optical axis are equal. The EGSM beam is split by a beam splitter to generate two beam paths: one beam path is sent out as the signal beam

Fig. 4.30 Experimental-setup diagram for generating a partially polarized EGSM beam

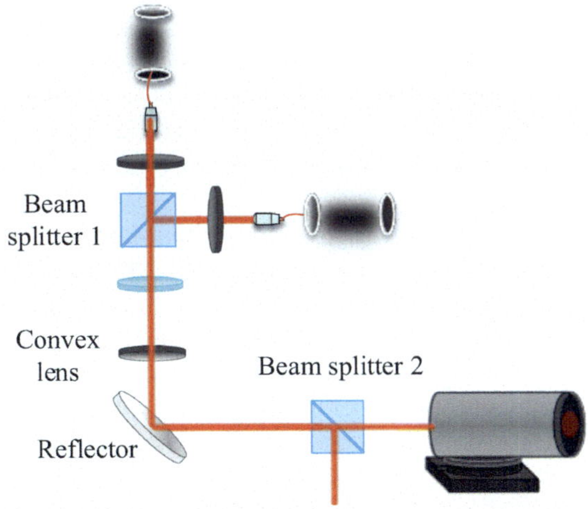

through the transmitting antenna, and the other beam path is coupled into the optical fiber as a local-oscillator beam through the coupling lens and then mixed with the signal beam received by the receiving antenna through 90°.

The mixed-frequency beam is processed by the balanced detector and then input to the oscilloscope to record the value of the intermediate-frequency (IF) signal. Then, polarizer 3 and a reflector are added between the convex lens and beam splitter, respectively, to obtain a fully polarized EGSM beam and a partially polarized EGSM beam, and to record the intermediate-frequency signal values of the oscilloscope in different polarization states.

4.7.3 Analysis of Results

In a completely polarized beam, the beam waves vibrate in only one direction, and all the beam waves on the polarization plane have the same direction of vibration. In an unpolarized beam, the beam waves vibrate uniformly in all directions and the beam waves on the vibrating plane do not exhibit any particular vibration direction. In a partially polarized beam, the vibration of the beam wave is only along a certain direction; however, vibration components exist in other directions. Therefore, the beam-wave intensity of a partially polarized beam is not completely fixed on the polarization plane but changes to a certain extent.

The intensity and phase difference of these vibration components affect the degree of polarization and beam-intensity distribution of the partially polarized beam. Therefore, the performance of a partially polarized beam is between that of a completely polarized beam and an unpolarized beam because it has both the directional characteristics of a polarized beam and the random characteristics of an unpolarized beam.

Figure 4.31 shows the IF signal waveform collected by the oscilloscope when the EGSM beam is in different polarization states. It can be seen from the figure that when the beam is unpolarized, the peak-to-peak value of the IF signal is 544.1 mV. When the beam is partially polarized, the peak-to-peak value is 740.1 mV. When the beam is completely polarized, the peak-to-peak value of the intermediate-frequency signal is 1.1 V. The system performance was verified to be between completely polarized and non-polarized when a partially polarized EGSM beam was used for coherent detection.

In this experiment, the sending terminal is used to control the sending of 109-bit signals, and the bits are synchronized at the receiving terminal. The received data are compared with the sent symbol signal to calculate the system bit error under different polarization states at a rate of 10^{-8}, and the specific result is shown in Fig. 4.32.

Figure 4.33 shows the measured optical power of the input signal when the bit-error rate is 10^{-8} under different polarization states of the EGSM beam.

It can be seen that when the signal EGSM beam is in a fully polarized state, its input-signal optical power fluctuates around -51 dBm, and when it is in a non-polarized state, its input-signal optical power fluctuates around -45 dBm. When the

Fig. 4.31 IF-signal waveform diagrams

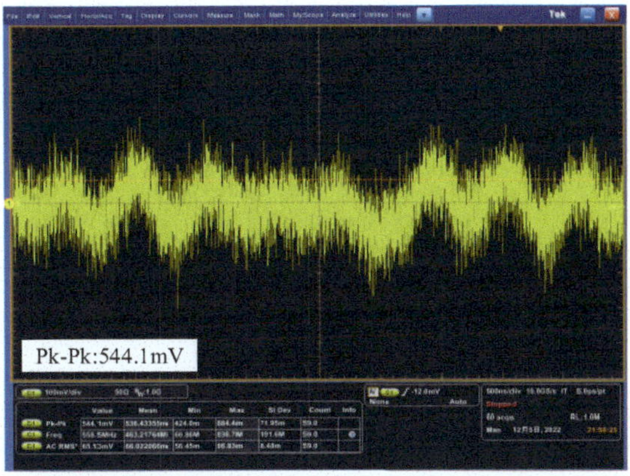

(a) Unpolarized EGSM beam

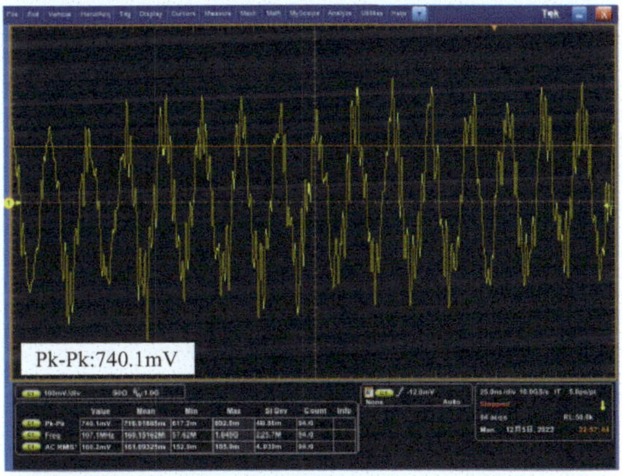

(b) Partially polarized EGSM beam

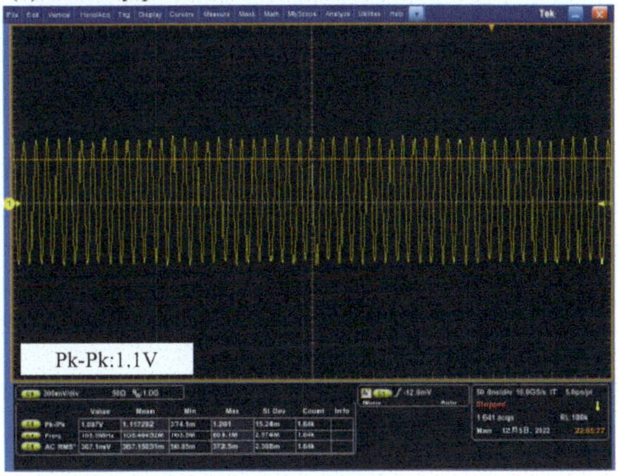

(c) Completely polarized EGSM beam

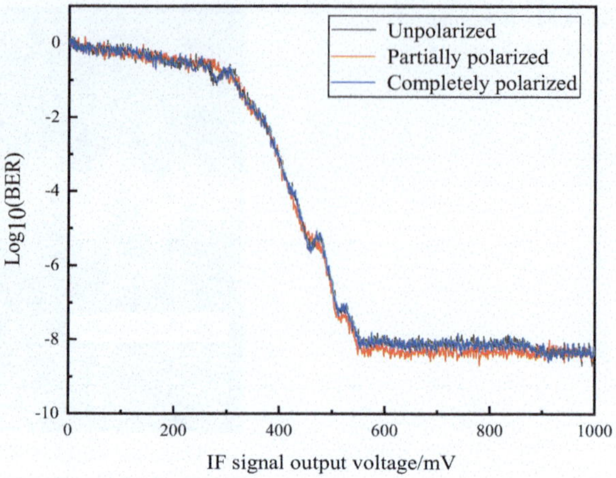

Fig. 4.32 Bit-error rate of the heterodyne-detection system varies with the voltage of the intermediate-frequency signal

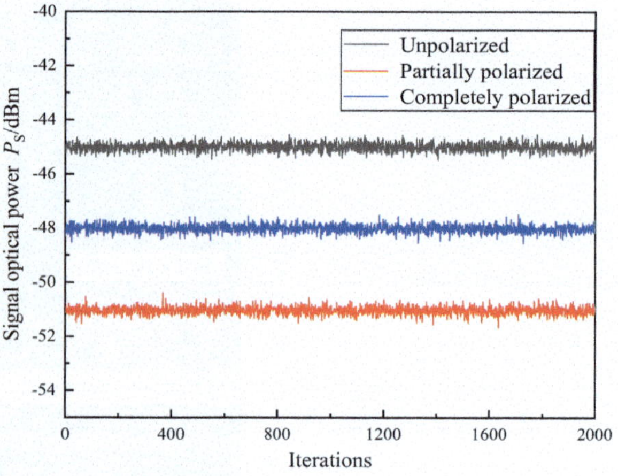

Fig. 4.33 Experimental curves of the impact of different EGSM-beam polarization states on the input-signal optical power

signal EGSM beam is partially polarized, its input signal beam power fluctuates around -48 dBm. The system performance is verified to be between fully polarized and non-polarized when a partially polarized EGSM beam is used for coherent detection.

According to the theoretical calculations in Sect. 4.6.2.2, the detection sensitivity of the fully polarized EGSM beam used for heterodyne detection is 3 dBm greater than that of the nonpolarized beam. It can be seen from the experimental results that the sensitivity of the fully polarized beam is lower than that of the non-polarized beam. The sensitivity of the non-polarized beam is 3 dBm larger but very close to 3 dBm.

The first reason for this result is errors caused by the experimental measurement and the second is the non-strict structure of the nonpolarized beam source. The powers of the two laser beams may not be completely equal, or they may not be completely orthogonal before combining. There may be an s-beam deviation when adjusting the polarizer, resulting in factors such as $J_{xy} \neq 0$ and $J_{yx} \neq 0$, which will lead to errors in the experimental results.

Nevertheless, the experiment achieved the expected purpose to a certain extent and effectively verified the difference in the influence of partially coherent EGSM beams on the system sensitivity under different polarization states, which has certain guiding significance.

References

1. Hongyang L, Yange Z, Yong A et al (2014) Design and implementation of balance detector used in coherent optical communication system. Laser Optoelectron Progress 51(7):1–7
2. Yajun P, Long G, Chunhui W (2012) Analysis of IQ demodulation and signal noise ratio for 2 μm dual-balanced heterodyne detection. Chin J Lasers 39(1):1–5
3. Xiangyang L, Zongfeng M, Dele S (2015) Effect of Gaussian fields distribution on mixing efficiency for coherent detection. Infrared and Laser Eng 44(2):539–543
4. Hongzhan L, Yuefeng J, Nan X et al (2011) Effect of amplitude profile difference of signal and local oscillator wave on heterodyne efficiency in the inter-satellite coherent optical communication. Acta Optica Sinica 31(10):71–76
5. Haitao C, Huajun Y, Tuohui L et al (2011) Influence of fiber offset on coupling efficiency of spatial beam-single mode fiber. Laser Infrared 7(1):77–80
6. Zengfeng Q (2010) Study on the SNR capability optimization of fiber lidar system. Harbin Institute of Technology, Harbin
7. Chengqiang L (2016) Investigating the effect of optics factors on the performance of laser heterodyne detection system. Graduate School of Chinese Academy of Sciences, Chang Chun (Changchun Institute of Optics, Fine Mechanics and Physics)
8. Gori F, Santarsiero M et al (1998) Beam coherence-polarization matrix. Pure and Appl Opt J Eur Opt Soc Part A 7(5):941
9. Gallion P, Debarge G (1984) Quantum phase noise and field correlation in single frequency semiconductor laser systems. IEEE J Quantum Electron 20(4):343–349
10. Wolf E (2003) Unified theory of coherence and polarization of random electromagnetic beams. Phys Lett A 312(5–6):263–267
11. Roychowdhury H, Ponomarenko SA, Wolf E (2005) Change in the polarization of partially coherent electromagnetic beams propagating through the turbulent atmosphere. J Mod Opt 52(11):1611–1618
12. Roychowdhury H (2006) Changes in the spectral, coherence and polarization properties of stochastic electromagnetic beams on propagation. University of Rochester
13. Wei L, Liren L, Jianfeng S et al (2007) Change in degree of coherence of partially coherent electromagnetic beams propagating through atmospheric turbulence. Opt Commun 271(1):1–8
14. Salem M, Dogariu A (2004) Optical heterodyne detection of random electromagnetic beams. Optica Acta Int J Opt 51(15):2305–2313
15. Gori F, Santarsiero M, Piouero G et al (2001) Partially polarized Gaussian Schell-model beams. J Opt A: Pure Appl Opt 3(1):1–9
16. Chengqiang L, Tingfeng W, Heyong Z et al (2014) Effect of source parameters on polarization characteristics of electromagnetic beam propagating in atmospheric turbulence. Acta Phys Sin 63(10):118–125
17. Meiling D, Jinhong L, Jilin W (2013) Influence of different propagation paths on the propagation of laser in atmospheric turbulence. Optoelectron Lett 9(6):477–480
18. Chao Z, Liying T, Siyuan Y et al (2015) Fiber coupling efficiency for a Gaussian-beam wave propagating through non-Kolmogorov turbulence. Opt Express 23(12):15242–15255

19. Series P (2015) Propagation data and prediction methods required for the design of Earth-space telecommunication systems. Recommendation ITU-R, pp 618–712
20. Chao L, Shanqiu C, Xinyang L et al (2014) Performance evaluation of adaptive optics for atmospheric coherent laser communications. Opt Express 22(13):15554–15563
21. Kai C (2011) Experiment research of correlation between coherence and polarization of beam. National University of Defense Technology, Chang Sha

Chapter 5
Influence of the Beam Spatial Distribution on the Mixing Efficiency

Compared with traditional free-space optical communication, coherent detection has the advantages of a long communication distance and flexible modulation mode, and has gradually become the research focus of many scholars. Research on the factors affecting the mixing efficiency of coherent-detection systems can provide a basis for improving their performance.

This chapter mainly studies the mixing performance of different combination models when the signal beam and local-oscillator beam have different spatial distributions, and the performance of a vortex-beam coherent-detection system with oblique-path transmissions.

5.1 Beams with Different Spatial Distributions and Coherent Detection

In this section, plane waves and vortex beams with different spatial distributions are introduced, including several plane waves with different amplitude distributions as well as the common vortex beam and Laguerre–Gaussian (LG) beam. Subsequently, these beams will be used as the signal and local-oscillator beams to study the coherent-detection performance of the pairwise-combination model. Finally, the principle of coherent detection and the structure of the detection system are introduced, and the evaluation index of coherent detection is derived.

5.1.1 Beams with Different Spatial Distributions

The properties of several differently distributed plane waves and vortex beams are analyzed. Gaussian, Airy, and uniformly distributed plane waves are listed. The

X. Ke and C. Ke, *Performance-Improvement Limits in Optical Wireless Coherent Communication*, Optical Wireless Communication Theory and Technology,
https://doi.org/10.1007/978-981-96-2000-5_5

beam-field expression, amplitude-distribution curve, and beam-intensity distribution image are presented. The most commonly used Laguerre–Gaussian beam is selected for the vortex-beam research, and its expression, beam intensity, and phase distribution are introduced.

5.1.1.1 Plane Waves

Several plane waves with different amplitude distributions are described below.

Gaussian Distribution

$$U_s = \exp\left(-\frac{r^2}{\omega_0^2}\right), \tag{5.1}$$

where ω_0 is the waist radius (Fig. 5.1).

Airy Distribution

$$U_s = 2J_1\left(\frac{\pi r}{F\lambda}\right) / \left(\frac{\pi r}{F\lambda}\right), \tag{5.2}$$

where f is the F-number of the lens, J_1 is the first-order Bessel function of the first kind, and λ is the beam wavelength (Fig. 5.2).

Uniform Distribution

$$U_s = 1 \tag{5.3}$$

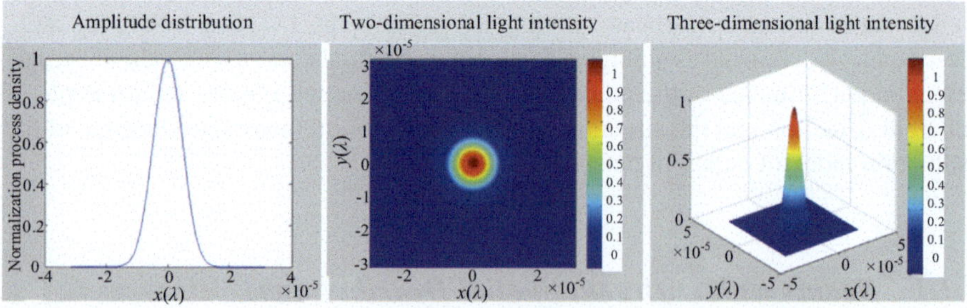

Fig. 5.1 Amplitude and intensity distribution of a Gaussian distributed beam

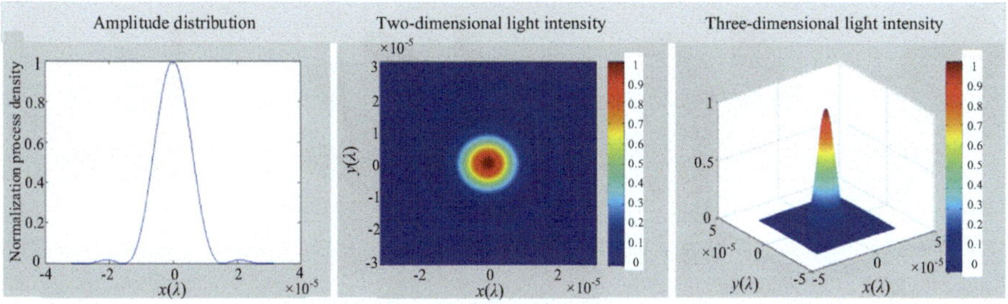

Fig. 5.2 Amplitude and intensity distributions of an Airy distributed beam

5.1.1.2 Vortex Beams

The vortex beam has a spiral phase factor $\exp(il\theta)$; l is the topological charge of the vortex beam and its symbol represents the rotation direction of the spiral phase. Vortex beams can be distinguished by the value of their orbital angular momentum (OAM). OAM beams with different modes are orthogonal to each other. Therefore, vortex beams with different OAM values transmitted through the same channel do not interfere with each other, and can be used for multiplexing communication. Hence, they have attracted considerable attention from researchers worldwide.

The beam-field expression of a vortex beam is [1]

$$u(r, \theta, z) = u_0(r, z) \exp(il\theta) \exp(-ikz), \tag{5.4}$$

where z is the propagation direction, u_0 is the amplitude distribution, l is the topological charge, and k is the wavenumber.

In Eq. (5.4), when θ is equal to 0, it represents the expression of a Gaussian beam. This implies that a vortex beam can only be generated by adding a phase factor. It can be observed that the vortex beam not only has spin angular momentum but also orbital angular momentum.

Figure 5.3 shows the three-dimensional phase distribution of ordinary Gaussian beams and vortex beams with different topological charges obtained through simulation. Among them, Fig. 5.3a shows that the phase of an ordinary Gaussian beam with a topological charge of 0 exhibits a planar distribution without a spiral distribution. Figure 5.3b–d show vortex beams with topological-charge numbers of 1, 3, and 5, respectively; the number of spirals in their phase distribution is consistent with the topological-charge value. By adding a phase factor of $\exp(il\theta)$ to the expression of a Gaussian beam, it can be transformed into a vortex beam with different topological charges by controlling the value of the topological charge l.

The Laguerre–Gaussian (LG) beam is a type of vortex beam; its beam field with propagation distance is expressed as [1]

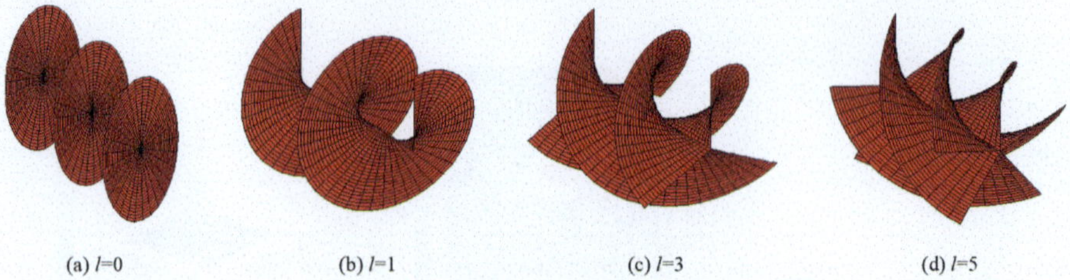

Fig. 5.3 Phase-distribution structure of Gaussian beams and vortex beams [2]

$$u_p^l(r,\theta,z)=\sqrt{\frac{2p!}{\pi(|l|+p)!}\frac{1}{w(z)}}\left[\frac{\sqrt{2}r}{w(z)}\right]^{|l|}L_p^{|l|}\left[\frac{2r^2}{w^2(z)}\right]\exp\left[\frac{-r^2}{w^2(z)}\right]$$
$$\exp\left(\frac{ikr^2z}{2(z^2+z_R^2)}\right)\exp\left[-i(2p+|l|+1)\arctan\left(\frac{z}{z_R}\right)\right]\exp(-il\theta), \tag{5.5}$$

where p is the radial index of the LG beam, r is the distance from a point in space to the central axis, and θ is the azimuth; w_0 is the waist radius of the beam, $w(z) = w_0[1 + (z / z_R)^2]^{1/2}$ is the beam radius of the LG beam propagating to z, $L|l|\ p(\cdot)$ is the associated Laguerre polynomial, $z_R = kw_0 / \lambda$ is the Rayleigh distance, and $k = 2\pi/\lambda$ is the wavenumber.

When the radial index $p = 0$ and $L|l|\ p(\cdot) = 1$ at a transmission distance $z = 0$, Eq. (5.5) can be expressed as [2]

$$u^l(r,\theta)=\sqrt{\frac{2}{\pi(|l|)!}\frac{1}{w_0}}\left(\frac{\sqrt{2}r}{w_0}\right)^{|l|}\times\exp\left(-\frac{r^2}{w_0^2}\right)\exp(-il\theta). \tag{5.6}$$

We simulate the expression of the LG beam in Eq. (5.6) with the following parameters: wavelength $\lambda = 1550$ nm and $\omega_0 = 0.03$ m. The intensity and phase distributions of a single LG beam with different topological charges and a transmission distance of 0 are shown in Fig. 5.4. Figure 5.4a1 and a2 respectively show the intensity and phase distribution of LG beams with a topological charge $l = 1$. Similarly, Fig. 5.4b1 and b2 respectively show the intensity and phase distributions of an LG beam with a topological charge of $l = 3$.

By comparing Fig. 5.4a1 and b1, it can be observed that the central intensity of the LG beam is 0, and the intensity distribution is circular, with a dark spot in the center. In addition, as the topological charge increases, the spot radius of the LG beam also increases. Comparing Fig. 5.4a2 and b2, it can be observed that in the phase distribution of the LG beam, a series of equal phase lines are emitted from the center, with a number equal to the topological charge. As the topological charge increases, the number of equiphase lines also increases.

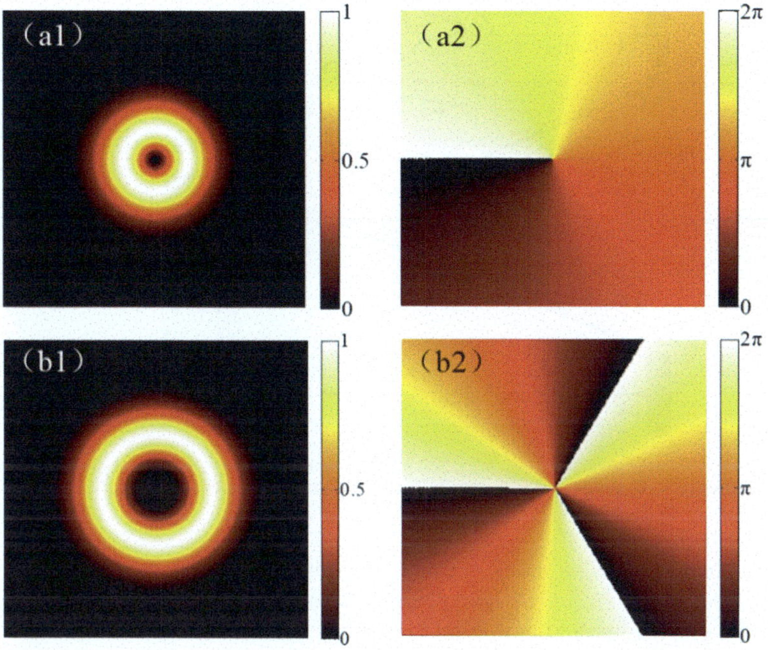

Fig. 5.4 Intensity and phase distribution of an LG beam. **a1** l = 1, intensity; **a2** l = 1, phase; **b1** l = 3, intensity; **b2** l = 3, phase

5.1.2 Mixing Efficiency Affected by Atmospheric Turbulence

As shown in Fig. 5.5, a common coherent-detection system combines a signal beam transmitted through the atmosphere with a local-oscillator (LO) beam at the receiving end to form a beam, which is then converged onto the surface of the detector through a focusing lens for frequency mixing [3, 4]. The essence of frequency mixing is the interference between the signal beam and local-oscillator beam on the surface of the photodetector. The necessary conditions for the interference of two beam waves are a fixed phase difference and the same polarization direction when the two beam waves are superimposed [5, 6].

The intermediate-frequency (IF) signal output by the detector contains various types of noise, and the mixing efficiency is a key indicator for evaluating the performance of coherent-detection systems. It reflects the degree of coherent matching between the signal beam and local-oscillator beam and affects the signal-to-noise ratio of the communication system. Changes in the superimposed beam-field characteristics on the detector surface directly affect the mixing efficiency. Therefore, an evaluation index for mixing efficiency is introduced to evaluate the mixing performance of the system [7].

The phase difference between the signal beam and LO beam discussed above is constant; however, in practice, when the signal beam passes through the atmosphere,

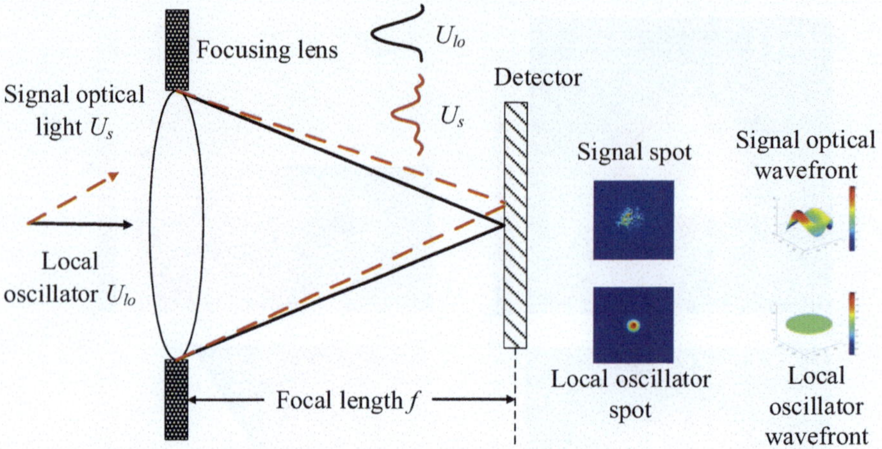

Fig. 5.5 Schematic diagram of a coherent-detection system

the atmospheric refractive index is uneven, resulting in a distortion of the phase of the signal-beam wavefront [8]. The phase difference caused by atmospheric turbulence leads to a decline in the mixing efficiency and affects the performance of the coherent optical communication system. The expression for a signal beam affected by turbulence is

$$E_{s1}(r, \varphi, t) = U_{s1}(r, \varphi) \exp[i(\omega_{s1} t + \varphi_{s1})] = U_s(r, \varphi, t) \exp(\Phi), \tag{5.7}$$

where U_s and U_{s1} are the amplitudes before and after the influence of the optical signal receiving turbulence, respectively; ω_{s1} and φ_{s1} are the angular frequency and initial phase after the turbulence, respectively; and Φ are the joint expressions of the amplitude and phase fluctuations caused by atmospheric turbulence. $U_s(r, \varphi, t)$ is the relationship between the complex amplitude of the signal beam at any point on the detector and time; $U_s(r, \varphi, t) = U_s(r, \varphi) \exp[i(\omega_s t + \varphi_s)]$, where ω_s and φ_s are the original angular frequency and initial phase of the signal beam, respectively. The expression for the Φ is given as follows:

$$\Phi = \ln \frac{U_{s1}(r, \varphi)}{U_s(r, \varphi)} + i[\Phi_{s1} - \Phi_s] = \chi + iS, \tag{5.8}$$

where χ is the change in beam intensity, ln is the natural logarithm of base e, S is the wavefront phase distortion, and Φ_s and Φ_{s1} are, respectively, the joint expressions of the amplitude and phase fluctuations of the signal beam before and after being affected by atmospheric turbulence.

Substituting Eq. (5.8) into Eq. (5.7) yields

$$U_{s1} = U_s(r, \varphi) \exp[i(\omega_s t + \varphi_s)] \exp(\chi + iS). \tag{5.9}$$

Equation (5.9) as the signal light, the mixing efficiency can be expressed as:

$$\eta = \frac{\left| \int_0^{r_0} U_s(r) U_{lo}(r) \exp(-iS) r dr \right|^2}{\int_0^{r_0} [U_{lo}(r)]^2 r dr \int_0^{\infty} [U_{lo}(r)]^2 r dr}, \tag{5.10}$$

where $U_s(r)$ and $U_{lo}(r)$ represent the amplitude distribution of the detector-surface signal beam and local-oscillator beam, respectively.

Equation (5.10) shows that the mixing efficiency of coherent detection is affected by atmospheric turbulence. The Gaussian signal transmitted over a long distance reaches the detector surface as a plane wave [9]. The phase fluctuations under the influence of atmospheric turbulence can be described using the Kolmogorov statistical theory. The phase fluctuation of a plane wave is [10]

$$S = 2.914 C_n^2 L k^2 \rho^{5/3}, \tag{5.11}$$

where C_n^2 is the atmospheric refractive-index structure constant, L is the transmission distance of the signal beam, k is the wavenumber, and ρ is the two-dimensional space-vector diameter, which is larger than the internal turbulence scale and smaller than the external scale. The mixing efficiency of a coherent optical communication system affected by atmospheric turbulence is

$$\eta = \frac{\left| \int_0^{r_0} U_s(r) U_{lo}(r) \exp(-i(2.914 C_n^2 L k^2 \rho^{5/3})) r dr \right|^2}{\int_0^{r_0} [U_{lo}(r)]^2 r dr \int_0^{\infty} [U_{lo}(r)]^2 r dr}. \tag{5.12}$$

5.2 Mixing Performance of a Vortex-Beam and Plane-Wave Combination Model

The previous section introduced the principle of coherent detection and beams with different spatial distributions. This section analyzes the mixing efficiency of coherent detection when the signal beam has Gaussian and Airy distributions under ideal and turbulent conditions, and when the local-oscillator beam has Gaussian, Airy, and uniform distributions. By selecting different combinations of the signal beam and local-oscillator beam, the mixing efficiency is calculated to determine the best combination. The possibility of using Gaussian, Airy, and uniformly distributed beams for the coherent detection of a local-oscillator beam at the receiving end when the signal beam is a vortex beam is discussed.

5.2.1 *Signal Beam Is a Plane Wave*

5.2.1.1 Signal Beam Has a Gaussian Distribution

Coherent optical communication systems typically employ Gaussian signal beams. The Gaussian beam becomes larger after a long-distance transmission. The proportion of the signal spot received by the detector is very small, and can be approximately regarded as a plane wave. After the Gaussian beam is focused by the lens and reaches the detector surface, the amplitude remains Gaussian; however, the coefficient of the Gaussian distribution function changes.

Because the constant coefficient of the expression of the local-oscillator and signal beams will appear simultaneously in the numerator and denominator during the calculation of the mixing efficiency, it can be reduced. Hence, the expression of the Gaussian signal beam is $U_s(r) = \exp(r^2 / \omega_0{}^2)$, and the local oscillator can have a Gaussian, Airy, or uniform distribution. J_1 is the first kind of first-order Bessel function, and F is the F-number of the receiving optical system.

In the following analysis, the following substitutions are performed: $X = \pi r / F\lambda$, $X_O = \pi r_O/F\lambda$, $Q = r/\omega$, and $Q_0 = r/\omega_0$. In the following calculation, the wavelength $\lambda = 1550$ nm and the waist radius of the Gaussian beam $\omega_0 = 1.6\lambda$.

The following presents an analysis of the mixing efficiency and receiving optical-system parameters under ideal conditions when the signal beam has a Gaussian distribution. Equations (5.13)–(5.15) are the theoretical formulas for the mixing efficiency under ideal conditions [9].

When the signal beam has a Gaussian distribution and the local-oscillator beam is uniformly distributed,

$$\eta_{G-U} = \frac{\left|\int_0^{r_0} \exp(-Q^2) r dr\right|^2}{\int_0^{r_0} r dr \int_0^{\infty} [\exp(-Q^2)]^2 r dr} = \frac{2\left|1 - \exp(-Q_0^2)\right|^2}{Q_0^2}, \tag{5.13}$$

where $X = \pi r/F\lambda$, $X_O = \pi r_O/F\lambda$, $Q = r/\omega$, $Q_0 = r/\omega_0$, r_O is the detector radius, and ω_0 is the waist radius.

When the local oscillator has a Gaussian distribution,

$$\eta_{G-G} = \frac{\left|\int_0^{r_0} \exp(-Q^2) r dr\right|^2}{\int_0^{r_0} [\exp(-Q^2)]^2 r dr \int_0^{\infty} [\exp(-Q^2)]^2 r dr} = 1 - \exp\left(-2Q_0^2\right). \tag{5.14}$$

When the local-oscillator beam has an Airy distribution,

$$\eta_{G-A} = \frac{\left|\int_0^{r_0} \left[J_1(X)/\left(\frac{\pi r}{F\lambda}\right)\right] \exp(-Q^2) r dr\right|^2}{\int_0^{r_0} \left[J_1(X)/(X)\right]^2 r dr \int_0^{\infty} \left[\exp(-Q^2)\right]^2 r dr} = \frac{\frac{8}{\omega_0^2}\left|\int_0^{r_0} J_1(X) \exp(-Q^2) dr\right|^2}{\left[1 - J_0^2(X_0) - J_1^2(X_0)\right]}. \tag{5.15}$$

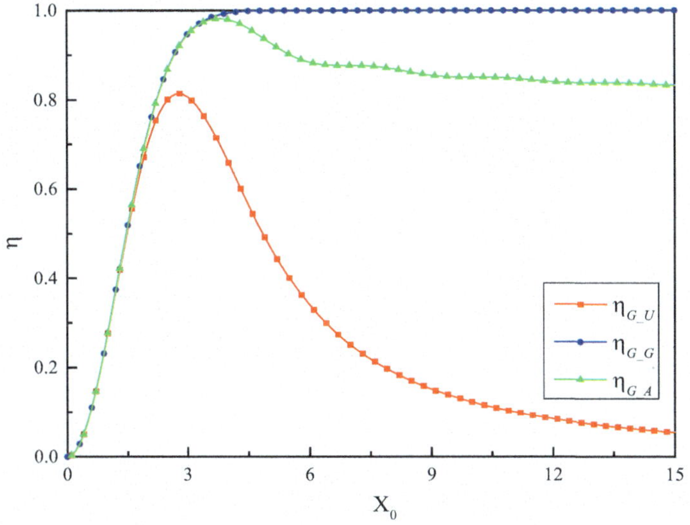

Fig. 5.6 Mixing efficiency of a signal beam with a Gaussian distribution and a local-oscillator beam with different distributions

From Fig. 5.6, it can be observed that when the signal beam is Gaussian and the local-oscillator beam is uniformly distributed, the mixing efficiency reaches a maximum value of 81.45% with an increase in the receiving optical-system parameters. When the receiving optical-system parameters exceed this value, the mixing efficiency rapidly decreases as they increase, indicating that this model is sensitive to them and they need to be adjusted based on the system parameters to achieve a higher mixing efficiency.

When the signal beam and local-oscillator beam both have Gaussian distributions, the mixing efficiency increases linearly to the maximum value of 100% with the increase of the receiving optical-system parameters. When an appropriate detector radius is selected, the mixing efficiency when the signal beam has a Gaussian distribution and the local-oscillator beam has an Airy distribution can approach 100%. However, as the receiving optical-system parameters continue to increase, the mixing efficiency decreases but can still be maintained at approximately 80%.

Following is a numerical analysis of the mixing efficiency under the influence of atmospheric turbulence when the signal beam has a Gaussian distribution, and the local-oscillator beam has different spatial distributions.

Local-Oscillator Beam Is Uniformly Distributed

When the local-oscillator beam is uniformly distributed, the mixing efficiency$\eta_{G\text{-}U}$ under the influence of atmospheric turbulence can be obtained from Eq. (5.12), as follows:

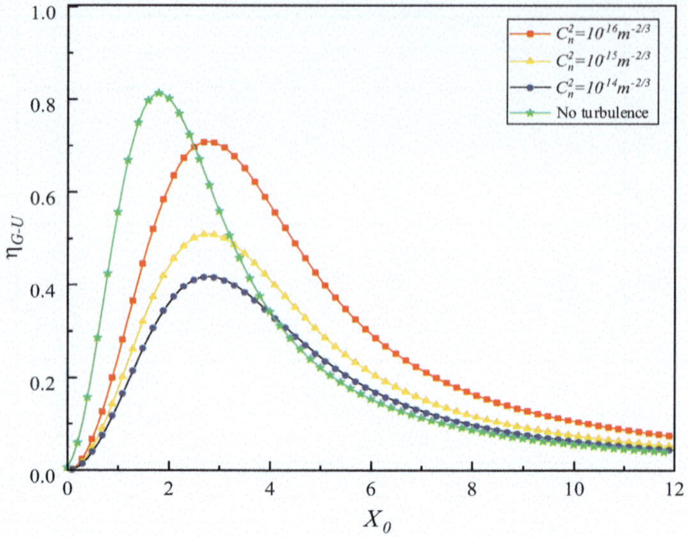

Fig. 5.7 Mixing efficiency of a Gaussian–uniform distribution model

$$\eta_{G-U} = \frac{\left|\int_0^{r_0} \exp(-Q^2)\exp(-i(2.914C_n^2Lk^2\rho^{5/3}))rdr\right|^2}{\int_0^{r_0} rdr \int_0^{\infty}[\exp(-Q^2)]^2 rdr}. \quad (5.16)$$

The variation curve of the mixing efficiency is shown in Fig. 5.7, when the transmission distance $L = 500$ m, the two-dimensional space-vector length is $\rho = 0.1$ m, and C_n^2 takes different values under the conditions of no turbulence and turbulence in Eq. (5.16).

From Fig. 5.7, it can be observed that the mixing efficiency reaches a maximum value of 81.45% with an increase in the receiving optical-system parameters. Subsequently, the mixing efficiency rapidly decreases with an increase in the receiving optical-system parameters, indicating that the mixing efficiency of this model is more sensitive to them, and that it is necessary to adjust their values to achieve a higher mixing efficiency.

As shown in the figure, the mixing efficiency gradually decreases with an increase in C_n^2. When $C_n^2 = 10^{-14}$ m$^{-2/3}$, the mixing efficiency is less than 40%.

Local-Oscillator Has a Gaussian Distribution

When the local-oscillator beam has a Gaussian distribution, it can be obtained from Eq. (5.12) that the mixing efficiency under the influence of atmospheric turbulence $\eta_{G\text{-}G}$ is

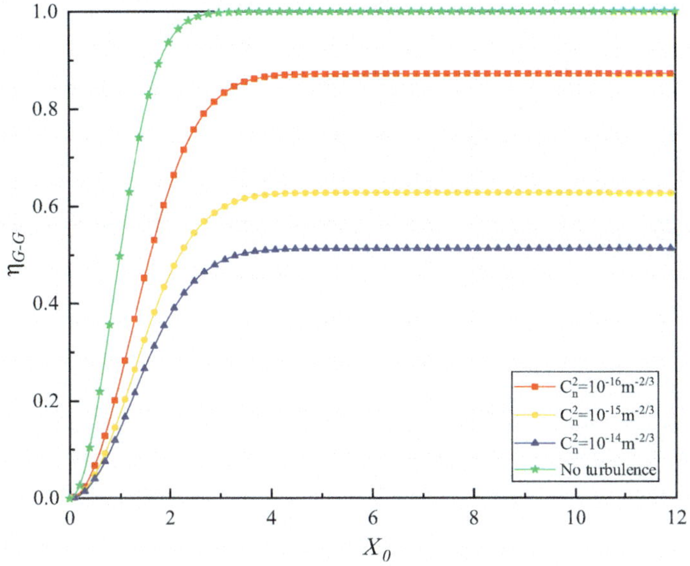

Fig. 5.8 Mixing efficiency of a Gaussian–Gaussian distribution model

$$\eta_{G-G} = \frac{\left|\int_0^{r_0} \exp(-Q^2)\exp(-i(2.914C_n^2Lk^2\rho^{5/3}))rdr\right|^2}{\int_0^{r_0}[\exp(-Q^2)]^2 rdr \int_0^{\infty}[\exp(-Q^2)]^2 rdr}. \tag{5.17}$$

The variation curve of the mixing efficiency under turbulent conditions with different values of C_n^2, obtained by combining Eq. (5.17), is shown in Fig. 5.8. In the absence of turbulence, the mixing efficiency rapidly increases to a maximum of 100% with an increase in the receiving optical-system parameters and is gradually reduced by the influence of atmospheric turbulence. When $C_n^2 = 10^{-16}\ \text{m}^{-2/3}$, the mixing efficiency exceeds 80%, indicating that the mixing performance of this model is less affected by atmospheric turbulence.

Local-Oscillator Beam Has an Airy Distribution

When the local-oscillator beam has an Airy distribution, it can be obtained from Eq. (5.12) that the mixing efficiency $\eta_{G\text{-}A}$ under the influence of atmospheric turbulence is

$$\eta_{G-A} = \frac{\left|\int_0^{r_0} [J_1(X)/(X)]\exp(-Q^2)\exp(-i(2.914C_n^2Lk^2\rho^{5/3}))rdr\right|^2}{\int_0^{r_0}[J_1(X)/(X)]^2 rdr \int_0^{\infty}[\exp(-Q^2)]^2 rdr}. \tag{5.18}$$

The variation curves of the mixing efficiency under ideal and turbulent conditions with different C_n^2 values are shown in Fig. 5.9. In the absence of turbulence, when the

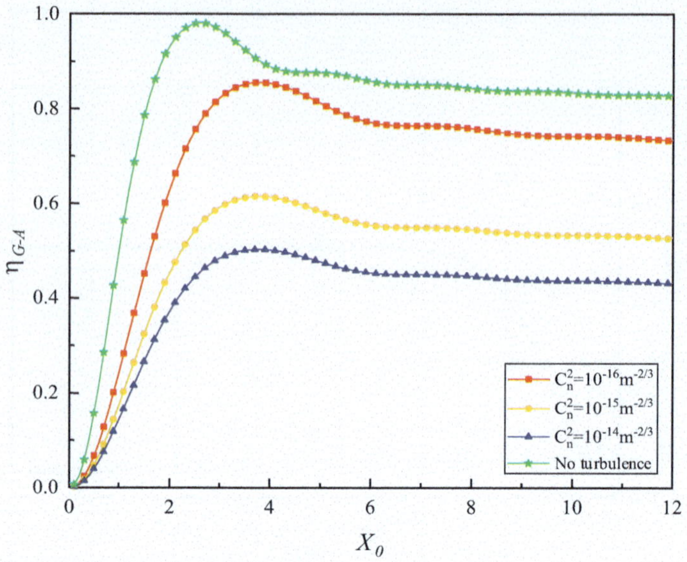

Fig. 5.9 Mixing efficiency of a Gaussian–Airy distribution model

appropriate receiving optical-system parameters are selected, the mixing efficiency can approach 100% in the case of a Gaussian-distribution signal and Airy-distribution local oscillator. However, when the receiving optical-system parameters continue to increase, the mixing efficiency decreases but can still be maintained at approximately 80%. As the turbulence intensity increases, the mixing efficiency gradually decreases. When $C_n^2 = 10^{-14}\ \text{m}^{-2/3}$, the mixing efficiency is less than 50%, which is 50% lower than that when there is no turbulence.

5.2.1.2 Signal Beam is Uniformly Distributed

The uniform signal beam is diffracted through the lens shown in Fig. 5.7, and when focused on the detector surface, the spatial distribution becomes an Airy distribution [9]. The local oscillator beam can be Gaussian, Airy, or uniformly distributed, and the mixing efficiency can be calculated using Eq. (5.33), yielding Eqs. (5.19)–(5.21) [9].

When the local-oscillator beam is an Airy distribution,

$$\eta_{A-A} = \frac{\left|\int_0^{r_0} [J_1(X)/(X)]^2 rdr\right|^2}{\int_0^{r_0} [J_1(X)/(X)]^2 rdr \int_0^{\infty} [J_1(X)/(X)]^2 rdr} = \frac{\left|\int_0^{r_0} [J_1(X)/(X)]^2 rdr\right|^2}{\int_0^{\infty} [J_1(X)/(X)]^2 rdr}. \\ = 1 - J_0^2(X_0) - J_1^2(X_0) \tag{5.19}$$

When the local oscillator is Gaussian distributed,

$$\eta_{A-G} = \frac{\left|\int_0^{r_0}[J_1(X)/(X)]\exp(-Q_0^2)rdr\right|^2}{\int_0^{\infty}[J_1(X)/(X)]^2 rdr\int_0^{r_0}[\exp(-Q_0^2)]^2 rdr} = \frac{\frac{8}{\omega_0^2}\left|\int_0^{r_0}J_1(X)\exp(-Q_0^2)dr\right|^2}{[1-\exp(-2Q_0^2)]}. \tag{5.20}$$

When the local-oscillator beam is uniformly distributed,

$$\eta_{A-U} = \frac{\left|\int_0^{r_0}[J_1(X)/(X)]rdr\right|^2}{\int_0^{r_0}rdr\int_0^{\infty}[J_1(X)/(X)]^2 rdr} = \frac{\left|\int_0^{X_0}J_1(X)dX\right|^2}{\frac{1}{2}(X_0)^2\cdot\frac{1}{2}} = 4[1-J_0(X_0)]^2/(X_0)^2. \tag{5.21}$$

Based on the mixing-efficiency equations obtained for the different combination models above, it is observed that when $\lambda = 1550$ nm, $\omega_0 = 1.6\lambda$, and F uses the most suitable value for each model, the mixing-efficiency curve with the receiving optical-system parameter X_0 is as shown in Fig. 5.10.

As shown in the figure, when the local-oscillator beam is uniformly distributed, the mixing efficiency is greatly affected by the parameters of the receiving optical system. First, it increases linearly. Then, as the parameters of the receiving optical system increase, it reaches a maximum value near $X_0 = 2.9$, and then decreases. There is a serious mismatch between the signal beam and the local-oscillator beam, and this model is not suitable.

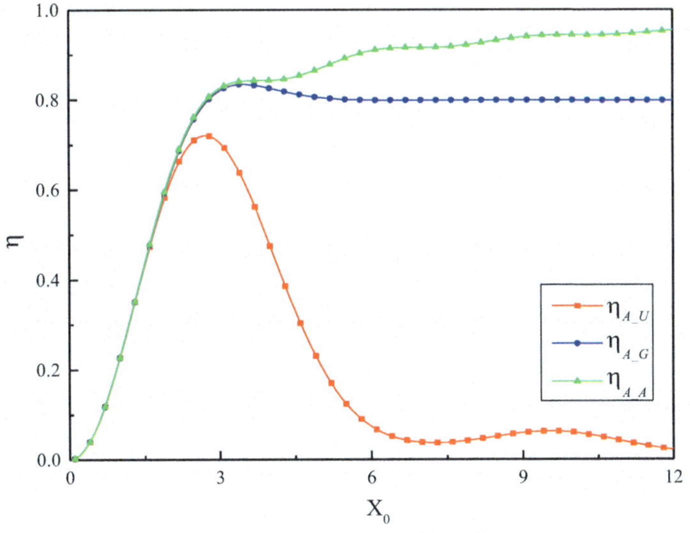

Fig. 5.10 Mixing efficiency when the signal beam has an Airy distribution and the local-oscillator beam has different distributions

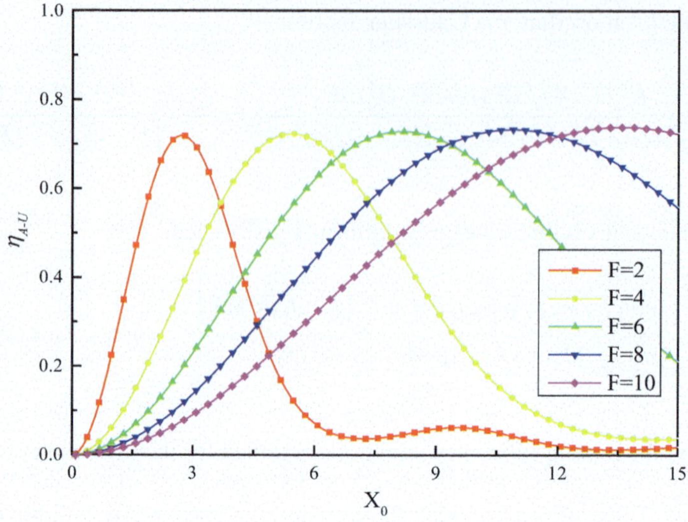

Fig. 5.11 Mixing efficiency of an Airy–Uniform distribution model with different F-numbers

When the local oscillator is Gaussian distributed, the mixing efficiency rapidly increases with the increase of the receiving optical-system parameters, and then remains at a constant value, indicating that this model has strong adaptability to the receiving optical system and can maintain a high mixing efficiency. When the local oscillator also has an Airy distribution, the mixing efficiency increases linearly and then slowly. If the receiving aperture of the receiving optical system is sufficiently large, the mixing efficiency can reach 100%.

Following is a numerical analysis of the mixing efficiency after an atmospheric-turbulence transmission when the signal beam is Airy distributed and the local-oscillator beam has different spatial distributions.

Local-Oscillator Beam is Uniformly Distributed

When the local-oscillator beam is uniformly distributed, the variation in the mixing efficiency with the parameters of the receiving optical system at different F values is shown in Fig. 5.11. The figure shows that the optimal mixing efficiency of this model is 70%, which is lower than those of the other two models. The conditions for achieving the maximum mixing efficiency are relatively strict. As the F-number increases, the receiving optical-system parameters required to achieve the maximum mixing efficiency increase.

When the local-oscillator beam is uniformly distributed, it can be obtained from Eq. (5.12) that the mixing efficiency $\eta_{A\text{-}U}$ under the influence of atmospheric turbulence is

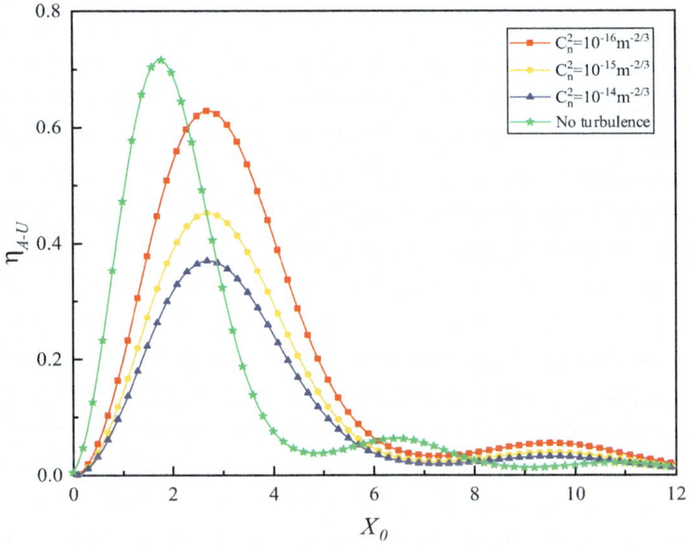

Fig. 5.12 Mixing efficiency of an Airy–uniform distribution model

$$\eta_{A-U} = \frac{\left|\int_0^{r_0} [J_1(X)/(X)] \exp(-i(2.914C_n^2 Lk^2 \rho^{5/3})) r dr\right|^2}{\int_0^{r_0} r dr \int_0^{\infty} [J_1(X)/(X)]^2 r dr}. \tag{5.22}$$

According to Eq. (5.22), the variation curve of the mixing efficiency with different values of C_n^2 is as shown in Fig. 5.12. When the signal beam is Airy distributed and the local-oscillator beam is uniformly distributed, the mixing efficiency quickly reaches a maximum value of 72.1% and then decreases, indicating that this model is greatly affected by the parameters of the receiving optical system. The conditions for reaching the maximum value are relatively strict, and when atmospheric turbulence occurs, the receiving optical-system parameters required to reach the maximum mixing efficiency increase. When $C_n^2 = 10^{-14}\ \mathrm{m}^{-2/3}$, the mixing efficiency is less than 40%.

Local-Oscillator Beam Has an Airy Distribution

The mixing efficiencies of the different models with different values of F when the local-oscillator beam has an Airy distribution are shown in Fig. 5.13. After a linear increase, the mixing efficiency slowly approached 100%, and the larger the F-number, the larger the receiving optical-system parameters required to achieve the maximum mixing efficiency. This model has a weak adaptability to the receiving system.

When the local-oscillator beam has an Airy distribution, it can be obtained from Eq. (5.12) that the mixing efficiency $\eta_{A\text{-}A}$ under the influence of atmospheric

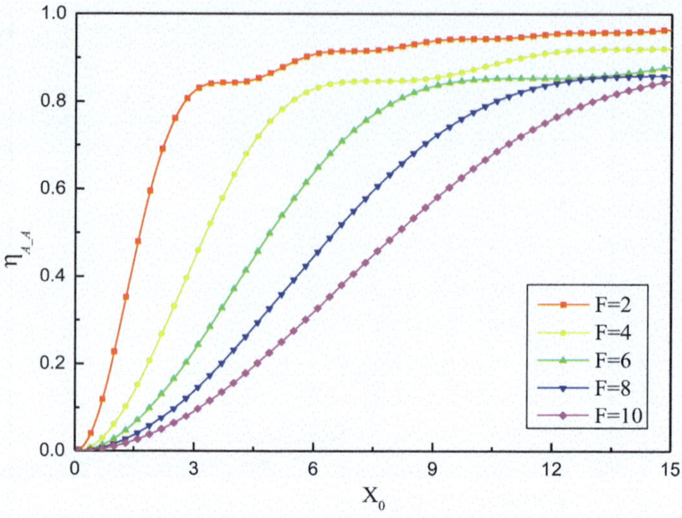

Fig. 5.13 Mixing efficiency of an Airy–Airy distribution model with different F-numbers

turbulence is

$$\eta_{A-A} = \frac{\left|\int_0^{r_0} [J_1(X)/(X)]^2 \exp(-i(2.914C_n^2Lk^2\rho^{5/3}))rdr\right|^2}{\int_0^{r_0} [J_1(X)/(X)]^2rdr \int_0^{\infty} [J_1(X)/(X)]^2rdr}. \tag{5.23}$$

The variation curve of the mixing efficiency with different values of C_n^2 obtained by combining Eq. (5.23), is shown in Fig. 5.14. When both the signal beam and local-oscillator beam have an Airy distribution, the mixing efficiency first rapidly increases to approximately 80% with an increase in the receiving optical-system parameters, then slowly increases and gradually approaches 100%, and then gradually decreases with an increase in C_n^2.

Local-Oscillator Has a Gaussian Distribution

The F-number of the receiving optical system was set to different values and the detector radius was set to a fixed value to obtain the change curve of the mixing efficiency, as shown in Fig. 5.15. The mixing efficiency of the Airy–Gaussian model increases with the increase in the receiving optical-system parameters, and then gradually stabilizes.

The change rule of the mixing efficiency of different receiving optical systems with different F-numbers is also different. The maximum mixing efficiency of the receiving system with $F = 4$ is 82%, whereas the change in the mixing efficiency with $F = 2$ is similar to that of the Airy–uniform combination model. This is because the radius of the Airy-disk spot is smaller than that of the Gaussian spot; hence, the

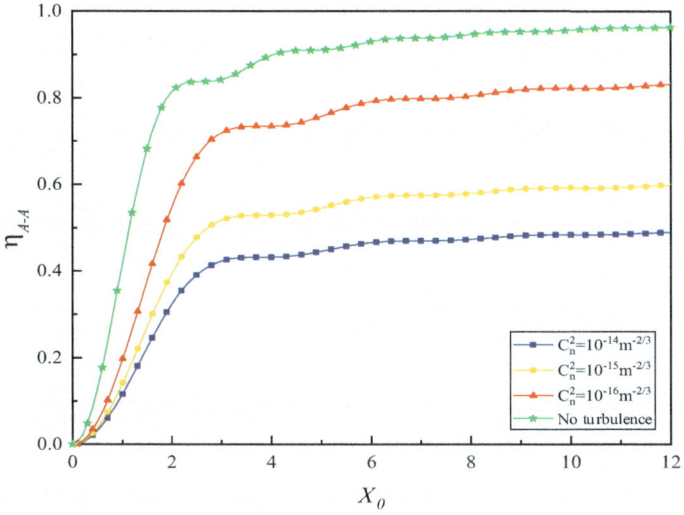

Fig. 5.14 Mixing efficiency of an Airy–Airy distribution model

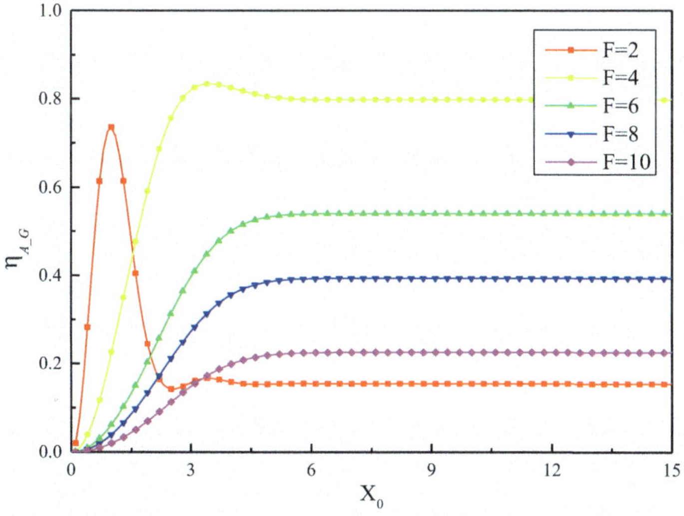

Fig. 5.15 Mixing efficiency of an Airy–Gaussian distribution model with different F-numbers

superposition part of the Gaussian-distributed LO beam is mainly in the center, and the Gaussian-distributed beam can be regarded as an approximately uniform distribution. Therefore, the trend of change is similar to that of the Airy–uniform combination model.

When the local oscillator is Gaussian distributed, it can be obtained from Eq. (5.12) that the mixing efficiency $\eta_{A\text{-}G}$ under the influence of atmospheric turbulence is

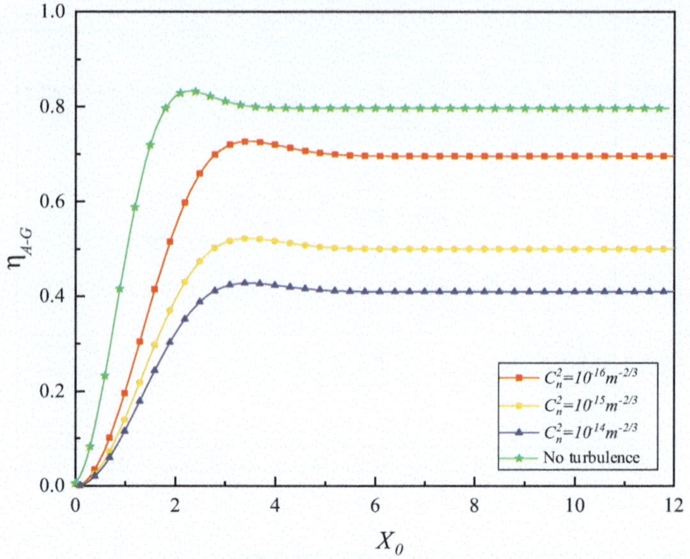

Fig. 5.16 Mixing efficiency of an Airy–Gaussian distribution model

$$\eta_{A-G} = \frac{\left|\int_0^{r_0} [J_1(X)/(X)] \exp(-Q^2) \exp(-i(2.914C_n^2 Lk^2 \rho^{5/3}))rdr\right|^2}{\int_0^{\infty} [J_1(X)/(X)]^2 rdr \int_0^{r_0} [\exp(-Q^2)]^2 rdr}. \quad (5.24)$$

According to Eq. (5.24), the variation curve of the mixing efficiency with different values of C_n^2 is as shown in Fig. 5.16. When the signal beam has an Airy distribution and the local-oscillator beam has a Gaussian distribution, the mixing efficiency reaches 83.42% with the most suitable receiving optical-system parameters. As the receiving optical-system parameters increase, the mixing efficiency increases linearly and reaches a maximum value of 83.42%. Subsequently, it remains stable, indicating that the model has strong adaptability to the receiving optical system. When affected by atmospheric turbulence, the mixing efficiency gradually decreases with an increase in C_n^2.

From the above analysis, it can be seen that the signal beam has an Airy distribution and the local-oscillator beam has a Gaussian distribution, which is the optimal and most commonly used model in practical systems. Below, a detailed analysis of the factors affecting the mixing efficiency of this model is conducted. As shown in Eq. (5.12), in the absence of turbulence, the mixing efficiency is related to the F-number of the receiving optical system and the waist radius of the Gaussian beam. These two factors are discussed below.

Equation (5.24) shows that the mixing efficiency of the Airy–Gaussian model is affected by the parameters of the receiving optical system and the radius of the Gaussian beam. The corresponding mixing efficiencies for different X_0 and ω_0 are shown in Fig. 5.17.

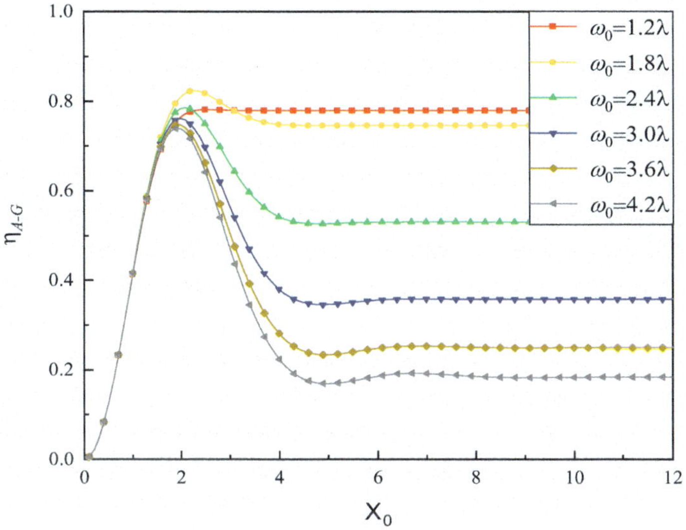

Fig. 5.17 Mixing efficiency of an Airy–Gaussian distribution model with different ω_0

From Fig. 5.17, it can be observed that the mixing efficiency in the initial stage increases linearly with an increase in the receiving optical-system parameters. As the parameters increase, the mixing efficiency also increases. However, as the parameters continue to increase, the mixing efficiency either remains unchanged or decreases. As they increase to a certain extent, the detector receives more background noise, resulting in a decrease in the mixing efficiency. The mixing efficiency cannot continue to increase with an increase in the receiving optical-system parameters.

The collimation mismatch refers to the angle between the optical axis of the signal beam and the axis of the photosensitive surface of the detector. This angle is referred to as the mismatch angle θ. The mismatch angle affects the mixing efficiency of the system. The formula for the mixing efficiency, including the mismatch angle, is

$$\eta = \frac{8}{Q} \frac{\left| \int_0^{X_0} \exp\left(-\frac{X^2}{Q^2}\right) J_1(X) J_0(k\theta r) dX \right|}{1 - \exp\left(-Q^2\right)}, \tag{5.25}$$

where θ is the mismatch angle and J_0 is the zero-order Bessel function.

For research convenience, the relationship between the loss of the mismatch angle and mixing efficiency is discussed under the most suitable condition, which is $F = 2$, $\lambda = 1550$ nm, and $\omega_0 = 1.6\lambda$. It can be seen from Fig. 5.18 that an increase in the mismatch angle will seriously reduce the mixing efficiency, and a micron-level mismatch angle will lead to a serious mismatch between the signal and LO beams, which will reduce the system performance. Therefore, the mismatch angle should be avoided as far as possible in the design and construction of a coherent-detection system to ensure the collimation of the signal beam.

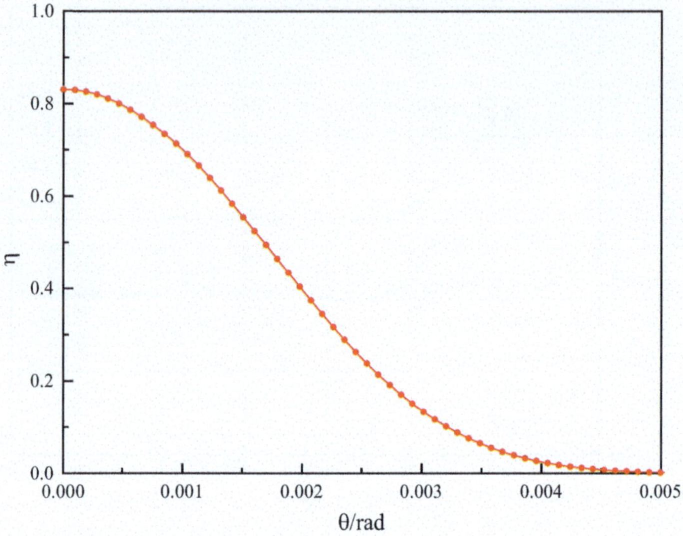

Fig. 5.18 Influence of the mismatch angle on the mixing efficiency of an Airy–Gaussian model

When the F-number of the receiving optical system differs, the relationship curve between the mixing efficiency and the waist radius of the Gaussian beam is as shown in Fig. 5.19. This figure shows that the mixing efficiency of the system can be improved by selecting an appropriate F-number. When an appropriate F-number is selected, the mixing efficiency can reach the ideal maximum value. From the trend of the curve changes, the larger the waist radius, the greater the F-number required for the receiving optical system.

The phase difference between the signal and local-oscillator beams can cause a mismatch in their fields, resulting in a decrease in the mixing efficiency. The following section analyzes the change in mixing efficiency caused by aberrations of the receiving antenna.

A Zernike polynomial is used to represent the wavefront of the transmitted signal beam. The Zernike polynomial containing the primary aberration is [11]

$$\begin{cases} W_{tilt}(\rho, \varphi) = W_{111}\rho\cos\varphi \\ W_{defocus}(\rho, \varphi) = W_{020}\varphi^2 \\ W_{spherical\ aberration}(\rho, \varphi) = W_{040}\rho^4, \\ W_{coma}(\rho, \varphi) = W_{131}\rho^3\cos\varphi \\ W_{astigmatism}(\rho, \varphi) = W_{222}\rho^2\cos\varphi \end{cases} \tag{5.26}$$

where ρ is the normalized radius, $\rho = r/R$, $r = 100$ mm is the radius of the receiving antenna, and W is the magnitude of the wavefront aberration. Therefore, the mixing efficiency formula with an aberration influence is [11]

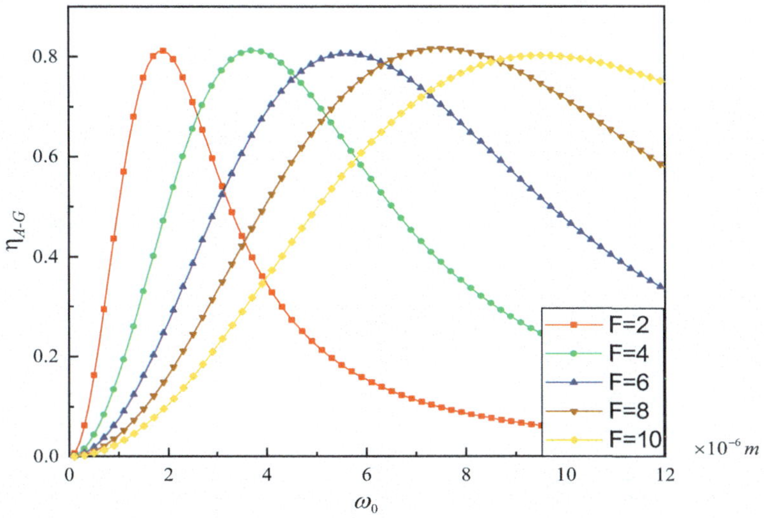

Fig. 5.19 Relationship between the mixing efficiency, ω_0, and F-number with an Airy optical signal and Gaussian local-oscillator signal

$$\eta = \frac{\left| \int_0^{2\pi} \int_0^{r_0} U_s(r,\varphi) U_{lo}(r,\varphi) \exp[ikW_X(\rho,\varphi)] r dr d\varphi \right|^2}{\int_0^{2\pi} \int_0^{r_0} [U_{lo}(r,\varphi)]^2 r dr d\varphi \int_0^{2\pi} \int_0^{\infty} [U_s(r,\varphi)]^2 r dr d\varphi}, \tag{5.27}$$

where $kW_X(\rho, \varphi)$ is the phase difference between the signal beam introduced by the antenna aberration and the local-oscillator beam, and W is the magnitude of the wavefront aberration.

A spherical aberration is caused by the interceptions of different beam rays with different angular apertures emitted by the axial object points in the image side. The difference caused by the loss of symmetry in the image, caused by a beam emitted by an off-axis object point that is symmetrical to the main beam, is called a coma. Astigmatism is caused by the inconsistency between the meridional and sagittal rays in a beam emitted by an off-axis object at the point where they converge in the image. Field curvature refers to the curvature of the image field, which is the aberration of the outer point of the main axis. The magnification on the conjugate surface changes with the increase of the field of view, and there are different magnifications on the axis and edges. This deformation aberration is called a distortion.

The relationship between the size of the above aberrations and the mixing efficiency is shown in Fig. 5.20. Primary aberrations can decrease the mixing efficiency to various degrees, and the tilt difference has the most significant effect. When the aberration is 0.2λ, the mixing efficiency decreases by approximately 10%. Coma and spherical aberrations have the smallest impact on the mixing efficiency, and defocusing has a greater impact than astigmatism.

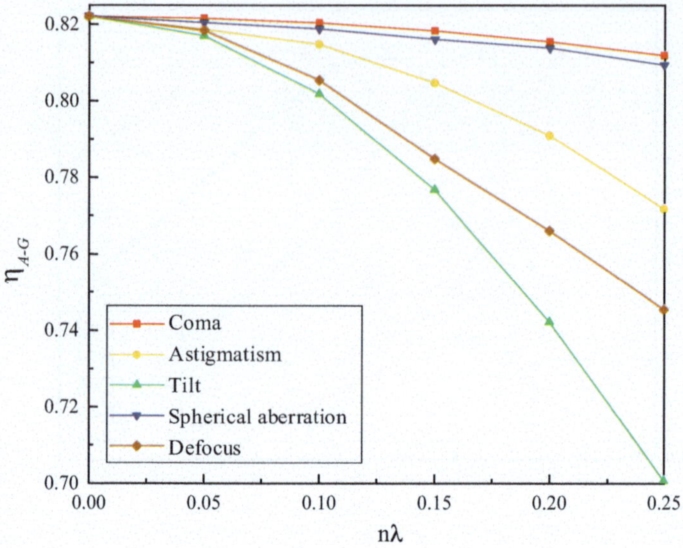

Fig. 5.20 Relationship between an antenna aberration and the mixing efficiency

5.2.2 *Signal Beam Is a Vortex Beam*

An LG beam is a common vortex beam used in optical communication systems. A Gaussian beam is often used as the detection medium when detecting vortices in a communication system. Owing to the particularity of the vortex-beam phase, the mixing efficiency of vortex optical heterodyne detection will be studied from the perspectives of amplitude and phase matching, and the feasibility of optical heterodyne detection for vortex optical communication systems will be discussed. The changes in the mixing efficiency are obtained when the signal beam is an LG beam and the local-oscillator beam is Gaussian, Airy, and uniformly distributed.

When the radial index is 0, the amplitude of the LG beam carrying orbital angular momentum at a transmission distance of 0 can be expressed as [1]

$$U(r,\theta) = \sqrt{\frac{2}{\pi}}\frac{1}{\omega_0}\exp\left(-\frac{r^2}{\omega_0^2}\right)\sqrt{\frac{1}{|l!|}}\left[\frac{r\sqrt{2}}{\omega_0}\right]^{|l|}\exp(il\theta). \tag{5.28}$$

5.2.2.1 Local-Oscillator Beam Has a Gaussian Distribution

When the local-oscillator beam has a Gaussian distribution, assuming that the two phases match, the effect of the amplitude on the mixing efficiency is studied, and the mixing efficiency $\eta_{G\text{-}L}$ is obtained as

$$\eta_{G-L} = \frac{\left| \int_0^{r_0} \sqrt{\frac{2}{\pi}\frac{1}{\omega_0}} \exp(-Q^2) \sqrt{\frac{1}{|l!|}} \left[\sqrt{2}Q\right]^{|l|} \exp(-Q^2) r dr \right|^2}{\int_0^{\infty} \left[\sqrt{\frac{2}{\pi}\frac{1}{\omega_0}} \exp(-Q^2) \sqrt{\frac{1}{|l!|}} \left[\sqrt{2}Q\right]^{|l|} \right]^2 r dr \int_0^{r_0} \left[\exp(-Q^2)\right]^2 r dr}. \quad (5.29)$$

When the local oscillator has a Gaussian distribution and the signal beam is an LG beam, the relationships between the mixing efficiency and the parameters of the receiving optical system under different topological-charge conditions are obtained according to Eq. (5.29), with topological charges of 1, 2, and 3. The variation curve is shown in Fig. 5.21.

From the graph, it can be seen that when the topological charge is 1, the mixing efficiency reaches approximately 80%. However, when the topological charge is 3, the mixing efficiency is approximately 30% and when the topological charge is 5, the mixing efficiency is only 10%. The mixing efficiency gradually decreases with an increase in the topological charge, and the receiving optical-system parameters required for stable mixing efficiency gradually increase. After reaching the maximum value, the mixing efficiency remains stable and no longer changes with an increase in the receiving optical-system parameters (Fig. 5.22).

Assuming that the amplitude of the signal beam matches that of the local-oscillator beam, the effect of the phase on the mixing efficiency is discussed. The mixing-efficiency expression calculated using Eq. (5.17) is used to set the receiving optical-system parameters to fixed values. The relationship between the topological charge and mixing efficiency is studied, as shown in Fig. 5.23.

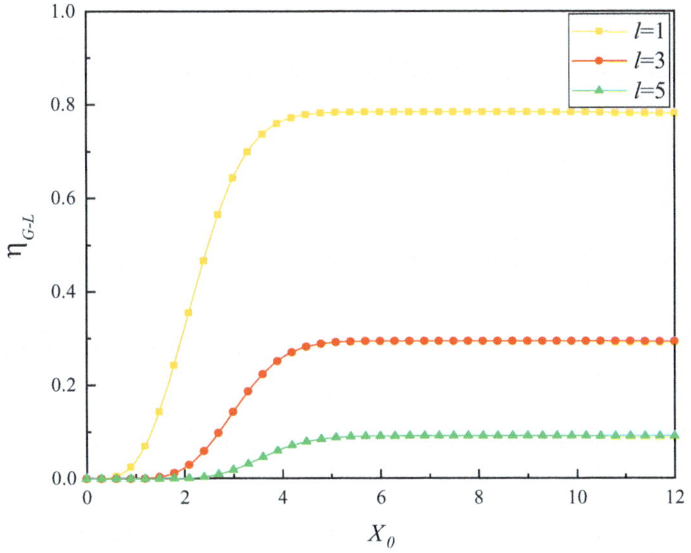

Fig. 5.21 Mixing efficiency when the signal beam is an LG beam and the local-oscillator beam has a Gaussian distribution

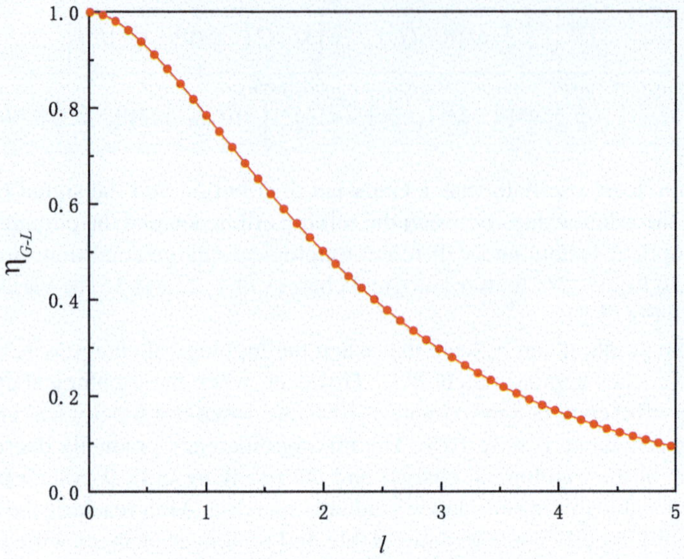

Fig. 5.22 Variation of the mixing efficiency with the topological charge

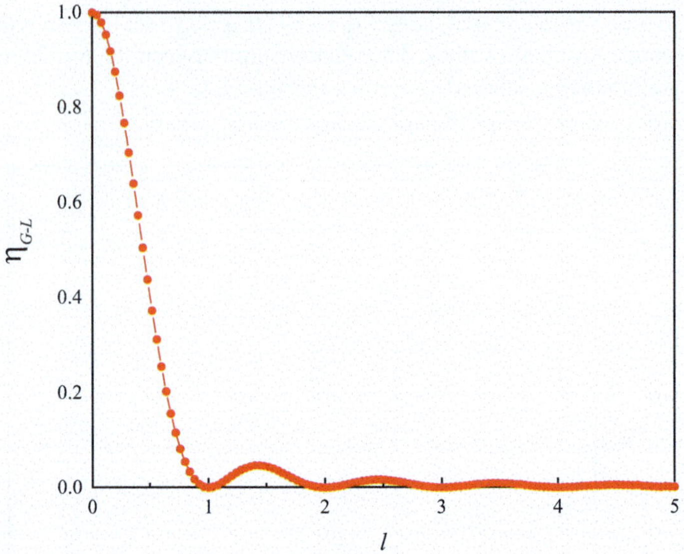

Fig. 5.23 Mixing efficiency when the local-oscillator signal is Gaussian and the signal is an LG beam

From the figure, it can be seen that when the topological charge is 0–1, the mixing efficiency rapidly decreases from 1 to 0, and when the topological charge is greater than 1, the mixing efficiency is less than 10%. When the topological charge is an integer, the mixing efficiency is almost 0. The mixing efficiency between two integer topological charges first increases and then decreases. In practical applications, the topological charge is typically an integer. Therefore, from the perspective of phase matching, Gaussian and LG beams cannot be mixed.

5.2.2.2 Local-Oscillator Beam Has an Airy Distribution

When the local oscillator has an Airy distribution, assuming phase matching, the relationship between the amplitude distribution and mixing efficiency is studied. The formula for the mixing efficiency $\eta_{A\text{-}L}$ is obtained as follows:

$$\eta_{A-L}=\frac{\left|\int_0^{r_0}\sqrt{\frac{2}{\pi}}\frac{1}{\omega_0}\exp\left(-Q^2\right)\sqrt{\frac{1}{|l!|}}\left[\sqrt{2}Q\right]^{|l|}\left[\frac{J_1(X)}{(X)}\right]rdr\right|^2}{\int_0^{\infty}\left[\sqrt{\frac{2}{\pi}}\frac{1}{\omega_0}\exp\left(-Q^2\right)\sqrt{\frac{1}{|l!|}}\left[\sqrt{2}Q\right]^{|l|}\right]^2rdr\int_0^{r_0}\left[\left[\frac{J_1(X)}{(X)}\right]\right]^2rdr}. \tag{5.30}$$

The relationship between the mixing efficiency and the parameters of the receiving optical system was calculated using Eq. (5.30). By changing the topological charge of the LG beam, the mixing-efficiency curves with the parameters of the receiving optical system at different topological charges were obtained, as shown in Fig. 5.24. When the local oscillator has an Airy distribution and the topological-charge number of the LG beam is 1, the mixing efficiency first increases to a maximum value of 72.6%, and then stabilizes at approximately 60%. When the topological charge is 3, the maximum mixing efficiency is only 20% and when the topological charge is 5, the mixing efficiency is less than 10%.

To more intuitively observe the relationship between the mixing efficiency and topological charge, the receiving optical-system parameters were set to fixed values. The independent variable was set to the topological charge to obtain the relationship curve between the topological charge and mixing efficiency, as shown in Fig. 5.25. The figure shows that the mixing efficiency decreases rapidly with the topological charge, approaching zero when the topological charge is 5.

5.2.2.3 Local-Oscillator Beam is Uniformly Distributed

Assuming phase matching, the effect of the amplitude distribution on the mixing efficiency is studied, and the mixing efficiency $\eta_{U\text{-}L}$ is obtained as

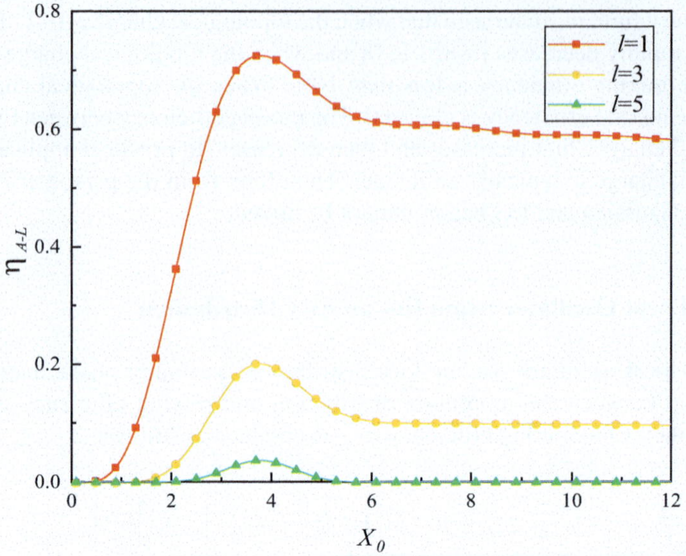

Fig. 5.24 Mixing efficiency when the signal is an LG beam and local-oscillator is an Airy beam

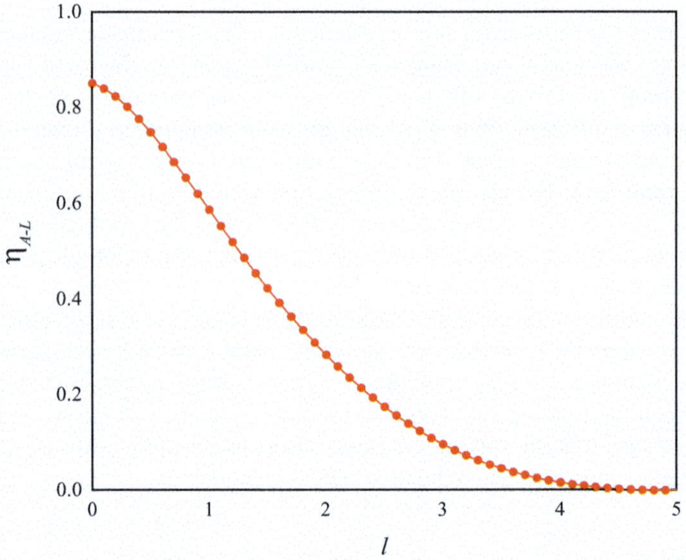

Fig. 5.25 Relation curve between the mixing efficiency and topological charge when the signal is an LG beam and the local oscillator is an Airy beam

$$\eta_{U-L} = \frac{\left| \int_0^{r_0} \sqrt{\frac{2}{\pi}\frac{1}{\omega_0}} \exp(-Q^2) \sqrt{\frac{1}{|l|!}} \left[\sqrt{2}Q\right]^{|l|} rdr \right|^2}{\int_0^{\infty} \left[\sqrt{\frac{2}{\pi}\frac{1}{\omega_0}} \exp(-Q^2) \sqrt{\frac{1}{|l|!}} \left[\sqrt{2}Q\right]^{|l|} \right]^2 rdr \int_0^{r_0} rdr}. \tag{5.31}$$

The relationship between the mixing efficiency and parameters of the receiving optical system when the uniformly distributed and LG beams have different topological charges was obtained using Eq. (5.31), and the mixing-efficiency variation curve is shown in Fig. 5.26. The mixing efficiency increases linearly to its maximum value, and then decreases with an increase in the receiving optical-system parameters. The trend of the mixing-efficiency change curve for different topological charges is the same; however, as the topological charge increases, the maximum value of the mixing efficiency gradually decreases, and the corresponding receiving optical-system parameters that reach the maximum value gradually increase.

The starting point of the linear increase also gradually shifts to the left, indicating that the larger the topological charge, the larger the receiving optical-system parameters required to achieve optimal mixing efficiency. This is because the radius of the central dark spot of the LG beam increases with an increase in the topological charge [12].

When the local-oscillator beam is uniformly distributed and the receiving optical system parameters are set to fixed values, the relationship between the mixing efficiency and topological charge is obtained according to Eq. (5.59), and the curve of the mixing efficiency with the topological charge is shown in Fig. 5.27. The graph

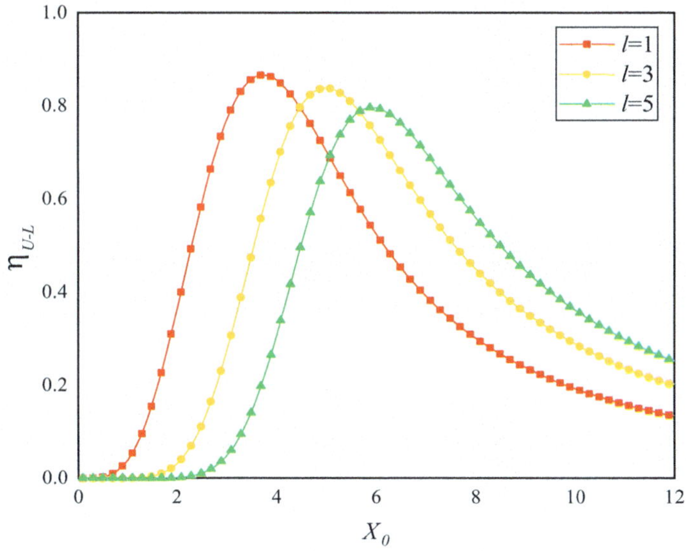

Fig. 5.26 Mixing efficiency when the signal optical is uniformly distributed and the signal is an LG beam

shows that the mixing efficiency gradually decreases with an increase in the topological charge. When the topological charge is 1, the mixing efficiency is approximately 60%; however, when the topological sum is 2, the mixing efficiency has already decreased to approximately 30%, the amplitude is seriously mismatched, and the mixing performance sharply decreases.

When the signal beam is a vortex beam, the most common LG beam is selected for research. When the local-oscillator beam is also a vortex beam, if the topological charges of the two beams are the same, the mixing efficiency can reach 100%. The larger the difference in topological charges, the lower the mixing efficiency. This characteristic can be used for information exchanges in vortex beams.

Assuming phase matching, the LG–Gaussian model has a mixing efficiency of 80% when the topological charge is 1, which matches the amplitude. However, as the topological charge increases, the mixing efficiency decreases linearly. The mixing efficiency of the LG–Airy model increases linearly to its maximum value with the parameters of the receiving optical system, and then decreases steadily and remains stable. The mixing efficiency decreases linearly with the increase of the topological charge.

When studying the effect of the phase on the mixing efficiency, assuming amplitude matching, it was found that the LG–Gaussian model signal-beam and local-oscillator beam phases could not match. When the topological charge is an integer, the mixing efficiency is approximately zero. Assuming phase matching, the mixing performance of the LG–uniform model is similar to that of the Gaussian–uniform model. As the topological charge increases, the mixing efficiency decreases. In

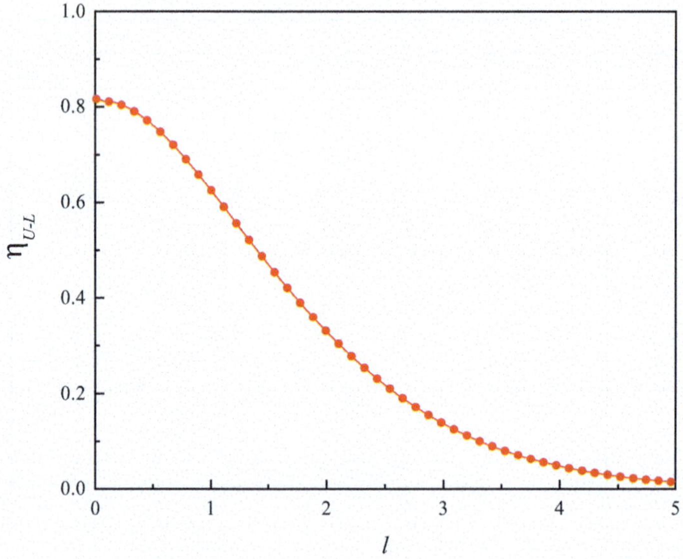

Fig. 5.27 Relation between the mixing efficiency and topological charge when the local-oscillator beam is uniformly distributed and the signal is an LG beam

summary, vortex beams cannot be combined with Gaussian, Airy, or uniformly distributed beams.

5.3 Analysis of the Impact of Atmospheric Turbulence on the Coherent-Detection Performance of Vortex Beams

Vortex coherent detection can achieve the coherent detection of a vortex beam at the receiving end and exchange information by changing the topological charge of the local-oscillator beam. This section first introduces the system structure of vortex-beam coherent detection and then analyzes the effects of the topological charge, altitude, and transmission distance on the performance of the vortex-beam coherent-detection system during an oblique signal-beam transmission. The same adaptive-optics system is used to correct the signal wavefront of the uplink and downlink. The effects of wavefront distortion pre- and post-correction are compared and analyzed.

5.3.1 Atmospheric Turbulence

Many atmospheric factors, such as pressure and temperature, can cause atmospheric changes. For example, changes in pressure can cause gas flow, whereas changes in temperature can affect the gas density and pressure. The gas flow caused by these changes constitutes a random motion of the atmosphere, which is called atmospheric turbulence. When a signal beam is transmitted in atmospheric turbulence, the fluctuation in atmospheric refraction caused by the atmospheric turbulence changes the wavefront of the transmitted signal beam, which affects the performance of the coherent communication system. Therefore, studying atmospheric factors is particularly important in communication systems, and the primary task is to study atmospheric turbulence.

5.3.1.1 Atmospheric-Refraction Structure Constants

The change in a signal-beam transmission under atmospheric turbulence is essentially caused by atmospheric refraction [13]. Atmospheric-refraction structure constants are important parameters for describing the intensity of atmospheric turbulence.

The empirical formula given by Hufnagel is as follows [14]:

$$C_n^2(h) = [2.2 \times 10^{-53} h^{10} (W/27) e^{-h/1000} + 10^{-16} e^{-h/1500}] e^{J(h,t)}, \tag{5.32}$$

where W is the wind speed at the test site and J is a uniform Gaussian random variable with a mean of zero, which can be expressed as

$$\langle J(h+h_1,t+\tau)\rangle = A\left(\frac{h_1}{100}\right)e^{-\frac{\tau}{5}} + A\left(\frac{h_1}{100}\right)e^{-\frac{\tau}{80}}. \tag{5.33}$$

In Eq. (5.33),

$$A\left(\frac{h_1}{100}\right) = \begin{cases} 1-\dfrac{h}{100}, h< 100 \\ 0,\ h \geq 100 \end{cases}. \tag{5.34}$$

In Eqs. (5.32), (5.33), and (5.34), h represents the altitude of the experimental site and τ is the time-interval constant. Function W indicates that the root-mean-square speed of the wind speed during the experiment is

$$W = \left[\frac{1}{15}\int_5^{20} v^2(h)dh\right]^{\frac{1}{2}}. \tag{5.35}$$

The modified Hufnagel–Valley model can be represented as [14]

$$\begin{aligned} C_n^2(h) &= 5.94\times 10^{-53}h^{10}\left(\frac{v}{27}\right)^2 \exp\left(-\frac{h}{1000}\right) \\ &+ 2.7\times 10^{-16}\exp\left(-\frac{h}{1500}\right) + C_0\left(-\frac{h}{100}\right), \end{aligned} \tag{5.36}$$

where v represents the local high-altitude wind-speed parameter, h represents the altitude, and C_0 represents the local atmospheric structure constant.

5.3.1.2 Atmospheric-Turbulence Refractive-Index Power-Spectrum Model

The atmospheric-turbulence refractive-index power-spectrum model is used to describe the statistical characteristics of atmospheric turbulence. The random flow of the atmosphere can form atmospheric-turbulence vortices of different sizes. Owing to the instability of the atmospheric environment, these turbulent vortices continuously decompose into smaller-order turbulent vortices. The size of the turbulence vortices is generally represented by internal and external scales, with l_0 representing small-scale vortices and L_0 representing large-scale turbulence vortices, as shown in Fig. 5.28.

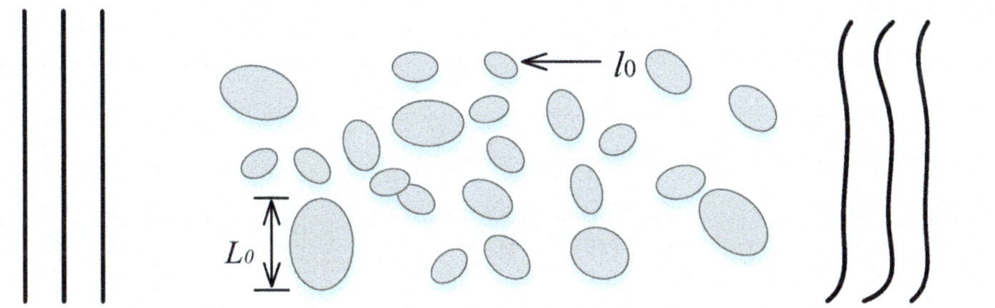

Fig. 5.28 Effect of atmospheric turbulence on a beam front [15]

Atmospheric turbulence can be divided into three regions based on the internal and external scales and their characteristics. The power-spectrum model characteristics of the different regions differ, as shown in Fig. 5.29.

Kolmogorov linked the random-fluctuation description of atmospheric refraction with atmospheric turbulence and expressed the power spectrum and C_n^2, namely, the Kolmogorov spectrum [16]:

$$\Phi_n(\kappa) = 0.033 C_n^2 \kappa^{-11/3}, \kappa_0 << \kappa << \kappa_m, \tag{5.37}$$

where κ is the number of beam waves, and C_n^2 represents the atmospheric-refraction structure constant. From Fig. 5.29, it can be seen that $\kappa \leq \kappa_0$ is the energy-input area, $\kappa_0 \leq \kappa < \kappa_m$ is the habitual sub-region, and $\kappa \geq \kappa_m$ is the dissipation zone.

The Kolmogorov spectrum is only applicable to the inertial sub-region of turbulence. To expand the applicable spatial-frequency range, Tatarskii used a Gaussian

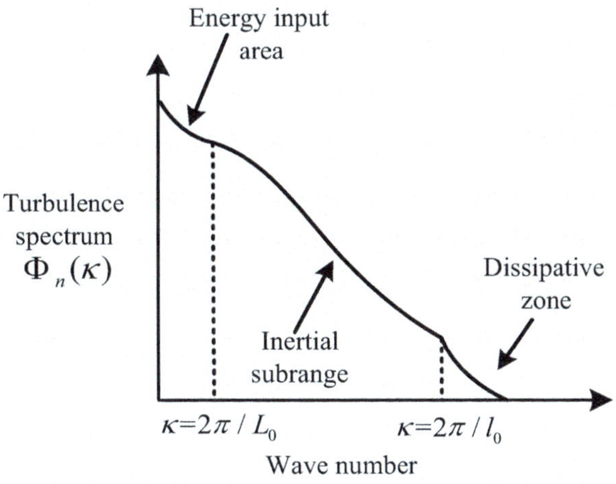

Fig. 5.29 Kolmogorov spectrum [15]

function to improve the dissipation region of the Kolmogorov spectrum and obtained the Tatarskii-spectrum model describing the turbulent-dissipation region, which is expressed as [16]

$$\Phi_n(\kappa) = 0.033C_n^2\kappa^{-11/3}\exp\left(-\frac{\kappa^2}{\kappa_m^2}\right), \kappa \gg \kappa_m, \tag{5.38}$$

where $\kappa_m = 5.92/l_0$. Von Karman modified the model and allowing it to be performed in all bands [16]:

$$\Phi_n(\kappa) = 0.033C_n^2\left(\kappa^2 + \kappa_0^2\right)^{-11/6}, \tag{5.39}$$

where $\kappa_0 = 2\pi/L_0$.

To compensate for the shortcomings of the Kolmogorov spectrum, internal and external scale parameters were introduced into the spectral model, and a modified Von Karman spectrum was proposed [16]:

$$\Phi_n(\kappa) = 0.033C_n^2\kappa^{-11/3} \cdot \left(\kappa^2 + \kappa_0^2\right)^{-11/6}\exp\left(-\frac{\kappa^2}{\kappa_m^2}\right). \tag{5.40}$$

Andrews established an approximate analysis of the Hill spectrum called the corrected atmospheric spectrum [17]:

$$\Phi_n(\kappa) = 0.033C_n^2(h) \cdot \left[1 + 1.802\left(\frac{\kappa}{\kappa_l}\right) - 0.254\left(\frac{\kappa}{\kappa_l}\right)^{7/6}\right] \cdot \frac{\exp\left(-\frac{\kappa^2}{\kappa_l^2}\right)}{\left(\kappa^2 + \kappa_0^2\right)^{11/6}}, \tag{5.41}$$

where $\kappa_l = 3.3/l_0$ and $\kappa_0 = 2\pi/L_0$.

5.3.1.3 Power-Spectrum Inversion Method for Simulating Turbulent Phase Screens

Below is an introduction to the steps for simulating atmospheric turbulence using the power-spectrum inversion method. Assuming that the atmospheric turbulence is locally homogeneous and isotropic, the Kolmogorov phase power-spectral density function $F_\varphi(k_x, k_y)$ can be expressed as [18]

$$F_\varphi(k_x, k_y) = 2\pi k^2 \Delta z \Phi(\kappa, z), \tag{5.42}$$

where k is the wavenumber and Δz is the thickness of the turbulent atmosphere. The refractive-index power-spectral density function $\Phi(\kappa, z)$ can be expressed as [19]

$$\Phi(\kappa, z) = 0.033C_n^2(z)\kappa^{-11/3}, \tag{5.43}$$

where z is the propagation direction of the beam.

To obtain the turbulence phase function $\varphi(x, y)$, the usual method is to filter the phase power-spectral density and then carry out a two-dimensional inverse Fourier transform. Here, a set of complex random matrices $H(k_x, k_y)$ subject to a Gaussian distribution is required, and the turbulence phase function obtained is [18]

$$\varphi(x, y) = \iint H(k_x, k_y)\sqrt{F_\varphi(k_x, k_y)}\exp[j(k_x x + k_y y)]dk_x dk_y. \tag{5.44}$$

To change it to a discretized expression [18],

$$\varphi(m, n) = \sum_{m=0}^{N_x}\sum_{n=0}^{N_y} H(m', n')\sqrt{F_\varphi(m', n')}\exp[2\pi j\left(\frac{m'm}{N_x} + \frac{n'n}{N_y}\right)]\Delta k_x \Delta k_y, \tag{5.45}$$

where $N_x = D_x/\Delta x$ and $N_y = D_y/\Delta y$ are the numbers of grid cells in the x and y directions, respectively, D_x and D_y are the length and width of the turbulent phase screen, respectively. Δk_x and Δk_y are the spatial-frequency intervals in the x and y directions, respectively.

However, this method lacks low-frequency components. A subharmonic method is used to compensate for the low-frequency phase and improve the authenticity of the simulated turbulence [19]; it can be expressed as

$$\varphi_{SH} = \sum_{p=1}^{N_p}\sum_{m'=-1}^{1}\sum_{n'=-1}^{1} H(m', n')f(m', n')\exp\left[j2\pi 3^{-p}\left(\frac{m'm}{N} + \frac{n'n}{N}\right)\right], \tag{5.46}$$

where P is the number of subharmonics and $f(m', n') = C \cdot 3^{-p}\left(f_{lx}^2 + f_{ly}^2\right)$ is the harmonic function.

We assume that the side length of the square turbulent phase screen is 4 m, the number of simulation points is 256, and the transmission distance is 1000 m. Figure 5.30 shows $C_n^2 = 1 \times 10^{-17}\text{m}^{-2/3}$ and the turbulent phase screens obtained using the power-spectrum inversion method. The comparison shows the two-dimensional and three-dimensional phase-distribution maps with added fifth-harmonic components and only high-frequency components.

By comparing Fig. 5.30a1 and a2 with Fig. 5.30b1 and b2, it can be observed that the turbulence phase-screen image is smoother after harmonics are added, and the tilted components of the three-dimensional image are more abundant. Therefore, it can be seen that combining the harmonic method with the power-spectrum inversion method produces a turbulence phase screen that is closer to real turbulence.

The equation of a beam wave propagating in a random medium can be expressed as [13]

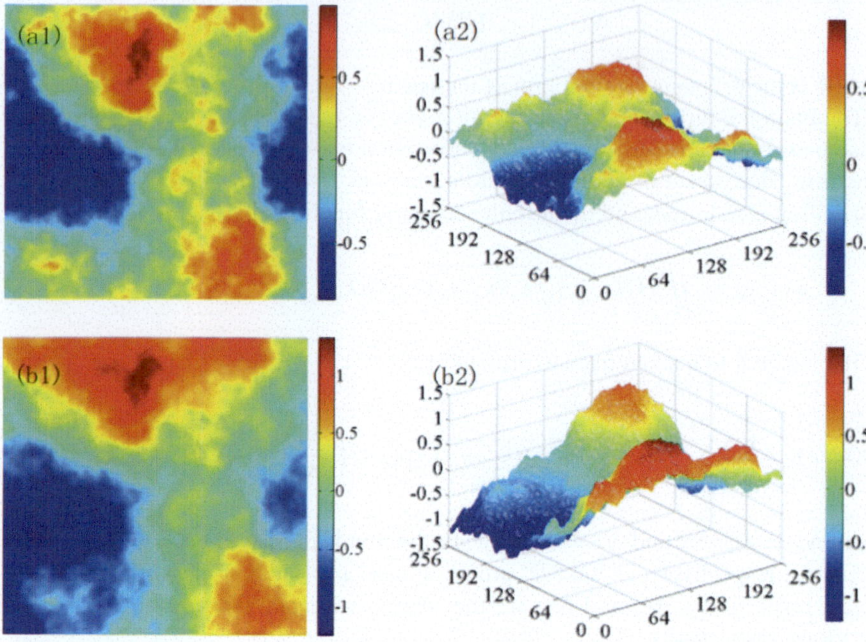

Fig. 5.30 Atmospheric-turbulence phase screen. **a1** Uncompensated, 2D distribution; **a2** uncompensated, 3D distribution; **b1** after compensation, 2D distribution; **b2** after compensation, 3D distribution

$$2ik\frac{\partial u}{\partial z} + \frac{\partial^2 u}{\partial x^2} + \frac{\partial^2 u}{\partial y^2} + 2k^2 n_1 u = 0, \tag{5.47}$$

where n_1 is the fluctuation value of atmospheric refraction and k is the wavenumber.

Consider an atmospheric-turbulent medium as several phase screens with varying thicknesses, as shown in Fig. 5.31. Each time the beam passes through the phase screen, it causes a change in phase and produces a new beam field. Therefore, a multiphase-screen model can be used to simulate the propagation characteristics of vortex beams in atmospheric media.

After the beam passes through the turbulent phase screen, only the phase changes and the current-beam phase screen is transmitted to the next phase screen. According to the Fresnel diffraction theory, the beam field can be expressed as [13]

$$u(r, z_{i+1}) = F^{-1}\left\{F\left[u(r, z_i)\exp[i\phi(r, z_i)]\right]\exp(ik\Delta z_{i+1} - i\frac{K_x^2 + K_y^2}{2k}\Delta z_{i+1})\right\}, \tag{5.48}$$

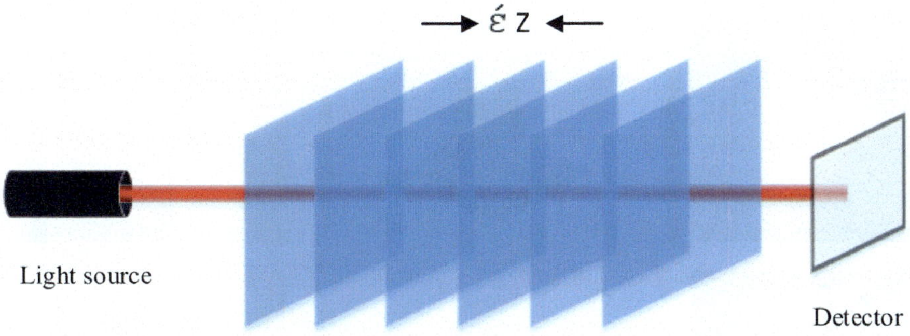

Fig. 5.31 Transmission diagram based on the principle of a multiphase screen

where F and F^{-1} represent the Fourier and inverse Fourier transforms, respectively; Δz represents the spacing between adjacent phase screens; and K_x and K_y are the spatial wavenumbers in the x and y directions, respectively.

5.3.1.4 Vortex-Mode Crosstalk

A vortex beam transmitted in free space may undergo wavefront distortion, owing to the influence of atmospheric turbulence, and crosstalk may occur between the different modes. In this section, the relevant content of the turbulence phase screen is used to simulate the transmission characteristics of an LG beam in turbulence.

Assuming the wavelength of the beam wave $\lambda = 1550\,\text{nm}$, waist radius $w_0 = 0.03\text{m}$, atmospheric structure constant $C_n^2 = 1 \times 10^{-15}\text{m}^{-2/3}$, and a transmission distance of 1000 m, the transmission characteristics of a single vortex beam with a topological charge of 2 can be obtained. Figure 5.32 shows the two-dimensional and three-dimensional beam intensity and phase distribution before and after the turbulent transmission. The changes in beam intensity and vortex-beam phase after transmission can be compared from multiple angles.

First, a horizontal comparison is conducted. Figure 5.32a1–a3 show the ideal vortex-beam images under ideal conditions. The intensity of a single vortex beam exhibits a hollow annular distribution. The phase distribution is divided into several parts by the isophase line emitted from the center, and the number of rays is the topological charge of the vortex beam.

Figure 5.32b1–b3 show images of the vortex beam transmitted through turbulence. From the figure, it can be seen that the intensity distribution of the vortex beam transmitted through turbulence is still a hollow ring structure; however, the energy on the ring is dispersed and weakened to varying degrees. The isophase line is no longer a smooth ray, and the phase distribution appears blurry and deformed. The beam transmitted through turbulence exhibits distortions in both intensity and phase

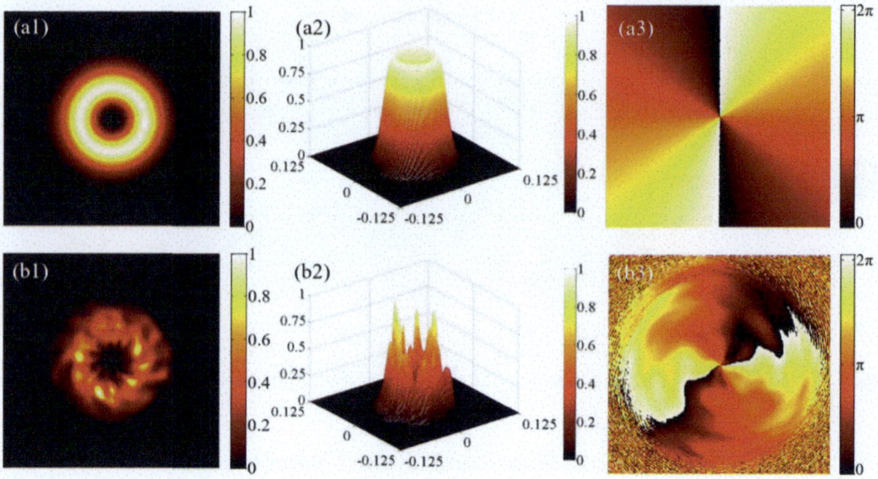

Fig. 5.32 Beam intensity and phase distribution of a vortex beam before and after an atmospheric-turbulent transmission

distribution, which can have a significant impact on the performance of a optical wireless communication system.

When a vortex beam propagates in atmospheric turbulence, its wavefront and beam intensity will be distorted, which affects the purity of the OAM mode. Moreover, crosstalk between different modes has a serious impact on the performance of a communication system. To quantitatively analyze the impact of crosstalk in this mode on the optical-signal transmission, the concept of a spiral spectrum is introduced. Theoretically, vortex beams with different OAM modes are orthogonal to each other and all modes can form a set of orthogonal bases that represent any vortex beam or spectral decomposition [2].

A vortex beam can be expanded, based on the harmonic function of the spiral spectrum [2]:

$$u(r, \varphi, z) = \frac{1}{\sqrt{2\pi}} \sum_{l=-\infty}^{+\infty} a_l(r, z) \exp(il\varphi), \tag{5.49}$$

where coefficient $a_l(r, z)$ is

$$a_l(r, z)=\frac{1}{\sqrt{2\pi}} \int_0^{2\pi} u(r, \varphi, z) \exp(-il\varphi) d\varphi. \tag{5.50}$$

By integrating $a_l(r, z)$ radially, the energy occupied by each mode can be obtained [2]:

$$R_l = \int_0^\infty |a_l(r, z)|^2 r dr. \tag{5.51}$$

According to Eq. (5.51), the relative power occupied by each mode of the vortex beam [2] is expressed as

$$P_l = \frac{R_l}{\sum_l R_l}. \tag{5.52}$$

The spiral-spectrum distribution before and after the turbulent propagation of an LG beam when $\lambda = 1550\text{nm}$, $w_0 = 0.03\text{m}$, $C_n^2 = 1 \times 10^{-15}\text{m}^{-2/3}$, and $z = 1000\text{m}$ is shown in Fig. 5.33. Figures 5.33a1 and a2 show the spiral-spectrum distribution of vortex-beam turbulence with a topological charge number of 2 at transmission distances of 0 and 1000 m. Figures 5.33b1 and b2 show the spiral-spectrum distributions of stacked beams with topological charges of 1 and 2 at propagation distances of 0 and 1000 m.

Figure 5.33a1 and b1 show vortex beams with topological charges of 2, 1, and 2, respectively, without a turbulent transmission. It can be seen that there is no crosstalk

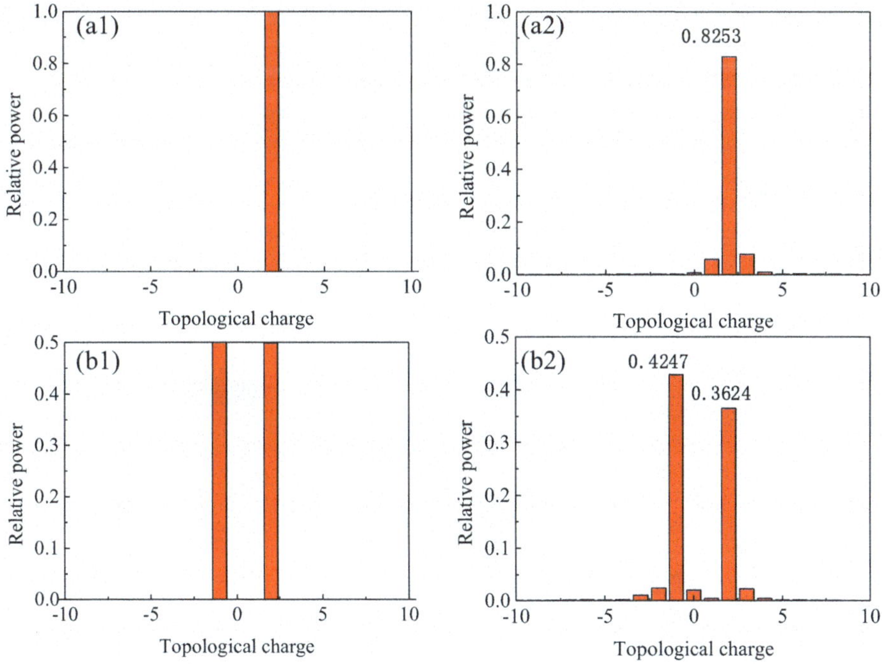

Fig. 5.33 Spiral distribution of an LG beam. **a1–a2** l = 2; **b1–b2** l = 0.1, 2; **a1–b1** z = 0, no turbulence; and **a2–b2** z = 1000, with turbulence

between adjacent modes. After being transmitted through atmospheric turbulence, it can be seen from Fig. 5.33a2 that the relative power of a single vortex beam has changed from 1 to 0.8253, and the remaining energy has transferred to adjacent modes, indicating the existence of mode crosstalk in the transmission of a single vortex beam in atmospheric turbulence.

Figure 5.33b2 shows the mode crosstalk of the superimposed beam, with the OAM-mode relative power of the transmitted beam decreasing from 0.5 to 0.4247 and 0.3624, respectively. The power spreads towards adjacent modes, and there may be overlapping crosstalk between different modes. For example, the relative power of the vortex beams at 0 for topological charges -1 and 2 both have values, so the relative powers of the two beams overlap.

5.3.2 System Structure

5.3.2.1 Transceiver System

The structure of a optical wireless communication OAM multiplexing coherent-detection system using bidirectional transmission is shown in Fig. 5.34. The transmission distance between the two communication terminals is z, the zenith angle of the transmission link is ψ, the transmitting end of the uplink (i.e., the receiving end of the downlink) is located at an altitude of h_0, and the receiving end of the uplink (i.e., the transmitting end of the downlink) is located at an altitude of H; $h_0 < H$. The adaptive-optics system is located at the receiving end of the uplink near the ground.

The signal laser at the transmitting end adopts a 1 × 4 coupler, which is divided into four beams, as shown in Fig. 5.35a. After the source signal is converted in series and parallel, each beam is modulated separately and converted into a vortex beam with topological charges of 1, 2, 3, and 4 using a spiral phase plate. It is then synthesized all the way through a 4 × 1 coupler and finally coaxially transmitted.

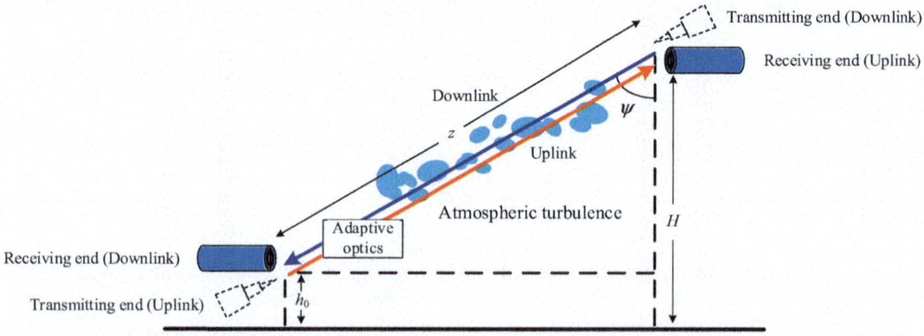

Fig. 5.34 Structure diagram of an orbital angular-momentum multiplexing coherent-detection system for optical wireless communication

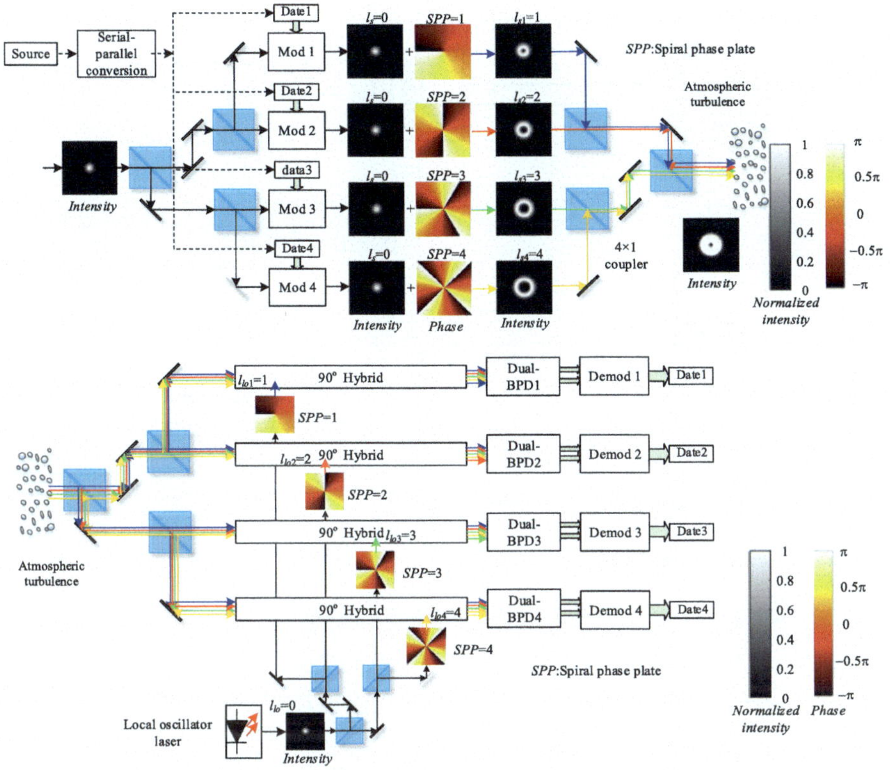

Fig. 5.35 Schematic diagrams of **a** transmitter and **b** receiver

The receiving end first adopts one-way coupling. The 1 × 4 coupler divides the beam into four channels and mixes the frequency with the local-oscillator vortex with topological charges of 1, 2, 3, and 4. After balanced detection, the electrical signal is recovered, demodulated, and converted into a single signal through a parallel–series conversion, thereby achieving a signal transmission from the source to the sink.

5.3.2.2 Adaptive-Optics System

From the analysis in Sect. 5.2, turbulence can cause beam-wavefront distortion, and atmospheric turbulence can cause mode crosstalk between vortex beams. Mode crosstalk can cause mutual interference between signals from different channels, reducing the detection accuracy at the receiving end, and seriously disturbing vortex optical multiplexing communication systems that rely on OAM states for information transmission.

An adaptive-optics system can dynamically compensate for a wavefront phase distorted by turbulence to alleviate the impact of atmospheric turbulence on the

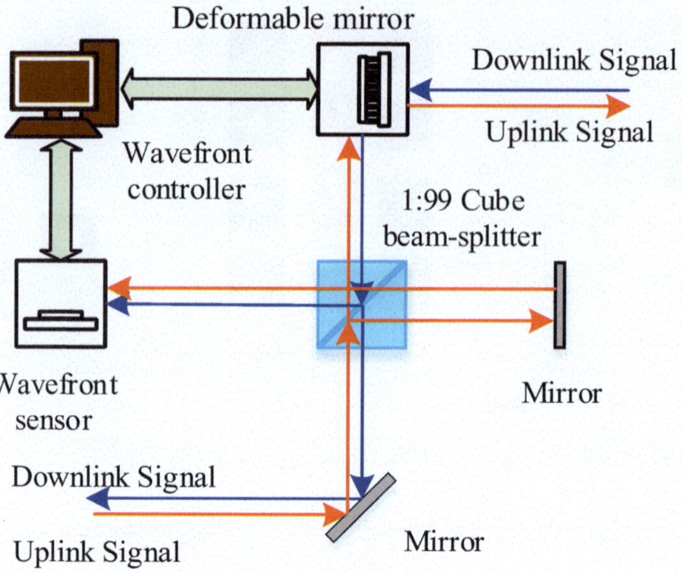

Fig. 5.36 Adaptive-optical system at the downlink receiving end

transmission signal. The adaptive-optics system includes three modules: wavefront sensor, wavefront control system, and wavefront corrector. The wavefront sensor measures the beam wavefront. The wavefront control system uses the distorted-wavefront information measured by the wavefront sensor to calculate and transmit the correct wavefront-control information to the wavefront corrector, thereby correcting the distorted wavefront.

The adaptive-optics system is located at the downlink receiving end (i.e., the uplink transmission end). The working-principle diagram is shown in Fig. 5.36. After the downlink optical signal transmitted through the atmosphere is reflected by the deformable mirror, the wavefront phase is collected by the wavefront sensor. Then, the calculated conjugate wavefront is applied to the deformable mirror to generate the conjugate wavefront, thus completing the correction. Owing to the reverse nature of the transmission link and reverse superposition effect of the wavefront, the optical signal of the uplink is pre-corrected at the transmitting end to suppress the turbulence.

5.3.3 Vortex-Beam Coherent Detection Gain

The following section uses the content and formulae related to the vortex beam, turbulent phase screen, and coherent detection discussed earlier to derive the coherent gain formula for vortex-beam coherent detection.

A Laguerre–Gaussian beam is a typical vortex beam and is expressed as [1]

$$U(\mathbf{r},z)=\sqrt{\frac{2p!}{\pi(p+|l|)!}}\frac{1}{w(z)}\left[\frac{r\sqrt{2}}{w(z)}\right]^{|l|}L_p^{|l|}\left[\frac{2r^2}{w^2(z)}\right]\exp\left[\frac{-r^2}{w^2(z)}\right],$$
$$\exp\frac{-i\kappa r^2 z}{2\left(z^2+z_R^2\right)}\exp\left(i(2p+|l|+1)\tan^{-1}\frac{z}{z_R}\right)\exp(-il\theta) \tag{5.53}$$

where i is the imaginary unit, $textbfr=(r,\theta)$, r and θ are the radial variable and angular-phase variable, respectively, z is the transmission distance, $z_R=\pi w_0^2/\lambda$, λ is the wavelength, w_0 is the initial beam waist radius, $w(z)=w_0\sqrt{1+\left(z/z_R\right)^2}$ is the beam waist radius at z, l is the topological charge, and $\kappa=2\pi/\lambda$ is the wave number. $L_p^{|l|}$ is the associated Laguerre polynomials:

$$L_p^{|l|}(x)=\sum_{m=0}^{p}(-1)^m\frac{(|l|+p)!}{(p-m)!(|l|+m)!m!}x^m, \tag{5.54}$$

where p is the radial exponent. To simplify this expression, we write Eq. (5.53) as follows:

$$U(\mathbf{r},z)=A(\mathbf{r},z)\exp(-il\theta). \tag{5.55}$$

For a modulated vortex with a topological charge of l_{s1}, the expression is

$$U_{s1}(\mathbf{r},z,t)=m_{s1}(t)\cos[\omega_s t+\pi\cdot n_{s1}(t)]\cdot A(\mathbf{r},z)\exp(-il_{s1}\theta), \tag{5.56}$$

where ω_s is the angular frequency of the signal beam, t is the time variable, $m_{s1}(t)$ and $n_{s1}(t)$ respectively map the information of the signal source modulated in the amplitude and phase of the signal beam, $m_{s1}(t)$ and $n_{s1}(t)=0$ or 1.

For a vortex beam with four different topological charges, the optical field expression after the coaxial multiplexing transmission is

$$U_s(\mathbf{r},z,t)=\sum_{k=1}^{4}U_{sk}(\mathbf{r},z,t)=\sum_{k=1}^{4}m_{sk}(t)\cos[\omega_s t+\pi n_{sk}(t)]\cdot A(\mathbf{r},z)\exp(-il_{sk}\theta). \tag{5.57}$$

Using the power-spectrum inversion method to generate a random phase screen, the high-frequency component can be represented as [19]

$$\phi_H(m,n)=\sum_{m'=-\frac{N_x}{2}}^{\frac{N_x}{2}-1}\sum_{n'=-\frac{N_y}{2}}^{\frac{N_y}{2}-1}a\left(m',n'\right)\sqrt{\Phi_\phi(m',n')}e^{i2\pi\left(\frac{m'm}{N_x}+\frac{n'n}{N_y}\right)}, \tag{5.58}$$

where N_x and N_y are the sampling points in the x and y directions, and $a(m', n')$ is a complex Gaussian random matrix; $\Phi_\phi(m', n')$ is the spectral density of the atmospheric phase power obtained by the Hill spectral-correction model [17]:

$$\Phi_\phi(\kappa_r) = 2\pi\kappa^2 \cdot 0.033C_n^2(h)\left[1 + 1.802\left(\frac{\kappa_r}{\kappa_l}\right) - 0.254\left(\frac{\kappa_r}{\kappa_l}\right)^{7/6}\right]\frac{\exp\left(-\frac{\kappa_r^2}{\kappa_l^2}\right)}{\left(\kappa_0^2 + \kappa_r^2\right)^{11/6}}, \tag{5.59}$$

where $\kappa_r = 2\pi f_r$ is the spatial wavenumber, $f_r = ({f_x}^2 + {f_y}^2)^{1/2}$, f_x and f_y are the spatial frequencies, that is, $f_x = 1/L_x$, $f_y = 1/L_y$, and L_x and L_y are the sizes of the phase screen in the x and y directions, respectively. $\kappa_l = 3.3/l_0$ and $\kappa_0 = 2\pi/L_0$, where l_0 and L_0 are the internal and external scales, respectively, and C_n^2 is the atmospheric refractive-index structure constant.

For an oblique transmission, the variation in C_n^2 with the altitude can be represented by the Hufnagel–Valley 21 model (wind speed of 21 m/s) as [14]

$$C_n^2(h) = 5.94 \times 10^{-53}h^{10}\left(\frac{21}{27}\right)^2 e^{\left(-\frac{h}{1000}\right)} + 2.7 \times 10^{-16}e^{\left(-\frac{h}{1500}\right)} + 1.7 \times 10^{-14} \cdot e^{\left(-\frac{h}{100}\right)}. \tag{5.60}$$

We place N_{screen} turbulent phase screens at equal intervals on the diagonal transmission link with a length of z. The C_n^2 of each phase screen can be calculated by Eq. (5.60):

$$C_n^2 = \frac{1}{\Delta h}\int_h^{h+\Delta h} C_n^2(\xi)d\xi. \tag{5.61}$$

The low-frequency component of the phase screen is [19]

$$\phi_L(m, n) = \sum_{p=1}^{N_p}\sum_{m'=-\frac{N_x}{2}}^{\frac{N_x}{2}-1}\sum_{n'=-\frac{N_y}{2}}^{\frac{N_y}{2}-1} a(m', n')\sqrt{\Phi_\phi(m', n')}e^{\left[i2\pi 3^{-p}\left(\frac{mm'}{N}+\frac{nn'}{N}\right)\right]}, \tag{5.62}$$

where N_p is the number of subharmonics. The signal beam after passing through a layer of the turbulent phase screen is expressed as

$$U_s(\mathbf{r}, z + \Delta z, t) = \mathcal{F}^{-1}\left[\mathcal{F}[U_s(\mathbf{r}, z, t)] \cdot \exp(i\phi_H(\mathbf{r}) + i\phi_L(\mathbf{r}))\right] \cdot e^{\left(\frac{-1(\kappa_X^2+\kappa_Y^2)\Delta Z}{2\kappa}\right)}, \tag{5.63}$$

where $\mathcal{F}$ and $\mathcal{F}^{-1}$ respectively represent the Fourier transform and inverse transform, $\phi_L(\mathbf{r})$ and $\phi_H(\mathbf{r})$ respectively represent the low-order and high-order phases of the phase screen, and $\Delta z = z/N_{screen}$ is the distance between the two phase screens; κ_x

$= 2\pi f_x$ and $\kappa_y = 2\pi f_y$ are the spatial wavenumbers. The phase compensation of adaptive-optics can be expressed as

$$U_{s-re}(\mathbf{r}, z, t) = U_s(\mathbf{r}, z, t) \cdot \exp\left[-i \sum_{n=1}^{N_{screen}} \mathcal{F}^{-1}\left[\mathcal{F}(\phi_{Hn}(\mathbf{r}) + \phi_{Ln}(\mathbf{r})) \cdot e^{\left(\frac{-i(\kappa_x^2+\kappa_y^2)\Delta z}{2\kappa}\right)}\right]\right], \tag{5.64}$$

where n is the sum variable and is taken as an integer.

The local-oscillator beam at the receiving end is a vortex beam with topological charge l_{lo}, expressed as

$$U_{lo}(\mathbf{r}, z, t) = A(\mathbf{r}, z)\cos(\omega_{lo}t)\exp(-il_{lo}\theta). \tag{5.65}$$

The signal beam $U_{s-re}(\mathbf{r}, z, t)$ and local-oscillator $U_{lo}(\mathbf{r}, z, t)$ beams are mixed in the mixer, which outputs the following four optical signals:

$$U_{hybrid1}(\mathbf{r}, z, t) = U_{s-re}(\mathbf{r}, z, t) + U_{lo}(\mathbf{r}, z, t) \tag{5.66}$$

$$U_{hybrid2}(\mathbf{r}, z, t) = U_{s-re}(\mathbf{r}, z, t) - U_{lo}(\mathbf{r}, z, t) \tag{5.67}$$

$$U_{hybrid3}(\mathbf{r}, z, t) = U_{s-re}(\mathbf{r}, z, t) + U_{lo}^*(\mathbf{r}, z, t) \tag{5.68}$$

$$U_{hybrid4}(\mathbf{r}, z, t) = U_{s-re}(\mathbf{r}, z, t) - U_{lo}^*(\mathbf{r}, z, t). \tag{5.69}$$

Equations (5.66)–(5.69) represent the four signals output by the mixer, and * represents the conjugation. The square of Eq. (5.69) is subtracted from the square of Eq. (5.68), and the sum and difference of the products are used to calculate the output current signal of the balanced detector, as follows:

$$\begin{aligned} I_1 &= 4\alpha \cdot \sum_{k=1}^{4} \langle U_{sk-re}(\mathbf{r}, z, t) \cdot U_{lo}(\mathbf{r}, z, t)\rangle \\ &= \alpha \cdot \sum_{k=1}^{4} 2m_{sk}(t) \cdot [\cos[(\omega_s + \omega_{lo})t + \pi n_{sk}(t)] + \cos[(\omega_s - \omega_{lo})t + \pi n_{sk}(t)]], \\ &\int_0^{2\pi}\int_0^R |A(\mathbf{r}, z)|^2 \cdot \exp[-i(l_{sk} - l_{lo})\theta] \cdot \exp[i\Delta\phi(\mathbf{r})] \cdot rdrd\theta \end{aligned} \tag{5.70}$$

where α is the photoelectric-conversion coefficient and R is the effective radius of the detector. $\Delta\phi(\mathbf{r})$ is the wavefront residual after correction for the adaptive beam. Owing to the high frequency corresponding to the sum-frequency term, photodetectors are generally unable to respond, and the frequency of the difference-frequency term changes much more slowly, compared to the beam field. Therefore, the output of the balanced detector is

$$I_1 = \alpha \cdot \sum_{k=1}^{4} 2m_{sk}(t) \cdot [\cos[(\omega_s - \omega_{lo})t + \pi n_{sk}(t)]]$$
$$\int_0^{2\pi} \int_0^R |A(\mathbf{r}, z)|^2 \cdot \exp[-i(l_{sk} - l_{lo})\theta] \cdot \exp[i\Delta\phi(\mathbf{r})] \cdot rdrd\theta. \tag{5.71}$$

The wavefront residual $\Delta\phi(\mathbf{r})$ disrupts the orthogonality between vortex beams with the same topological charge, and the mode expansion of each signal causes crosstalk between each signal.

The coherent gain G between the signal beam and local-oscillator beam can be expressed as

$$\begin{aligned} G &= \int_0^{2\pi} \int_0^R |A(\mathbf{r}, z)|^2 \cdot \exp[-i(l_{sk} - l_{lo})\theta] \cdot \exp[i\Delta\phi(\mathbf{r})] \cdot rdrd\theta \\ &= \int_0^{2\pi} \int_0^R \left[U_{sk}(\mathbf{r}, z, t) \cdot U_{lo}^*(\mathbf{r}, z, t)\right] rdrd\theta. \end{aligned} \tag{5.72}$$

The received signal $U_{sk}(\mathbf{r}, z, t)$ represents the vortex optical signal of the kth path affected by turbulence. Considering the computational complexity and symmetry of the topological charges of each path to be analyzed, each optical signal is expanded according to the spiral spectrum −5th–10th-order harmonic function, where the kth path signal is represented as

$$U_{sk}(\mathbf{r}, z, t) = \frac{1}{\sqrt{2\pi}} \sum_{l=-5}^{10} a_{kl}(\mathbf{r}, z) \exp(-il\theta), \tag{5.73}$$

where $a_{kl}(\mathbf{r}, z) = \frac{1}{\sqrt{2\pi}} \int_0^{2\pi} U_{sk}(\mathbf{r}, z, t) \exp(il\theta) d\theta$ and k represents the kth signal.

Similarly, the local oscillator $U_{lo}(\mathbf{r}, z, t)$ can be expanded as

$$U_{lo}(\mathbf{r}, z, t) = \frac{1}{\sqrt{2\pi}} \sum_{l=-5}^{10} b_{lo}(\mathbf{r}, z) \exp(-il\theta), \tag{5.74}$$

where $b_{lo}(\mathbf{r}, z) = \frac{1}{\sqrt{2\pi}} \int_0^{2\pi} U_{lo}(\mathbf{r}, z, t) \exp(il\theta) d\theta$.

The received signal beam $U_{sk}(\mathbf{r}, z, t)$ of Eq. (5.73) and the local-oscillator beam $U_{lo}(\mathbf{r}, z, t)$ of Eq. (5.74) are substituted into Eq. (5.72), and the coherent gain G is

$$G = \frac{1}{2\pi} \int_0^{2\pi} \int_0^R \left(\sum_{l=-5}^{10} a_{kl}(\mathbf{r}, z) \exp(-il\theta) \cdot \sum_{l=-5}^{10} b_{lo}(\mathbf{r}, z) \exp(il\theta) \right) rdrd\theta \tag{5.75}$$

Owing to the orthogonality of a vortex beam with different topological charges [20],

$$\int_0^{2\pi}\int_0^R |A(\mathbf{r}, z)|^2 \cdot \exp[-i(l_{sk} - l_{lo})\theta] \cdot rdrd\theta = \begin{cases} 1 & l_{sk} = l_{lo} \\ 0 & l_{sk} \neq l_{lo} \end{cases}. \tag{5.76}$$

Equation (5.76) can be simplified as

$$G = \frac{1}{2\pi}\int_0^{2\pi}\int_0^R \left(\sum_{l=-5}^{10} a_{kl}(\mathbf{r}, z) b_{lo}(\mathbf{r}, z)\right) rdrd\theta = \int_0^R \left(\sum_{l=-5}^{10} a_{kl}(\mathbf{r}, z) b_{lo}(\mathbf{r}, z)\right) rdr. \tag{5.77}$$

Owing to the fact that the local-oscillator beam does not pass through atmospheric turbulence without mode crosstalk, and therefore has value only when $l = l_{lo}$. At this point, the coherent gain G can be simplified as

$$G = \int_0^R [a_{kl}(\mathbf{r}, z) \cdot b_{lo}(\mathbf{r}, z)]|_{l=lo}\, rdr. \tag{5.78}$$

5.3.4 *Research on Factors Influencing the Coherence Performance*

5.3.4.1 Number of Topological Charges

Taking the transmission distance z = 10 km, the coherence gains of different topological charge signal light l_s = 1–4 after atmospheric transmission and different topological charge intrinsic oscillator light l_{lo} = 1–4 after mixing, respectively, in the uplink, downlink, uncorrected, and corrected are calculated by Eq. (5.78), and the results are shown in Fig. 5.37.

At the same transmission distance, the larger the topological charge of the vortex beam, the richer the vortex structure, the more vulnerable it is to turbulence, and the more severe the mode crosstalk. When the signal-beam topological charge $l_{s1} = 1$ and the local-oscillator beam topological charge $l_{lo} = 1$, the coherent gain in the case of an uncorrected uplink is calculated according to Eq. (5.78) as 0.867.

When the signal-beam topological charge $l_{s1} = 1$ and the local-oscillator beam topological charge $l_{lo} = 2$, the coherent gain in the uncorrected uplink case is calculated to be 0.156, according to Eq. (5.78). The distortion of the vortex-beam wavefront caused by atmospheric turbulence can reduce the coherence between the signal beam

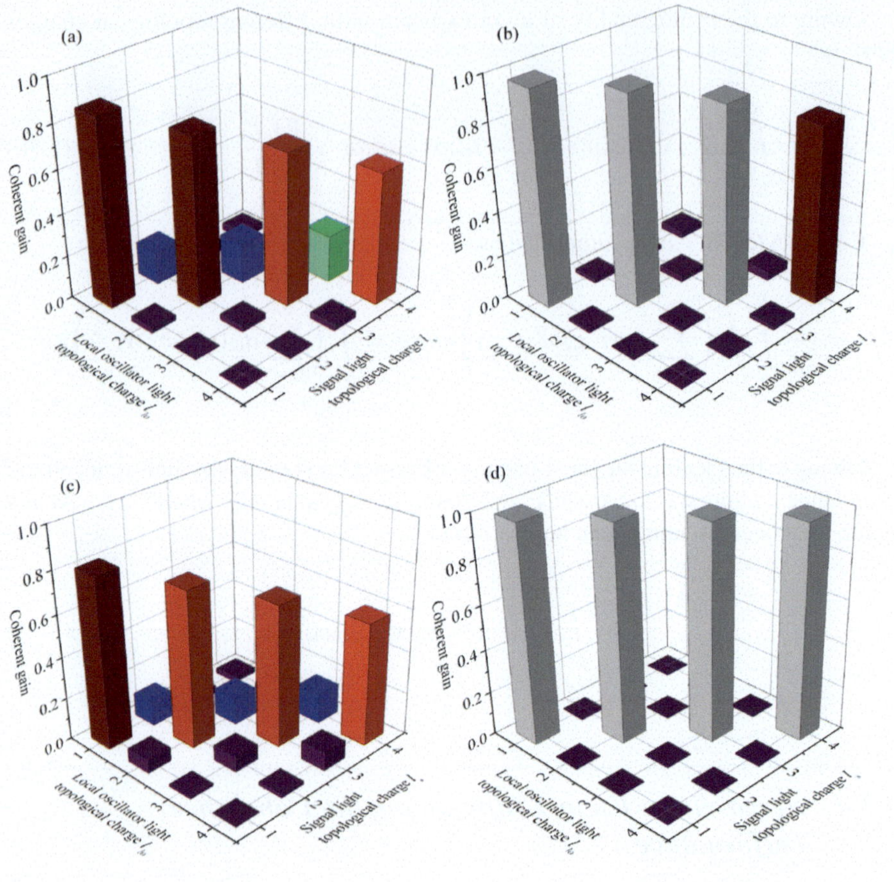

Fig. 5.37 Coherent gain of the signal beam and LO beam with different topological charges. **a** Uplink uncorrected; **b** uplink corrected; **c** downlink uncorrected; **d** downlink corrected

and the local-oscillator beam, while also affecting the orthogonality between signals from different paths.

The intensity of atmospheric turbulence is strongest near the surface [21], and the uplink beam propagates from a dense medium to a sparse medium, resulting in significant wavefront distortion and spot jitter of the beam. The downlink beam is transmitted from a sparse medium to a dense medium, and the wavefront distortion and spot jitter of the beam are relatively small; therefore, the wavefront distortion in the uplink changes significantly.

According to Eq. (5.78), when the local-oscillator topological charge $l_{lo} = 3$ and the transmission distance $z = 10$ km, the coherent gain after uplink pre-correction increases from 0.706 to 0.942. When the LO topological charge $l_{lo} = 3$ and the transmission distance $z = 10$ km, the coherence gain for the corrected downlink increases from 0.744 to 0.975, according to Eq. (5.78). The reversibility of an atmospheric-link transmission enables the use of the same adaptive-optics system to compensate for both uplink and downlink wavefront distortions.

5.3.4.2 Uplink-Receiver Altitude

We set the transmission distance to $z = 5$ km, change the height H of the uplink receiver from the ground (i.e., change the zenith angle ψ), and obtain the coherent gain of the signal beam with different topological charges after transmission, before and after uplink and downlink correction, as shown in Fig. 5.38. As H increases (i.e., the zenith angle ψ decreases), the coherent gain tends to increase, and the trends before and after the uplink and downlink corrections are not significantly different.

From Fig. 5.38, it can be observed that in the uncorrected case, when $H > 2$ km ($\psi < 1.16$ rad), the coherent gain after both downlink and uplink corrections tends toward a steady-state value, and the coherent gain remains stable with increasing transmission distance. This proves that, when the zenith angle $\psi < 1.16$ rad, the coherent gain is unaffected by changes in the zenith angle. After adaptive-optics is

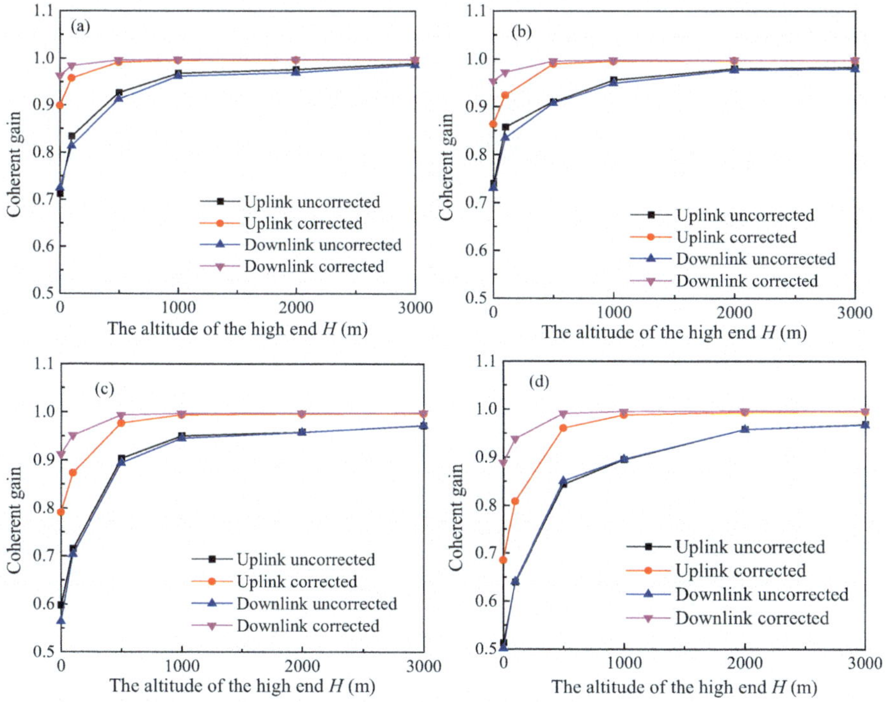

Fig. 5.38 Change of the signal-beam coherent gain with different topological charges transmitted at different altitudes. **a** $l_{s1} = 1$; **b** $l_{s2} = 2$; **c** $l_{s3} = 3$; **d** $l_{s4} = 4$

used to correct the wavefront distortion, the coherent gain remains relatively stable when $H > 1$ km ($\psi < 1.37$ rad).

From this, it can be concluded that, for the extreme case, the coherent gain of the system is most affected by turbulence when the signal beam is horizontally transmitted ($\psi = \pi/2$ rad) and least affected by turbulence when it is vertically transmitted ($\psi = 0$ rad) [22].

5.3.4.3 Transmission Distance

When the topological charges of the signal beam are 1, 2, 3, and 4, the variation in the coherent gain with the transmission distance of the signal beam is shown in Fig. 5.39, when the topological charges of the signal beam and local-oscillator beam are the same.

When the transmission distance $z = 8$ km and the local-oscillator topological charge number $l_{lo} = 1$–4, the coherent gains before the uplink correction are obtained according to Eq. (5.78) as 0.905, 0.743, 0.717, and 0.632, respectively. This indicates that when the transmission distance is the same, the larger the vortex beam topological-charge number, the more evident is the influence of turbulence and the

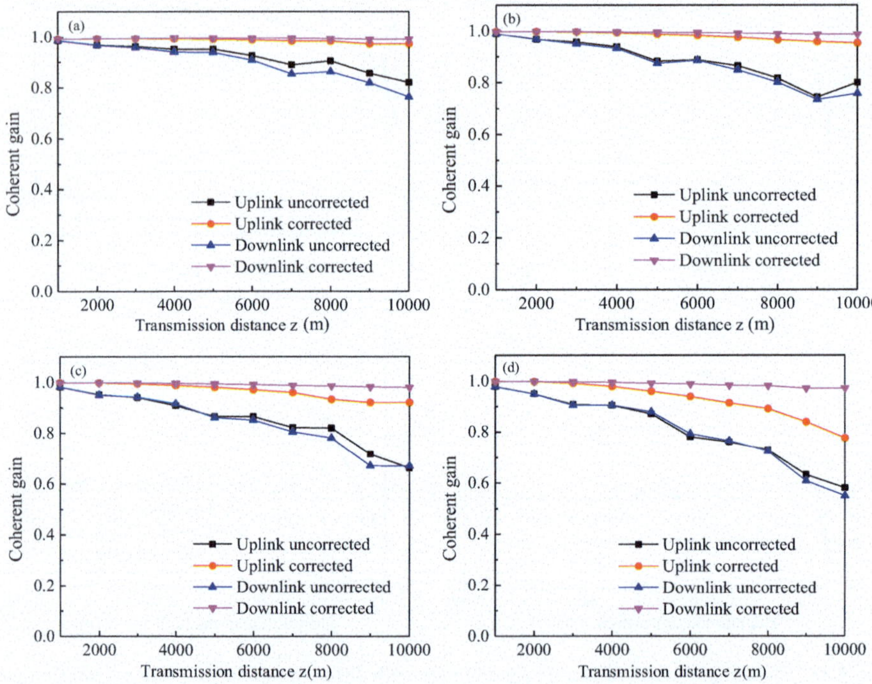

Fig. 5.39 Change of the coherent gain of the signal beam with different topological charges after different transmission distances. **a** $l_{s1} = 1$; **b** $l_{s2} = 2$; **c** $l_{s3} = 3$; **d** $l_{s4} = 4$

more severe is the mode degradation. Because adaptive-optics is usually suitable for wavefront correction in weak turbulence, the effect of wavefront correction using adaptive-optics in the downlink is more significant compared with the uplink.

When the transmission distance $z = 9$ km and LO topological charge $l_{lo} = 4$, the coherence gain corresponding to the downlink correction is 0.972, and the coherence gain corresponding to the uplink correction is 0.841, as calculated from Eq. (5.78). In addition, as the transmission distance and topological charge increase, the correction advantage of the downlink becomes more evident. Therefore, when designing a vortex-beam multiplexing coherent-detection communication system, the relationship between the transmission distance and number of multiplexing paths should be considered.

After transmitting the signal beam over a transmission distance of $z = 10$ km, an altitude of $H = 500$ m, and a topological charge of $l_s = 1$–4, the crosstalk between the uplink, downlink, uncorrected, and corrected signals is shown in Fig. 5.40.

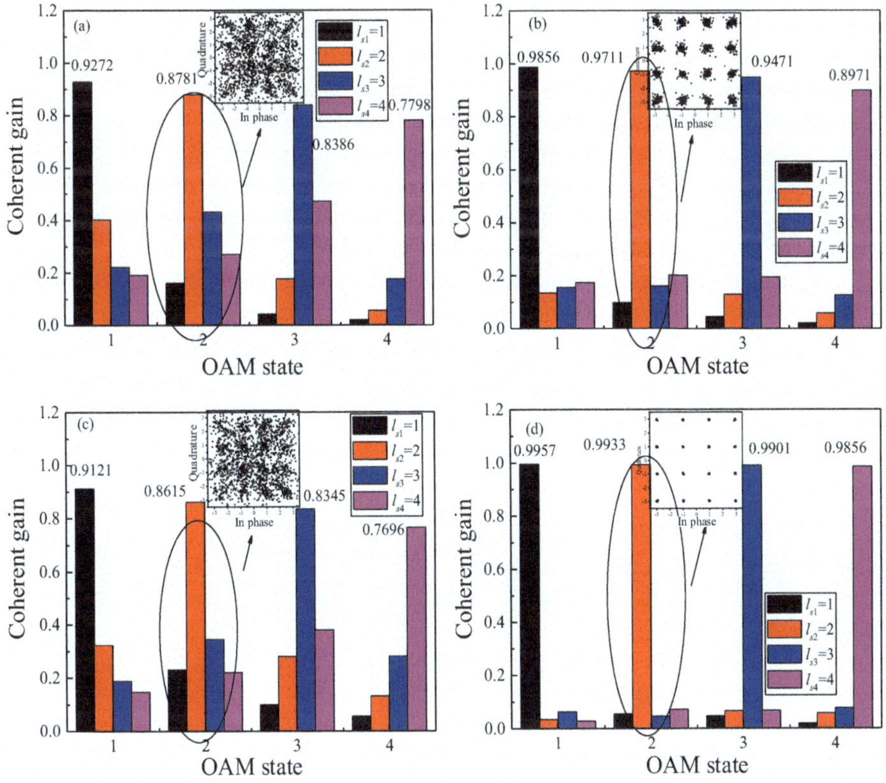

Fig. 5.40 Coherent gain of a signal beam with different topological charges after transmission and crosstalk between modes. **a** Uplink uncorrected; **b** uplink corrected; **c** downlink uncorrected; **d** downlink corrected

Taking the above chain as an example, when the transmission distance z = 10 km and the topological charge $l_{s2} = 2$ of the signal beam is transmitted, the coherent gain G_{22} (the first subscript represents the second transmission, and the second subscript represents the helical spectral decomposition topological charge 2) was calculated by Eq. (5.78), which increased from 0.8781 before correction to 0.9711. After the signal beam with topological-charge number $l_{s1} = 1$ is transmitted and calculated by Eq. (5.78), the coherent gain G_{12} decreases from 0.1581 before correction to 0.0949.

For the second channel of the receiving end, because the topological charge of the local-oscillator beam $l_{lo} = 2$, during the coherent-detection process, the spiral spectrum of the signal beam with topological charge $l_{s2} = 2$ carries effective information. However, the spiral spectrum of the signal beam with topological charge $l_{s1} = 1$ generates a spiral-spectrum component with the same topological charge as the signal beam with topological charge $l_{s2} = 2$, due to mode crosstalk, which causes noise and interference for the signal beam with topological charge $l_{s2} = 2$. Similarly, considering the interference of other signals, the signal constellation generated by one local oscillator with topological-charge number $l_{lo} = 2$ is dispersed, as shown in Fig. 5.40a.

While adaptive-optics can effectively suppress the crosstalk between various channels of vortex-beam multiplexing, the constellation shown in Fig. 5.40b is more concentrated at the corresponding modulation constellation points than that shown in Fig. 5.40a. In addition, for transmission distance $z = 10$ km, the constellation in Fig. 5.40d is more concentrated than that in Fig. 5.40b, indicating that the correction effect of the downlink is better than that of the uplink. Therefore, the convergence effect of the downlink constellation is more evident.

Using the Monte Carlo method, the total code number for the calculation is 106. Figure 5.41 shows the error rate of the uncorrected and corrected channels and the change curve of the system bit-error rate with the transmission distance. The bit-error rate (BER) gradually increases with an increase in the transmission distance, and the larger the number of topological charges, the greater the bit-error rate under the same transmission distance.

When the transmission distance z is less than 5 km, there is no significant difference in the correction effect between the downlink and uplink. As the transmission distance increases, the impact of turbulence on the downlink is less than that on the uplink. Comparing the changes in the error-rate curve in Figure 5.41b and d, it can be observed that when the transmission distance $z > 5$ km, the correction effect of the downlink is better than that of the uplink. Therefore, when the transmission distance $z > 5$ km, the error rates of each channel and the overall system in the downlink are lower than those in the uplink.

Taking the vortex-beam transmission with a signal-beam topological-charge number $l_{s3} = 3$ as an example, when the transmission distance is 10 km, the beam intensity and phase distribution of the signal beam are shown in the small figure in Fig. 5.41, which shows the uncorrected and corrected beam intensity and phase distribution after uplink and downlink transmissions, respectively. The corrected spot diameter was larger than the uncorrected spot diameter, owing to the use of pre-correction processing in the uplink. Wavefront correction does not affect the

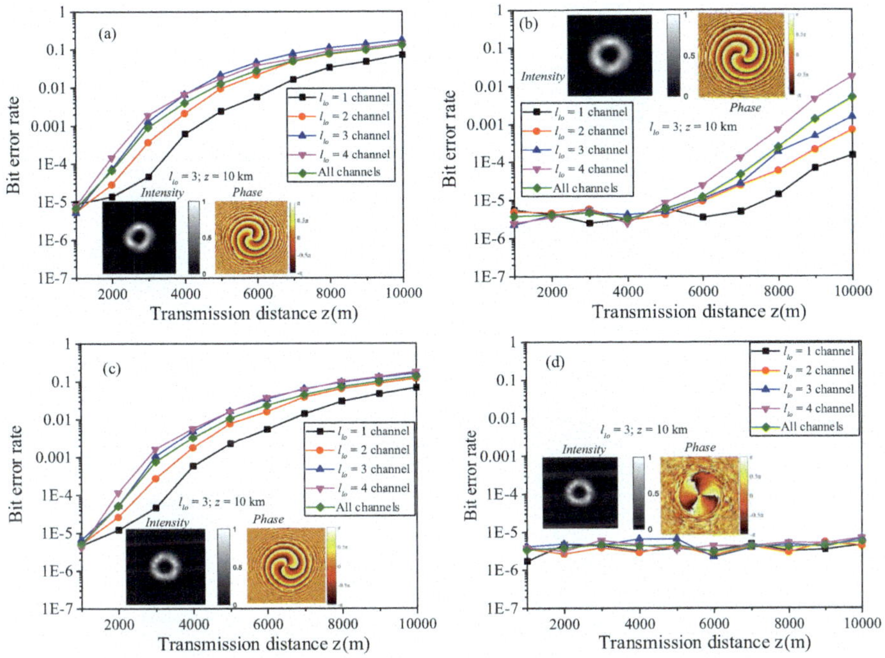

Fig. 5.41 Change of system error rate with transmission distance. **a** Uplink uncorrected; **b** uplink corrected; **c** downlink uncorrected; **d** downlink corrected

beam-intensity distribution; therefore, there was no difference in the beam-intensity distribution before and after the downlink correction.

Vortex-beam multiplexing can improve the channel capacity. The channel capacity is calculated when the number of multiplexed channels differs. Figure 5.42 shows the change in channel capacity with the transmission distance under a vortex-beam single-channel transmission and multiplexed transmission. Under the same conditions, the channel capacity of a multiplexed transmission is greater than that of a single-channel transmission. The channel capacity is based on the result of the system bit-error rate. The channel stochastic matrix is calculated, the conditional-information amount and the source-information amount are calculated using a Bayesian formula, and the channel capacity is finally calculated; thus, the change trends of the two are the same.

It can be seen from Fig. 5.42 that the more multiplexed the channels, the greater the channel capacity of the system. The channel capacity before uplink and downlink correction tends to decrease with an increase in transmission distance; however, after correction by the adaptive-optics system, the channel capacity of the uplink and downlink can recover the decrease in the channel capacity caused by the impact of atmospheric turbulence. An adaptive-optics system is used to correct the optical distortion of the signal, which can reduce the bit-error rate of the system while improving its channel capacity.

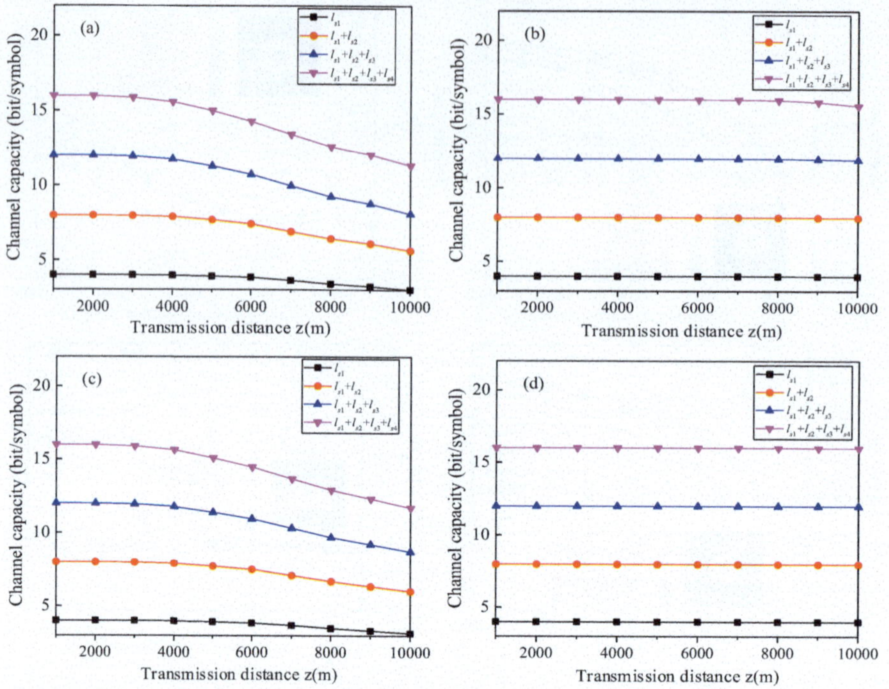

Fig. 5.42 Change of channel capacity with transmission distance. **a** Uplink uncorrected; **b** uplink corrected; **c** downlink uncorrected; **d** downlink corrected

5.4 Experimental Study on the Performance of Vortex Coherent Detection

The mixing performance of different combination models with different spatial distributions of the signal and local-oscillator beams was studied earlier, and the influence of the topological charges of the signal and local-oscillator beams on the mixing efficiency was analyzed. This section presents two sets of test optical paths, one to verify whether the vortex and Gaussian beams can mix, and the other to change the topological charge of the signal and LO beams, analyze the coherence performance of one vortex signal beam and LO beam under turbulent conditions, and verify the change in the IF signal.

5.4.1 Static Experimental Research on Vortex Coherent Detection

When the signal beam is a vortex beam and the local oscillator is a Gaussian beam, the mixing efficiency of the vortex beam with different topological charges and the Gaussian beam is observed. The experimental optical path was designed to verify whether the vortex and Gaussian beams described in Sect. 5.3 can be mixed.

The experimental schematic is shown in Fig. 5.43. The signal beam received by the receiving antenna is incident onto a spatial beam modulator to generate a vortex beam with a topological charge of 1. The desired vortex beam is selected through the aperture, and the splitter divides the beam into two paths. One path is focused by Lens 3 and is then incident onto the coherent receiving module for mixing. The root-mean-square value of the intermediate-frequency voltage is measured using an oscilloscope. The other path is collimated and reaches the wavefront sensor. The front of the vortex beam wave was observed to ensure experimental accuracy.

The root-mean-square values of the intermediate-frequency voltage measured twice through the above experiments are 55.64 mV and 49.87 mV, respectively, and the calculated mixing efficiency is 5.058% and 4.534%, respectively. The measured intermediate-frequency voltage also contains the noise power of the detector, and the actual mixing efficiency is significantly lower than the calculated mixing efficiency. Therefore, the Gaussian and vortex beams measured in the experiment cannot be mixed, which is consistent with the theoretical calculation results.

To investigate the mixing characteristics of a vortex beam, an experimental study in an indoor static environment, as shown in Fig. 5.44, was conducted. After the collimator outputs the signal beam, spiral phase screens with different topological charges are used to generate vortex beams with different topological charges, and the aperture is used to filter out other beam modes, retaining only the desired vortex-beam mode.

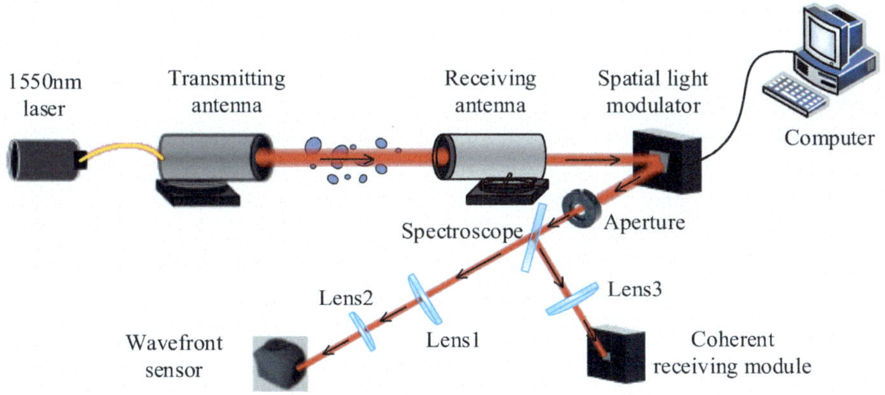

Fig. 5.43 Schematic diagram of the mixing experiment with an LG signal beam and Gaussian local-oscillator beam

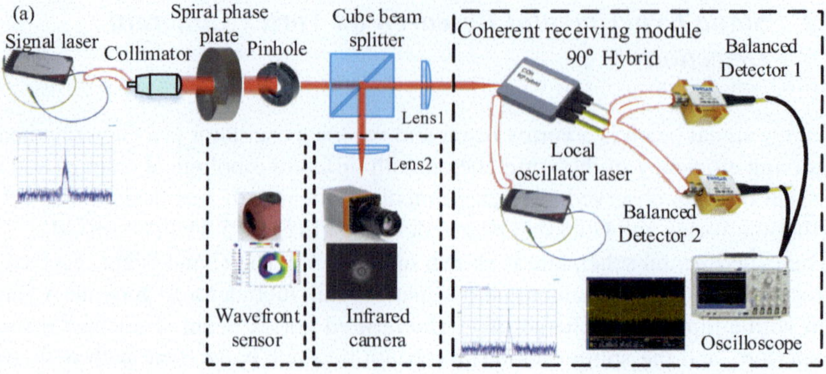

Fig. 5.44 Experimental diagram of mixing a vortex signal beam with a vortex local-oscillator beam

After the transmitted beam is processed by a splitting prism, one path uses a wavefront sensor to measure the wavefront phase or collects beam-intensity distribution information through an infrared camera. The other path of the optical signal is coupled into an optical mixer and mixed with the local-oscillator beam. The generation method for the local-oscillator beam is the same as that for the signal beam, and is not shown in detail in Fig. 5.44. The topological numbers of the signal beam and the local-oscillator beam are the same.

Vortex beams with different topological charges were obtained using different spiral phase screens. A balance detector was then used to convert the mixed optical signal into an intermediate-frequency current signal, and the oscilloscope displays and collects the corresponding intermediate-frequency current signal.

Using different spiral phase screens to generate vortex signal beams with different topological charges, Fig. 5.45 shows the signal spot information collected by an infrared camera and the beam-intensity distribution of the signal beam collected by a wavefront sensor. As the topological charge of the signal beam increases, the dark-spot radius at the center of the vortex beam also gradually increases. When the topological charges are $l = 1, 2, 3, 4$, and 5, the proportions of the vortex dark-spot diameter to the entire spot diameter were 19.66%, 30.13%, 36.67%, 44.37%, and 51.61%, respectively.

Figure 5.46 shows the wavefront phase of the vortex beam collected using a Shack–Hartmann wavefront sensor under different topological charges. When the topological charges of the signal beam are $l = 1, 2, 3, 4$, and 5, the peak-to-valley values of the wavefront phase are 3.05 μm, 3.19 μm, 3.49 μm, 2.71 μm, and 2.80 μm, respectively. From the reconstructed wavefront phase, it can be seen that the defocused terms of the wavefront occupy the vast majority of the overall wavefront.

Owing to the different sizes of dark spots in the vortex beam with different topological charges and the inherent aberration factors of different spiral phase screens, there are differences in the measured wavefront phase, and the wavefront phase that affects the subsequent mixing and coherent-detection processing is spatially distributed.

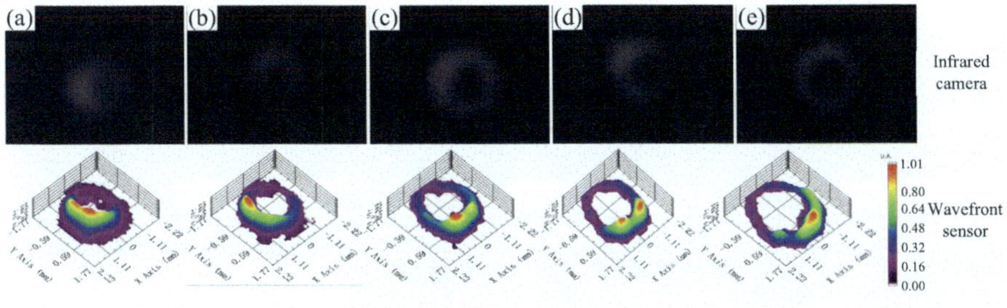

Fig. 5.45 Vortex optics spot with different topological charges and beam-intensity distributions. **a** $l = 1$; **b** $l = 2$; **c** $l = 3$; **d** $l = 4$; **e** $l = 5$

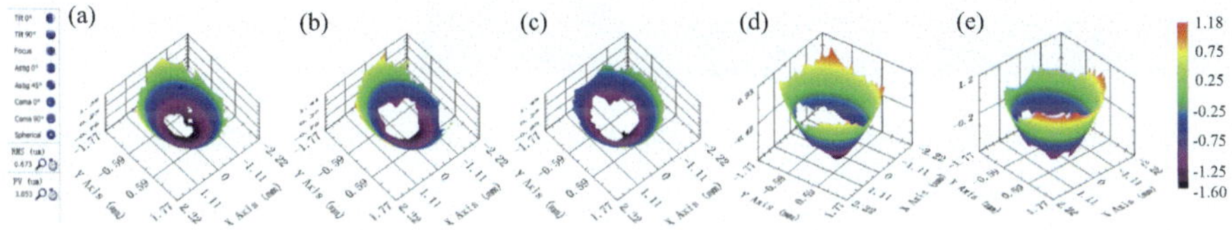

Fig. 5.46 Wavefront phase of vortex beams with different topological charges. **a** $l = 1$; **b** $l = 2$; **c** $l = 3$; **d** $l = 4$; **e** $l = 5$

Figure 5.47 shows the intermediate-frequency signal output by the vortex beam with different topological charges after coherent detection under the same transmission power. When the topological charges of the signal and local-oscillator beams are $l = 1, 2, 3, 4$, and 5, the average amplitudes of the intermediate-frequency signal are 1.15 V, 1.10 V, 1.08 V, 1.06 V, and 0.98 V, respectively. When the topological charges of the signal and local-oscillator beams are the same, the matching degree of the amplitude and phase is relatively high; thus, the mixing efficiency is higher.

However, as the topological charge of the signal beam and local-oscillator beam increases, the quality of the vortex beam generated by the spiral phase plate also deteriorates. Therefore, during the experimental process, the larger the topological charge, the smaller the actual measured intermediate-frequency current value; thus, the mixing efficiency of the system decreases. Figure 5.47 shows the amplitude of the intermediate-frequency current in the oscilloscope.

5.4.2 Experiment on the Effect of Atmospheric Turbulence on the Performance of Vortex Coherent Detection

Based on the above experiments, experiments were conducted in turbulent environments. The transmitting end was equipped with an erbium-doped fiber amplifier to ensure a sufficient power output of 15 dBm. Transmitting and receiving antennas

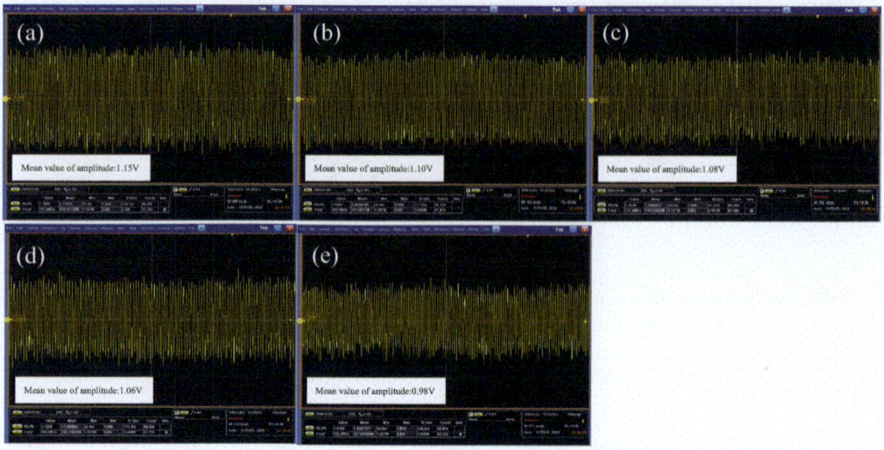

Fig. 5.47 Intermediate-frequency signal of the signal beam under different topological charges. **a** $l = 1$; **b** $l = 2$; **c** $l = 3$; **d** $l = 4$; **e** $l = 5$

were added between the spiral phase screen and the small hole to enable the vortex beam to be transmitted over long distances in turbulent atmospheric environments.

The experimental link was from the sixth floor of Xi'an University of Technology to Kaisen Fujing Yayuan. The communication distance was 1.3 km. The experimental period was December 3, 2022. The weather was cloudy and a northerly wind was at level 2 Fig. (5.48).

Figure 5.49 shows the beam spot and intensity distribution captured by an infrared camera after transmission of the vortex signal beam with topological charges of 1–5 in a turbulent environment. The dark spots of the vortex beam in the turbulent environment increase with an increase in the topological charge, and the radius of the central dark spot gradually increases. Moreover, the beam spots exhibit an irregular fragmentation trend, compared to static environments.

Figure 5.50 shows the wavefront phase of a vortex beam with different topological charges collected in a turbulent environment. When the topological charges are $l =$

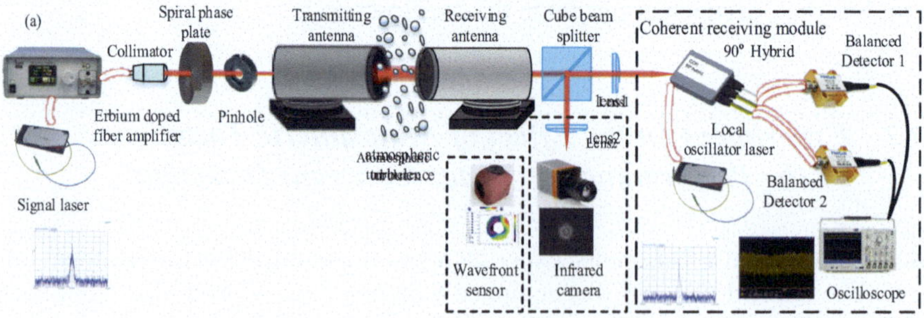

Fig. 5.48 Schematic diagram of a field experiment in a turbulent environment

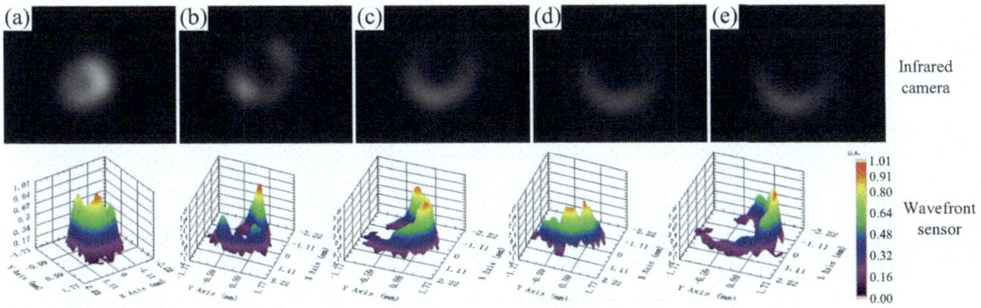

Fig. 5.49 Vortex optics spots with different topological charges and beam-intensity distributions in a turbulent environment. **a** $l = 1$; **b** $l = 2$; **c** $l = 3$; **d** $l = 4$; **e** $l = 5$

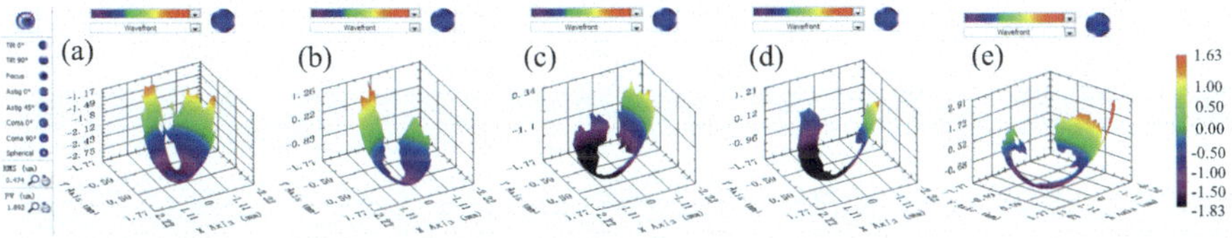

Fig. 5.50 Vortex-beam wavefront phase with different topological charges in a turbulent environment. **a** $l = 1$; **b** $l = 2$; **c** $l = 3$; **d** $l = 4$; **e** $l = 5$

1, 2, 3, 4, and 5, the peak-to-valley values of the wavefront phase are 1.89 μm, 3.14 μm, 2.88 μm, 3.25 μm, and 4.79 μm, respectively. As the topological charge increases, the degree of change in the wavefront phase in turbulent environments becomes significantly greater than that in static environments.

We calculated and compared the normalized intensity distribution and wavefront variance of a vortex beam with different topological charges in static and turbulent environments, as shown in Fig. 5.51. The variance of the normalized beam intensity in turbulent environments is greater than that in static environments, indicating that the beam intensity flickers more significantly in turbulent environments. Increasing the topological charge increases the variance of the normalized beam intensity and enhances the flickering of the beam intensity.

The wavefront-phase variance increases with an increase in the topological charge in turbulent environments, indicating that the larger the topological charge of the vortex beam, the weaker its ability to suppress turbulence and the more significantly it is affected by turbulence.

Vortex-beam signals with different topological charges in turbulent environments are mixed and processed to obtain the corresponding intermediate-frequency signals, as shown in Fig. 5.52. Compared to static environments, a vortex beam with the same topological charge has a more significant fluctuation in the envelope amplitude of the obtained intermediate-frequency signal after mixing in turbulent environments. As the topological charge of the signal beam increases, the amplitude and mixing efficiency of the intermediate-frequency signal decrease.

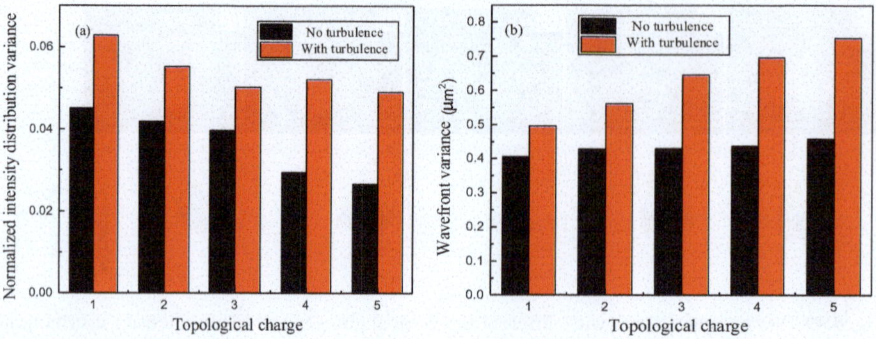

Fig. 5.51 Vortex beams with different topological charges in static and turbulent environments. **a** Normalized beam-intensity distribution variance; **b** wavefront variance

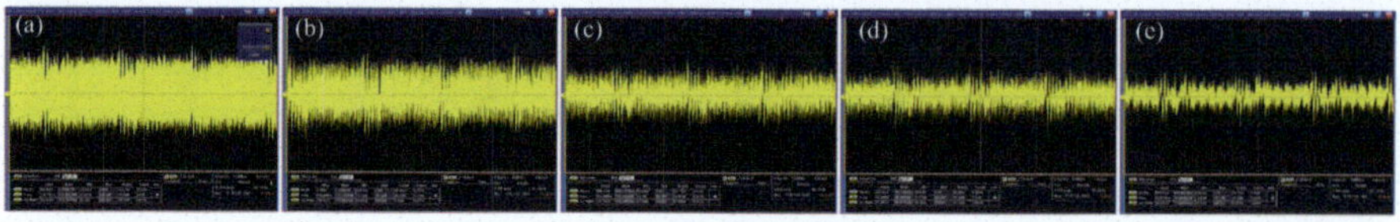

Fig. 5.52 Intermediate-frequency signal of a signal beam with different topological charges in a turbulent environment. **a** $l = 1$; **b** $l = 2$; **c** $l = 3$; **d** $l = 4$; **e** $l = 5$

References

1. Yao AM, Padgett MJ (2011) Orbital angular momentum: origins, behavior and applications. Adv Opt Photon 3(2):163–204
2. Ke XZ, Wang J (2018) Generation, transmission, detection, and application of vortex beams. Science Press, Beijing, pp 120–121
3. Tang M, Li WH, Zhang MM et al (2022) Homogeneous self zero difference coherent optical transmission technology. China Laser 49(12):1206002
4. Li G (2009) Recent advances in coherent optical communication. Adv Opt Photon 1(2):279–307
5. Li C, Wang Y, Wang J et al (2022) Convolutional neural network. Aided DP.64 QAM coherent optical communication systems. J Lightwave Technol 40(9):2880–2889
6. Liu W, Jin D, Shi W et al (2022) Performance analysis of coherent optical communication based on hybrid algorithm. Opt Laser Technol 149:107878
7. Modeling the heterodyne efficiency of array detector systems in the presence of target speckle. IEEE Photon J 11(4): 3–9
8. Guo Z (2012) Analysis of signal to noise ratio and heterodyne efficiency for reference beam laser Doppler velocimeter. Opt Laser Technol 44(1):108–113
9. Liu HZ, Ji YF, Xu N et al (2011) The influence of signal and local oscillator amplitude distribution on the mixing efficiency of inter satellite coherent optical communication systems. J Opt 31(10):1006001
10. Kong YX, Ke XZ, Yang Y (2015) Research on the influence of atmospheric turbulence on spatial coherent optical communication. Progress in Laser Optoelectron 52(08):080601
11. Xiang JS, Pan LC (2009) The heterodyne efficiency of spatial coherent optical communication and the influence of antenna aberration. Optoelectronics 36(11):53–57

12. Ren Y, Xie G, Huang H et al (2014) Adaptive optics based simultaneous pre and post turbulence compensation of multiple orbital angular momentum beams in a bidirectional free space optical link. Optica 1(6):376–382
13. Rao RZ (2005) The propagation of light in turbulent atmosphere. Anhui Science and Technology Press, Hefei pp 3–6
14. Hufnagel RE, Stanley NR (1964) Modulation transfer function associated with image transmission through turbulent media. JOSA 54(1):52–61
15. Fried DL, Mevers GE, Keister MP (1967) Measurements of laser beam scintillation in the atmosphere. JOSA 57(6):787–797
16. Andrews LC, Phillips RL (2005) Laser beam propagation through random media. Oxford University Press, Washingdon, pp 68–69
17. Andrews LC, Miller WB (1995) The mutual coherence function and the backscatter amplification effect for a reflected Gaussian.beam wave in atmospheric turbulence. Waves Random Media 5(2):167–182
18. Tatarski VI, Silverman RA, Chako N (1961) Wave propagation in a turbulent medium. Phys Today 14(12):46–47
19. Yang HB, Xu H (2019) Numerical simulation of atmospheric turbulence phase screen based on power spectrum inversion method. Appl Optoelectronic Technol 34(4):73–76
20. Yang C, Xu C, Ni W et al (2017) Turbulence heterodyne coherent mitigation of orbital angular momentum multiplexing in a free space optical link by auxiliary light. Opt Express 25(21):25612–25624
21. Duan MY, Dan X, Ai Y (2014) Research and progress on laser atmospheric turbulence simulation devices. Opt Commun Technol 38 (01):49–52
22. Ke XZ, Tan ZK (2018) Effect of angle of arrival fluctuation on heterodyne detection in slant atmospheric turbulence. Appl Opt 57(5):1083–1090

Chapter 6
Mechanism and Experimental Study of Wavefront Distortion on Mixing Efficiency

6.1 Multi-beam-Transmission Coherent-Detection Principle and Wavefront-Distortion Generation Mechanism

This chapter considers a optical wireless coherent-detection system as the research object, and studies wavefront distortion caused by atmospheric turbulence and other environmental factors, as well as its correction. The principle mechanism of wavefront distortion in Kolmogorov and non-Kolmogorov atmospheric turbulence is expounded.

Mathematical models of multi-beam-transmission wavefront distortion and the optical wireless coherent-detection system's mixing efficiency, signal-to-noise ratio, and bit-error rate in different atmospheric-turbulence models are established. The influence of wavefront distortion and its correction on the performance of a optical wireless coherent-detection system is analyzed; and indoor and outdoor wavefront-distortion correction experiments are established.

6.1.1 Multi-beam-Delivery Coherent Detection

Owing to its advantages, traditional coherent-detection technology is widely used at the receiving end of optical wireless coherent communication systems to detect weak signals. However, considering the performance degradation of coherent-detection systems at the receiving end caused by long-distance signal-beam transmissions, communication systems usually adopt multiple-input multiple-output (MIMO) technology. It can effectively suppress the fading and fluctuation of a signal beam during transmission and improve the channel capacity [1, 2]. On this basis, a mathematical model of a multi-beam transmission is established, and a optical wireless

X. Ke and C. Ke, *Performance-Improvement Limits in Optical Wireless Coherent Communication*, Optical Wireless Communication Theory and Technology,
https://doi.org/10.1007/978-981-96-2000-5_6

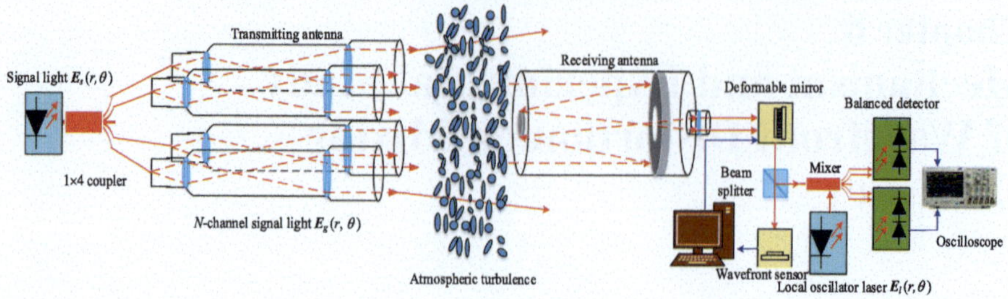

Fig. 6.1 Optical wireless coherent-detection system with a multi-beam transmission structure

coherent-detection system with a multi-beam transmission structure is shown in Fig. 6.1.

A laser-beam source is used as the signal beam, and the transmitting end uses N signal beams to transmit. After atmospheric turbulence, one beam is synthesized on the end face of the receiving antenna, which is expressed as

$$E_s(r,\theta)=\sum_{g=1}^{N}E_g(r,\theta)\times e^{\left[ik\left(r\sin\theta_g+r\cos\theta_g-\frac{r^2}{2L}\right)\right]},\tag{6.1}$$

where the spatial angular frequency of the wave is $k=2\pi/\lambda$, λ is the wavelength, and θ_g is the inclination angle of the gth signal beam $E_g(r,\theta)$. Assuming that all N signal beams are sent out by the same type of laser, the beam field of the gth signal beam at the transmitting end can be expressed as [3]

$$E_g(r,\theta)=A_g(r,\theta)e^{-i(\omega_g t+\varphi_g)},\tag{6.2}$$

where A_g, ω_g, and φ_g are the amplitude, frequency, and phase of the gth signal beam, respectively. According to the basic theory of optical wireless coherent detection, the signal-to-noise ratio SNR_{MIF} at the receiving end can be expressed as

$$SNR_{MIF}=\frac{2\eta K}{h\nu B}\cdot\frac{\left|\iint_S E_s(r,\theta)E_l(r,\theta)rdrd\theta\right|^2}{\iint_S|E_l(r,\theta)|^2rdrd\theta}.\tag{6.3}$$

When multiple beams are transmitted at the transmitting end, the system mixing efficiency for heterodyne detection can be expressed as

$$\gamma_{MRF}=\frac{\left|\iint_S E_s(r,\theta)E_l(r,\theta)rdrd\theta\right|^2}{\iint_S|E_s(r,\theta)|^2rdrd\theta\cdot\iint_S|E_l(r,\theta)|^2rdrd\theta}.\tag{6.4}$$

The system mixing efficiency of homodyne detection is expressed as

$$\gamma_{MZF} = \frac{\left[\iint_S |A_s(r,\theta)||A_l(r,\theta)|\cos(\Delta\varphi) r dr d\theta\right]^2}{\iint_S |E_s(r,\theta)|^2 r dr d\theta \cdot \iint_S |E_l(r,\theta)|^2 r dr d\theta}. \quad (6.5)$$

Furthermore, we provide the system bit-error rate of heterodyne detection in a multi-beam transmission [4]:

$$BER_{MRF} = \frac{1}{2} erfc\left(\sqrt{\eta N_p \gamma_{MRF}}\right). \quad (6.6)$$

The system bit-error rate for homodyne detection is expressed as

$$BER_{MZF} = \frac{1}{2} erfc\left(\sqrt{2\eta N_p \gamma_{MZF}}\right). \quad (6.7)$$

Combined with the mixing-efficiency expression, the correction gain is defined as

$$G_\gamma = 10\lg\left(\frac{\gamma_{corrected}}{\gamma_{uncorrected}}\right). \quad (6.8)$$

The correction gain measures the multiplier of the mixing-efficiency improvement after wavefront correction, and the unit is dB.

6.1.2 Atmospheric Turbulence

Changes in the atmospheric humidity and temperature gradients lead to atmospheric turbulence. The wavefront distortion caused by environmental factors, such as atmospheric turbulence, causes the beam to diverge beyond the diffraction phenomenon, which is reflected in the random change of the beam center of mass under beam drift, and thus, causes beam-amplitude and phase fluctuations [5].

In addition, atmospheric turbulence gradually destroys the spatial coherence of a propagating signal beam. This weakening of the spatial coherence limits the degree of collimation and focusing of the signal beam, which, in turn, leads to a significant decrease in system performance at the receiving end. Therefore, optical wireless coherent-detection systems are particularly sensitive to spatial-coherence loss. The characteristics of this part of atmospheric turbulence can be explained statistically, which also lays the theoretical foundation for the effect of atmospheric turbulence. Figure 6.2 shows a schematic diagram of atmospheric turbulence.

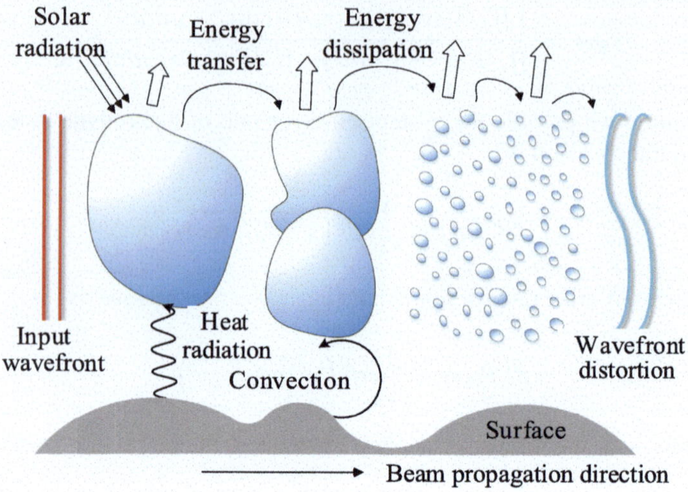

Fig. 6.2 Schematic diagram of atmospheric turbulence

6.1.2.1 Kolmogorov Turbulence Spectrum

Under the influence of the inertial force, atmospheric turbulence decomposes from chaotic large-scale vortex motion into smaller eddies, forming a continuum between the outer turbulent scale L_0 and inner turbulent scale l_0, called the inertial region. The famous Kolmogorov power spectrum obeys the -11/3 power theorem in the inertial region, which can be defined as [5]

$$\Phi_n(\kappa) = 0.033 C_n^2 \kappa^{-11/3} \quad 1/L_0 \ll \kappa \ll 1/l_0, \tag{6.9}$$

where κ is the amplitude of the wavenumber vector, and C_n^2 ($\mathrm{m}^{-2/3}$) is the atmospheric refractive-index structure constant. The research results show that the atmospheric refractive-index structure constant C_n^2 can be used to predict the intensity of atmospheric turbulence and is closely related to the altitude and location. In other words, the physical strength and regularity of atmospheric turbulence are fully reflected in C_n^2.

Physically, the turbulence intensity depends on the atmospheric-coherence length r_0, which is characterized by the atmospheric refractive-index structure constant C_n^2, which is an important parameter for measuring the turbulence intensity. The atmospheric-coherence length r_0 is expressed as follows:

$$r_0 = \left[0.423 \sec(\psi) k^2 \int_{h_0}^{H} C_n^2(h_w) dh_w \right]^{-\frac{3}{5}}, \tag{6.10}$$

where the secant function $\sec(x) = 1/\cos(x)$, ψ is the zenith angle, h_w is the height parameter, and h_0 represents the height of the uplink transmitter or downlink receiver from the ground. The link height H is defined as follows:

$$H = h_0 + L\cos(\psi), \tag{6.11}$$

where L is the transmission distance. The distribution of the atmospheric refractive-index structure constant C_n^2 is described by the Hufnagel–Valley turbulence model, and the atmospheric-turbulence structure parameter $C_n^2(h_w)$ at height h_w is calculated [6]:

$$\begin{aligned} C_n^2(h_w) = {} & C_n^2(0)e^{-h_w/100} + 5.94 \times 10^{-53}\left(\frac{v_w}{27}\right)^2 h_w^{10} e^{-h_w/1000} \\ & + 2.7 \times 10^{-16} e^{-h_w/1500}, \end{aligned} \tag{6.12}$$

where $C_n^2(0)$ is the ground refractive-index structure constant; generally, $C_n^2(0) = 1.7 \times 10^{-14}\ \mathrm{m}^{-2/3}$. The wind-speed model v_w is the root-mean-square wind speed along the vertical path:

$$v_w = \sqrt{v_g^2 + 30.69v_g + 348.91}. \tag{6.13}$$

When the ground wind speed is unknown, its approximate value $v_g = 2.8$ m/s is taken, and $v_w = 21$ m/s is obtained.

The Hufnagel-Valley night model can be obtained by correcting the parameters of the night experimental-measurement data. The expression of the atmospheric-turbulence structure parameter $C_n^2(h_w)$ is [5]

$$\begin{aligned} C_n^2(h_w) = {} & 8.16 \times 10^{-54} h_w^{10} e^{-h_w/1000} + 3.02 \times 10^{-17} e^{-h_w/1500} \\ & + 1.9 \times 10^{-15} e^{-h_w/100}. \end{aligned} \tag{6.14}$$

Davis provided a basis for quantitatively estimating the intensity of atmospheric turbulence and subsequent applications: When $C_n^2 < 6.4 \times 10^{-17}\ \mathrm{m}^{-2/3}$, the atmosphere is generally considered to be weakly turbulent. When $6.4 \times 10^{-17}\ \mathrm{m}^{-2/3} < C_n^2 < 2.5 \times 10^{-13}\ \mathrm{m}^{-2/3}$, it is considered to be medium turbulence; and when $C_n^2 > 2.5 \times 10^{-13}\ \mathrm{m}^{-2/3}$, it is strong turbulence. The greater the atmospheric refractive-index structure constant C_n^2, the greater the turbulence intensity; the smaller the atmospheric-coherence length r_0, the greater the turbulence intensity.

6.1.2.2 Non-Kolmogorov Turbulence Spectrum

An in-depth study of atmospheric turbulence found that experimental data obtained from scintillation measurements, lidar measurements, etc. indicate that some atmospheric-turbulence models deviate from the predictions of the Kolmogorov turbulence spectrum. The influence of the non-Kolmogorov turbulence spectrum on beam propagation has gradually attracted the attention of researchers. Many studies have shown that the non-Kolmogorov turbulence-spectrum model is more consistent with actual atmospheric turbulence. The non-Kolmogorov power spectrum is defined as [7]

$$\Phi_n(\kappa, \alpha) = A(\alpha)\tilde{C}_n^2 \kappa^{-\infty}, \tag{6.15}$$

where α is the spectral power-law exponent in non-Kolmogorov atmospheric turbulence, and $\tilde{C}_n^2$ is the refractive-index structural constant of non-Kolmogorov atmospheric turbulence, in units of $\mathrm{m}^{-\alpha+11/3}$. $A(\alpha)$ is the consistency function, which can be expressed as

$$A(\alpha) = \frac{\Gamma(\alpha - 1)}{4\pi^2} \cos\left(\frac{\alpha}{2}\pi\right) \quad 3 < \alpha < 4, \tag{6.16}$$

where $\Gamma(*)$ is a gamma function. When $\alpha = 11/3$ is satisfied, it degenerates into the refractive-index structure constant C_n^2 under Kolmogorov turbulence; that is, $\tilde{C}_n^2(11/3) = C_n^2$. Considering the influence of the turbulence-scale factors, the non-Kolmogorov turbulence spectrum is expressed as

$$\Phi_n(\kappa, \alpha) = A(\alpha)\tilde{C}_n^2 \left(\kappa^2 + \kappa_0^2\right)^{-\frac{\alpha}{2}} \exp\left(\frac{-\kappa^2}{\kappa_m^2}\right), \tag{6.17}$$

where $\kappa_0 = 2\pi/L_0$ and $\kappa_{\mathrm{m}} = c(\alpha)/l_0$. Function $c(\alpha)$ of the power spectral index is as follows:

$$c(\alpha) = \left[2\pi \frac{A(\alpha)}{3} \Gamma\left(\frac{5 - \alpha}{2}\right)\right]^{\frac{1}{(\alpha-5)}}. \tag{6.18}$$

According to the relationship between the vortex and the inner and outer turbulence scales, atmospheric turbulence is divided into three regions: when the vortex scale is larger than the outer scale L_0, it is the input area; when the vortex scale is between the inner and outer scales, it is the inertial area; and when the vortex scale is smaller than the inner scale, it is the dissipative area [8]. Among these, the inertial region is the focus of atmospheric-turbulence research and the Kolmogorov turbulence spectrum is valid only in the inertial region.

With the irregular change in the eddy scale, the atmospheric turbulence has an uneven spatial distribution, which no longer conforms to the Kolmogorov turbulence-spectrum model and limits the application scope of the Kolmogorov turbulence-spectrum model in measuring the performance of optical wireless communication systems. Therefore, non-Kolmogorov turbulence-spectral models have attracted considerable attention.

6.1.3 Wavefront Distortion of Optical Signals in Atmospheric Turbulence

6.1.3.1 Wavefront Distortion in Kolmogorov Atmospheric Turbulence

During the transmission of a signal beam, the phase mismatch becomes increasingly serious, owing to the influence of atmospheric turbulence [9–11]. Using the general Kolmogorov turbulence model to describe atmospheric turbulence, Eq. (6.2) degenerates to

$$E_g(r,\theta) = A_g(r,\theta)e^{-i(\omega_g t+\varphi_g+\Delta\varphi)}, \tag{6.19}$$

where $\Delta\varphi$ is the phase difference caused by atmospheric turbulence. In an optical system, the wavefront distortion caused by a phase mismatch is described by Zernike polynomials; that is, the polar-coordinate function $\Delta\varphi(r,\theta)$ is expressed as [12]

$$\Delta\varphi(r,\theta) = \sum_{i=1}^{N} a_i Z_i(r,\theta), \tag{6.20}$$

where N is the number of Zernike items; a_i is the coefficient of the i-th Zernike polynomial, where the mean-square parameter related to the wavefront coefficient is called the distortion amplitude; and Z_i is the i-th Zernike polynomial. In an actual communication system, when a sufficiently large N-order wavefront distortion is used, the residual term can be omitted.

Here, it is necessary to determine the Zernike coefficient a_i. The diagonalization algorithm is used to describe the process of simulating wavefront distortion, and a_i can be obtained from the covariance matrix. For the wavefront phase-coefficient vector $A = [a_1, a_2, \ldots, a_n]^T$, the covariance matrix C can be written as [13, 14]

$$C = E\left[A, A^T\right] = \begin{bmatrix} c_{11} & c_{12} & \cdots & c_{1n} \\ c_{21} & c_{22} & \cdots & c_{2n} \\ \cdots & \cdots & \cdots & \cdots \\ c_{n1} & c_{n2} & \cdots & c_{nn} \end{bmatrix} \cdot \left(\frac{D}{r_0}\right)^{5/3}, \tag{6.21}$$

where n is a positive integer and D is the receiving aperture of the antenna. The ratio D/r_0 of the receiving aperture D to the atmospheric-coherence length r_0 is used to measure the intensity of atmospheric turbulence. In the covariance matrix C, the covariance expressions of Zernike polynomials Z_i and Z_j are

$$E(a_i, a_j) = c_{ij}\left(\frac{D}{r_0}\right)^{5/3}, \tag{6.22}$$

where $E(\cdot)$ represents the covariance operator of random variables, and the covariance c_{ij} is expressed as [15]

$$c_{ij} = \frac{0.046}{\pi}\left[(n_i + 1)(n_j + 1)\right]^{1/2}(-1)^{\frac{n_i+n_j-2m}{2}} \times \delta_{m_i m_j}\int_0^\infty x^{-14/3}J_{n_i+1}(qx)J_{n_j+1}(qx)dx, \tag{6.23}$$

where n_i, m_i, n_j, and m_j are the Zernike polynomial order and angular frequency corresponding to a_i and a_j, respectively, q is the Bessel integral-function parameter, $J_{n+1}(*)$ is the first kind of $n + 1$-order Bessel function, and $x \in$ the real number set. The Kronecker function δ_{mimj} is expressed as

$$\delta_{m_i m_j} = \begin{cases} 1, & m_i = m_j \\ 0, & m_i \neq m_j \end{cases}, \tag{6.24}$$

where, according to the Bessel indefinite-integral function defined by Weber in *Journal Für Math*, the solution can be obtained [16]:

$$\int_0^\infty x^{-14/3}J_{n_i+1}(qx)J_{n_j+1}(qx)dx = \frac{(q)^{11/3}\Gamma(14/3)\Gamma[(n_i+n_j-5/3)/2]}{2^{14/3}\Gamma[(n_i-n_j+17/3)/2]\Gamma[(n_j-n_i+17/3)/2]\Gamma[(n_i+n_j+23/3)/2]}. \tag{6.25}$$

Substituting Eq. (6.25) into Eq. (6.23), c_{ij} is obtained as

$$c_{ij} = 2.2698\left[(n_i + 1)(n_j + 1)\right]^{1/2}(-1)^{\frac{n_i+n_j-2m}{2}} \times\delta_{m_i m_j}\frac{\Gamma[(n_i+n_j-5/3)/2]}{\Gamma[(n_i-n_j+17/3)/2]\Gamma[(n_j-n_i+\frac{17}{3})/2]\Gamma[(n_i+n_j+23/3)/2]}. \tag{6.26}$$

According to Eq. (6.21), when the covariance matrix is not zero, the Zernike polynomials do not satisfy statistical independence. Therefore, Karhunen–Loève polynomials are introduced to expand the wavefront and construct a random quantity with a specific variance [17]. The wavefront distortion is described by Karhunen–Loève polynomials as follows:

$$\Delta\varphi(r,\theta) = \sum_{j=1}^{N} b_j K_j(r,\theta), \tag{6.27}$$

where b_j is a statistically independent random coefficient and $K_j(r, \theta)$ is a Karhunen–Loève polynomial. Then, the Zernike polynomials can be expanded by Karhunen–Loève polynomials as

$$K_j(r,\theta) = \sum_{j=1}^{N} V_{ij} Z_j(r,\theta), \tag{6.28}$$

where V_{ij} represents the conversion matrix. According to Eqs. (6.28) and (6.27), the wavefront distortion can be described as

$$\Delta\varphi(r,\theta) = \sum_{i=1}^{N} b_i \sum_{j=1}^{N} V_{ij} Z_j(r,\theta). \tag{6.29}$$

Combined with Eq. (6.20), the wavefront-distortion coefficient vector A can be expressed as

$$A = V \cdot B, \tag{6.30}$$

where vector $B = [b_1, b_2, \ldots, b_n]^T$, V is the Karhunen–Loève polynomial transformation matrix, and the Zernike coefficient describing the wavefront distortion in Eq. (6.20) can be obtained using Eq. (6.30). Multiplying the obtained wavefront-coefficient vector A by $(D/r_0)^{5/3}$ yields the exact covariance Zernike polynomial weights estimated from the Kolmogorov turbulence spectrum to produce the desired wavefront distortion. In this chapter, all atmospheric-turbulence models are regarded as Kolmogorov turbulence models unless otherwise specified.

Through analysis, it can be concluded that the mean-square residual error can be defined as [18]

$$\Delta = \int d\rho W(\rho) \langle [\varphi(r,\theta) - \varphi_c(r,\theta)]^2 \rangle, \tag{6.31}$$

where $\varphi(r, \theta)$ is the wavefront phase of the signal beam, $\varphi_c(r, \theta)$ is the compensated first J-term wavefront distortion, and $\rho = r/R$. $W(\rho)$ is the polynomial-expansion parameter:

$$W(\rho) = \begin{cases} \frac{1}{\pi} & \rho \le 1 \\ 0 & \rho > 1 \end{cases}. \tag{6.32}$$

Equation (6.31) can be simplified as

$$\Delta_J = \langle \varphi^2 \rangle - \sum_{j=1}^{J} \left\langle |a_j|^2 \right\rangle, \tag{6.33}$$

where $\langle \varphi^2 \rangle$ is the wavefront-distortion phase error caused by atmospheric turbulence and J is the number of wavefront-distortion items to be compensated. When $J > 10$, the mean-square residual error can be expressed as

$$\Delta_J \approx 0.2944 J^{-\frac{\sqrt{3}}{2}} \left(\frac{D}{r_0}\right)^{\frac{5}{3}} \quad \text{rad}^2. \tag{6.34}$$

Through calculations, the corresponding relationship between the mean-square residual error of the wavefront distortion and the atmospheric-turbulence intensity in Table 6.1 can be obtained. As shown in Table 6.1, Δ_1 is the residual error of the piston item, which is ignored; $|\Delta_3 - \Delta_1|$ is the remaining error of the tilt items in two directions, and the calculation shows that the overall tilt component accounts for 86.99% of the entire wavefront distortion. $|\Delta_4 - \Delta_3|$ is the remaining error of the defocus item, which accounts for 2.23% of the entire wavefront distortion.

$|\Delta_6 - \Delta_4|$ is the residual error of astigmatism in two directions, which accounts for 4.49% of the entire wavefront distortion. $|\Delta_8 - \Delta_6|$ is the remaining error of the coma aberration in two directions, which accounts for 1.19% of the entire wavefront distortion. $|\Delta_{11} - \Delta_{10}|$ is the residual error of the spherical aberration, which accounts for 0.23% of the entire wavefront distortion. The remainder are errors of higher-order wavefront-distortion items, accounting for 4.87% of the entire wavefront distortion.

Table 6.1 Wavefront-distortion mean-square residual errors

Number of items J	Mean-square residual error	Number of items J	Mean-square residual error	Number of items J	Mean-square residual error
1	$1.0299(D/r_0)^{5/3}$	8	$0.0525(D/r_0)^{5/3}$	15	$0.0279(D/r_0)^{5/3}$
2	$0.582(D/r_0)^{5/3}$	9	$0.0463(D/r_0)^{5/3}$	16	$0.0267(D/r_0)^{5/3}$
3	$0.134(D/r_0)^{5/3}$	10	$0.0401(D/r_0)^{5/3}$	17	$0.0255(D/r_0)^{5/3}$
4	$0.111(D/r_0)^{5/3}$	11	$0.0377(D/r_0)^{5/3}$	18	$0.0243(D/r_0)^{5/3}$
5	$0.088(D/r_0)^{5/3}$	12	$0.0352(D/r_0)^{5/3}$	19	$0.0232(D/r_0)^{5/3}$
6	$0.0648(D/r_0)^{5/3}$	13	$0.0328(D/r_0)^{5/3}$	20	$0.022(D/r_0)^{5/3}$
7	$0.0587(D/r_0)^{5/3}$	14	$0.0304(D/r_0)^{5/3}$	21	$0.0208(D/r_0)^{5/3}$

6.1.3.2 Wavefront Distortion in Non-Kolmogorov Atmospheric Turbulence

In a non-Kolmogorov turbulent flow, the signal beam at the transmitter degenerates after being subjected to atmospheric turbulence [19]:

$$E_g(r,\theta) = A_g(r,\theta)e^{-i(\omega_g t+\varphi_g+\Delta\varphi)}. \tag{6.35}$$

Here, it is necessary to determine the Zernike coefficients in non-Kolmogorov atmospheric turbulence. Considering the orthogonality of the Zernike polynomials, the correlation function $\langle a_i\, a_{i'}\rangle$ of the coefficients a_i and $a_{i'}$ of any two terms of the corresponding Zernike polynomials is given:

$$\langle a_i a_{i'}\rangle = \frac{1}{\pi^2}\iint Z_i(r,\theta)Z_{i'}(r',\theta')\langle\phi(r)\phi(r'),\rangle dr dr' \tag{6.36}$$

where $\langle\phi(r)\,\phi(r')\rangle$ is the phase covariance function. A Fourier transform of Eq. (6.36) yields

$$\langle a_i a_{i'}\rangle = \iint Q_i(f)Q_{i'}(f')\Phi_n(f,f')df\,df', \tag{6.37}$$

where $Q_i(f)$ is the Fourier transform of $Z_i(r,\theta)$, $Q_{i'}(f')$ is the Fourier transform of $Z_{i'}(r',\theta')$, and f is the beam-wave spatial frequency. In a non-Kolmogorov turbulent transport, the normalized power-spectral density $\Phi_n(f,f')$ is expressed as [20, 21]

$$\Phi_n(f,f') = (4\pi)^{2-\alpha}\frac{\Gamma\left(\frac{\alpha}{2}\right)}{2^{2-\alpha}\pi\alpha\Gamma\left(-\frac{\alpha}{2}\right)}2\left(\frac{8}{\alpha-2}\Gamma\left(\frac{2}{\alpha-2}\right)\right)^{\frac{(\alpha-2)}{2}}\left(\frac{D}{\widetilde{r}_0}\right)^{\alpha-2}f^{-\alpha}\delta_{ff'}, \tag{6.38}$$

where $\widetilde{r}_0$ is the atmospheric-coherence length in a non-Kolmogorov turbulent flow, and $\delta_{ff'}$ is the Kronecker function. Substituting Eq. (6.38) into (6.37), the Zernike-polynomial coefficient correlation function under non-Kolmogorov turbulence conditions is obtained by an integral transformation:

$$\begin{aligned}\langle a_i a_{i'}\rangle &= (-1)^{\frac{(n+n'-2m)}{2}}2^{3-\alpha}\sqrt{(n+1)(n'+1)}\left(\frac{D}{\widetilde{r}_0}\right)^{\alpha-2}\\ &\cdot\frac{\left(\frac{8}{\alpha-2}\Gamma\left(\frac{2}{\alpha-2}\right)\right)^{\frac{(\alpha-2)}{2}}\Gamma\left(\frac{\alpha}{2}\right)\Gamma\left(\frac{n+n'-\alpha+2}{2}\right)\Gamma(\alpha+1)}{\alpha\Gamma\left(-\frac{\alpha}{2}\right)\Gamma\left(\frac{n-n'+\alpha+2}{2}\right)\Gamma\left(\frac{n'-n+\alpha+2}{2}\right)\Gamma\left(\frac{n+n'+\alpha+4}{2}\right)}\delta_{mm'},\end{aligned} \tag{6.39}$$

where n, m, n', and m' are the Zernike polynomial order and angular frequency corresponding to a_i and $a_{i'}$ respectively, and $\delta_{mm'}$ is the Kronecker function. The atmospheric-coherence length in non-Kolmogorov turbulence $\widetilde{r}_0$ is expressed as [22]

$$\tilde{r}_0 = \left[\frac{2^{-\alpha}\alpha\Gamma(\alpha-1)\Gamma\left(\frac{-\alpha}{2}\right)\sin\left[\frac{(\alpha-3)\pi}{2}\right]}{(\alpha-1)\Gamma\left(\frac{\alpha}{2}\right)} k^2 \tilde{C}_n^2 L \right]^{\frac{-1}{\alpha-2}}, \tag{6.40}$$

where the non-Kolmogorov atmospheric refractive-index structure constant $\tilde{C}_n^2$ can be expressed using the atmospheric refractive-index structure constant C_n^2 of Kolmogorov turbulence [23]:

$$\tilde{C}_n^2 = -\frac{\frac{1}{2}\Gamma(\alpha)(2\pi)^{-11/6+\alpha/2}(\lambda L)^{11/6-\alpha/2}}{\Gamma\left(1-\frac{\alpha}{2}\right)\left[\Gamma\left(\frac{\alpha}{2}\right)\right]^2\Gamma(\alpha-1)\cos\left(\frac{\alpha\pi}{2}\right)\sin\left(\frac{\alpha\pi}{4}\right)} C_n^2. \tag{6.41}$$

In Eq. (6.41), when the beam is transmitted obliquely, the structural constant C_n^2 of the atmospheric refractive index satisfies the Hufnagel–Valley model. The Karhunen–Loève polynomial is introduced to expand the wavefront distortion, and the obtained wavefront coefficient vector A is used to generate the wavefront distortion in non-Kolmogorov turbulence.

6.1.4 Comparative Analysis

Using the theoretical derivation relationship between wavefront-distortion and optical wireless coherent-detection performance indicators, we take two beams at the transmitting end as an example. Numerical simulations are conducted on the performance changes in a optical wireless coherent-detection system at the receiving end, caused by wavefront distortions in Kolmogorov and non-Kolmogorov atmospheric turbulence.

When multiple beams are transmitted at the transmitting end, as the spectral power-law exponent α increases, the mixing efficiency and bit-error rate of the receiving-end system are simulated and analyzed by combining Eqs. (6.4) and (6.6).

The main simulation conditions are as follows: wavelength $\lambda = 1550$ nm, electron charge $e = 1.6 \times 10^{-19}$ C, photodetector quantum efficiency $\eta = 0.8$, Planck's constant $h = 6.63 \times 10^{-34}$ J s, carrier frequency $v = 1.94 \times 10^{14}$ Hz, number of photons received in a single bit $N_p = 10$, and detector radius $R = 100$ μm. The tilt angle $\theta_g = 2°$, amplitude $A_l = 1000\,A_g$, initial phase $\varphi_g = \varphi_l = 0$, receiving aperture $D = 200$ mm, wind speed $v_w = 21$ m/s, effective noise bandwidth $\Delta f = 200$ MHz, receiving-antenna horizontal height $h_0 = 10$ m, refractive-index structure constant $C_n^2(0) = 1.7 \times 10^{-14}$ m$^{-2/3}$, transmission distance $L = 10$ km, and zenith angle $\psi = \pi/3$.

The simulation results of the system mixing efficiency and bit-error rate with the spectral power-law exponent α during an oblique transmission are shown in Fig. 6.3.

Figures 6.3a, b show the change curves of the spectral power-law exponent α, mixing efficiency, and bit-error rate of the optical wireless coherent-detection system at the receiving end, following a multi-beam transmission at the transmitting end.

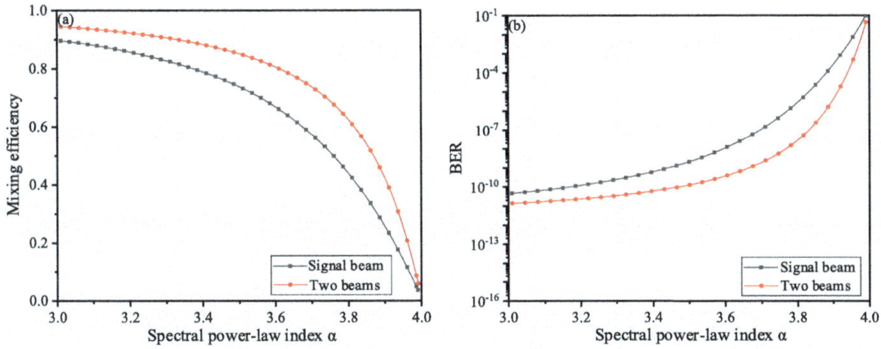

Fig. 6.3 Relationship between the spectral power-law exponent α and **a** the mixing efficiency and **b** the bit-error rate under different transmission beams

It can be seen that with an increase in the transmitted beam, the mixing efficiency of the system is improved, and the bit-error rate is effectively reduced. As the spectral power-law exponent α increases, the mixing efficiency of the system shows a downward trend.

In the spectral power-law exponent interval $3.6 < \alpha < 3.8$, the dual-beam transmission at the transmitting end improves the mixing efficiency of the system most significantly. In Kolmogorov atmospheric turbulence (spectral power-law exponent $\alpha = 11/3$), the mixing efficiency of a single-beam transmission is maintained at 0.61, and the bit-error rate tends to be on the order of 10^{-8}. The mixing efficiency of the double-beam transmission is maintained at 0.76, and the bit-error rate tends to be on the order of 10^{-9}.

It is concluded that an increase in the number of transmission beams at the transmitting end can partially compensate for the problems of the optical wireless coherent-detection system at the receiving end, in which the mixing efficiency decreases and the bit-error rate increases with an increase in the spectral power-law exponent α.

When the spectral power-law exponent $\alpha = 10/3$ in non-Kolmogorov atmospheric turbulence, the range of the zenith angle ψ is changed, and the values of the other parameters are the same as those given in Fig. 6.3. Combining Eqs. (6.4) and (6.6), in Kolmogorov and non-Kolmogorov atmospheric turbulence, the variation curves of the mixing efficiency and bit-error rate of a optical wireless coherent-detection system with zenith angle ψ during an oblique transmission are shown in Fig. 6.4.

Figures 6.4a, b show the change curves of the zenith angle ψ, mixing efficiency, and bit-error rate of the optical wireless coherent-detection system at the receiving end during a multi-beam transmission at the transmitting end. With an increase in the zenith angle ψ, the overall-frequency mixing efficiency of the system shows a downward trend, and the bit-error rate shows an increasing trend. As the number of transmission beams at the transmitting end increases, the mixing efficiency of the receiving-end system increases, and the bit-error rate gradually decreases.

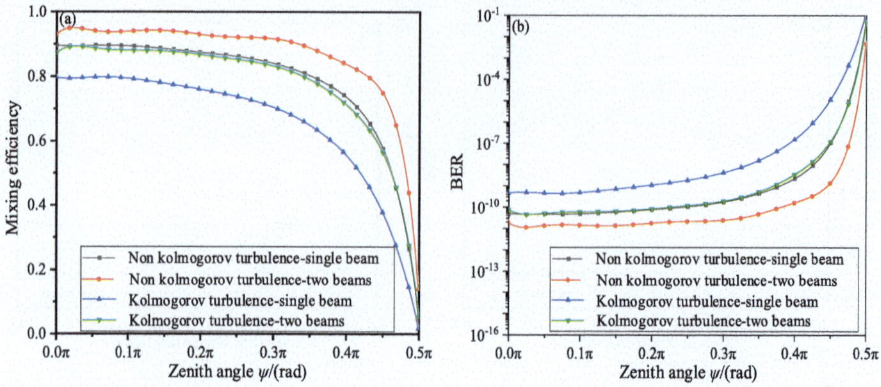

Fig. 6.4 Relationship between zenith angle ψ and **a** the mixing efficiency and **b** the bit-error rate under different transmission beams

In the cases of single-beam and double-beam transmissions at the transmitting end, when the zenith angle ψ in non-Kolmogorov atmospheric turbulence is 0.3π, the corresponding mixing efficiency of the single-beam receiving end is maintained at 0.84, and the bit-error rate is lower than 10^{-10} orders of magnitude. When the transmitting end has two beams, the mixing efficiency of the receiving end is 0.92, and the bit-error rate is lower than 10^{-11} orders of magnitude.

It is concluded that the increase in the number of transmission beams at the transmitting end can compensate for the problems of the optical wireless coherent-detection system at the receiving end, in which the mixing efficiency decreases and the bit-error rate increases with the increase in the zenith angle ψ.

With a spectral power-law exponent $\alpha = 10/3$ in non-Kolmogorov atmospheric turbulence, the range of the transmission distance L is changed, and the values of other parameters are in the same as those in Fig. 6.3. Combining Eqs. (6.4) and (6.6), in Kolmogorov and non-Kolmogorov atmospheric turbulence, the mixing efficiency and bit-error rate of the optical wireless coherent-detection system during an oblique transmission are shown in Fig. 6.5.

Figures 6.5a, b show that with an increase in the transmission distance L, the mixing efficiency in Kolmogorov atmospheric turbulence is lower than that in non-Kolmogorov atmospheric turbulence. As the number of transmission beams at the transmitting end increases, the mixing efficiency of the coherent-detection system at the receiving end increases, and the bit-error rate gradually decreases.

In non-Kolmogorov atmospheric turbulence, when the transmission distance is 6 km, the mixing efficiency of a single-beam transmission system at the corresponding transmitting end is maintained at 0.83, and the bit-error rate is lower than 10^{-10} orders of magnitude. For a dual-beam transmission at the transmitter, the mixing efficiency of the system is 0.91, and the bit-error rate tends to be on the order of 10^{-11}.

It is concluded that the increase in the number of transmission beams at the transmitting end can compensate for some of the problems in that, with an increase

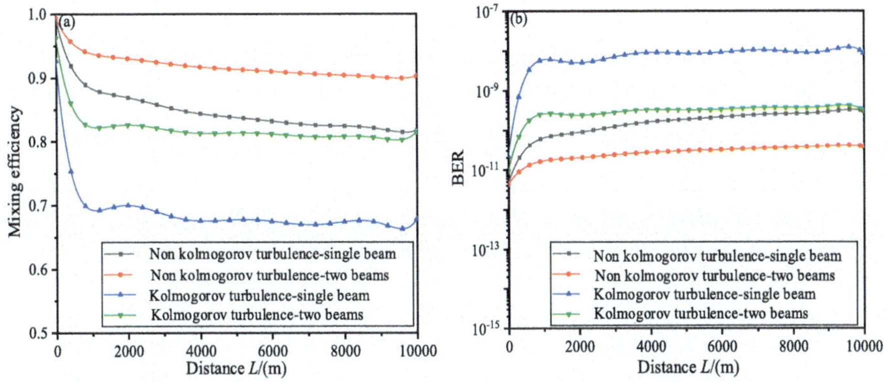

Fig. 6.5 Relationship between the transmission distance L and **a** the mixing efficiency and **b** the bit-error rate under different transmission beams

in the transmission distance L, the mixing efficiency of the optical wireless coherent-detection system at the receiving end decreases and the bit-error rate increases.

Combining Eqs. (6.22) and (6.24), in Kolmogorov and non-Kolmogorov atmospheric turbulence, the values of the parameters are consistent with those given by the simulation shown in Fig. 6.5. The mixing efficiency and bit-error rate of the optical wireless coherent-detection system during an oblique transmission are shown in Fig. 6.6 as a function of the Fresnel zone $(\lambda L)^{1/2}$.

Figures 6.6a, b show a multi-beam transmission at the transmitting end in different turbulence models and the relationship between the mixing efficiency and the bit-error rate of the optical wireless coherent-detection system at the receiving end as a function of the Fresnel zone $(\lambda L)^{1/2}$.

It can be seen that with the increase of the Fresnel zone $(\lambda L)^{1/2}$ in non-Kolmogorov atmospheric turbulence, the mixing efficiency gradually decreases and the bit-error

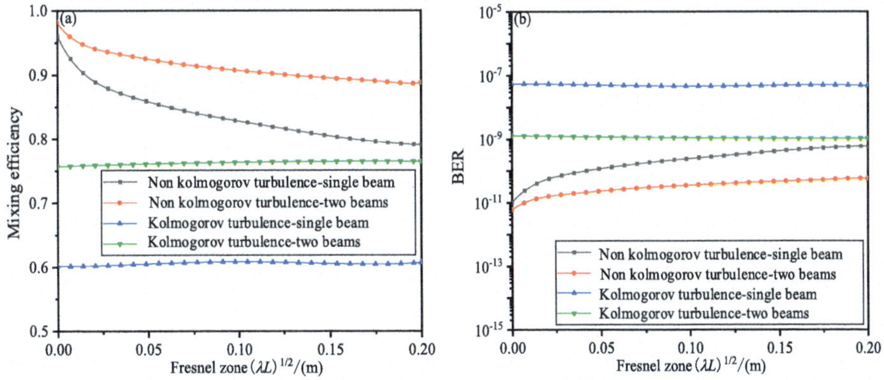

Fig. 6.6 Relationship between the Fresnel zone $(\lambda L)^{1/2}$ and **a** the mixing efficiency and **b** the bit-error rate under different transmission beams

rate gradually increases. When in Kolmogorov atmospheric turbulence, as the Fresnel zone $(\lambda L)^{1/2}$ increases, the mixing efficiency remains essentially unchanged. In Kolmogorov atmospheric turbulence, the mixing efficiency of the system is about 0.6 when the transmitting end transmits a single beam; when the transmitting end transmits two beams, the mixing efficiency is maintained at 0.76.

In non-Kolmogorov atmospheric turbulence, in the case of single-beam and double-beam transmissions at the transmitting end, when the Fresnel zone $(\lambda L)^{1/2}$ is 0.15 m, the mixing efficiency of the corresponding single-beam transmission system is maintained at 0.8, and the bit-error rate tends to be on the order of 10^{-10}. In the case of a dual-beam transmission at the transmitter, the mixing efficiency of the system is maintained at 0.9, and the bit-error rate is lower than 10^{-11} orders of magnitude.

It is concluded that an increase in the number of transmission beams at the transmitting end can partially compensate for the decrease in the mixing efficiency and increase in the bit-error rate of the optical wireless coherent-detection system at the receiving end, as the Fresnel zone $(\lambda L)^{1/2}$ increases.

6.2 Effect of Wavefront Distortion on the Beam-Transmission Characteristics

6.2.1 Wavefront Distortion Described by Mode Method

In research and analysis on wavefront distortion, it is necessary to fit the wavefront-measurement data for subsequent analysis. At present, the orthogonal polynomial combinations that can describe wavefront distortion using the mode method mainly include C-Zernike, Chebyshev, Legendre, and S-Zernike polynomials. The former is suitable for describing the wavefront distortion of a signal beam in a circular aperture because the C-Zernike polynomial basis is orthogonal to the unit circle, which can express the wavefront information well. The others are suitable for describing wavefront distortion in square apertures, such as cylindrical mirrors, gratings, etc. The outgoing wavefront of this type of optical system can be fitted with Chebyshev, Legendre, or S-Zernike polynomials to fit the wavefront distortion; such polynomials have orthogonality in a unit square area.

6.2.1.1 C-Zernike Polynomials

In the second section, the wavefront distortion $\Delta\varphi(r, \theta)$ caused by a phase mismatch can be decomposed into Zernike polynomials in the unit-circle domain for description. These Zernike polynomials (called C-Zernike polynomials) have continuous orthogonality in the unit-circle domain:

$$\Delta\varphi(r,\theta) = \sum_{i=1}^{N} a_i Z_i(r,\theta), \tag{6.42}$$

where N is the number of Zernike terms, a_i is the coefficient of the ith Zernike polynomial, and Z_i is the ith Zernike polynomial. The ith Zernike polynomial $Z_i(r, \theta)$ can be expressed as [24]

$$Z_n^m(r,\theta) = \begin{cases} Z_{odd\cdot i}(r,\theta) = \sqrt{2(n+1)}R_n^m(r)\sin m\theta, & m < 0 \\ Z_{even\cdot i}(r,\theta) = \sqrt{2(n+1)}R_n^m(r)\cos m\theta, & m > 0 \\ Z_i(r,\theta) = \sqrt{n+1}R_n^m, & m = 0 \end{cases}, \tag{6.43}$$

where n is the Zernike polynomial order, m is the angular frequency, and n and m are integers, satisfying $m \le n$, $n - |m| = even$. $Z_{odd\cdot i}$ corresponds to the mode item containing $\sin m\theta$ and $Z_{even\cdot i}$ corresponds to the pattern term with $\cos m\theta$. The radial polynomial $R_n^m(r)$ is expressed as

$$R_n^m(r) = \begin{cases} \sum_{s=0}^{\frac{(n-m)}{2}} \frac{(-1)^s (n-s)!}{s!\left[\frac{n+m}{2}-s\right]!\left[\frac{n-m}{2}-s\right]!} r^{n-2s}, & n-m are even \\ 0, & n-m are odd \end{cases} \tag{6.44}$$

The relationship between the ith term of the Zernike polynomial and its order n and angular frequency m is [25]

$$i = \frac{n(n+1)}{2} + |m| + \begin{cases} 0, & m > 0 \wedge n \equiv \{0,1\} \pmod 4 \\ 0, & m < 0 \wedge n \equiv \{2,3\} \pmod 4 \\ 1, & m \ge 0 \wedge n \equiv \{2,3\} \pmod 4 \\ 1, & m \le 0 \wedge n \equiv \{0,1\} \pmod 4 \end{cases}, \tag{6.45}$$

where |m| represents the modulo operation of m and ^ is the intersection symbol. The first 16-order C-Zernike polynomials are listed in Table 6.2.

Table 6.2 First 16-order C-Zernike polynomials

Number of items i	$Z_i(r, \theta)$	Number of items i	$Z_i(r, \theta)$
1	1	9	$\sqrt{8}\, r^3 \sin 3\theta$
2	$2r\cos\theta$	10	$\sqrt{8}\, r^3 \cos 3\theta$
3	$2r\sin\theta$	11	$\sqrt{5}(6r^4 - 6r^2 + 1)$
4	$\sqrt{3}(2r^2 - 1)$	12	$\sqrt{10}(4r^4 - 3r^2)\cos 2\theta$
5	$\sqrt{6}\, r^2 \sin 2\theta$	13	$\sqrt{10}(4r^4 - 3r^2)\sin 2\theta$
6	$\sqrt{6}\, r^2 \cos 2\theta$	14	$\sqrt{10}\, r^4 \cos 4\theta$
7	$\sqrt{8}(3r^3 - 2r)\sin\theta$	15	$\sqrt{10}\, r^4 \sin 4\theta$
8	$\sqrt{8}(3r^3 - 2r)\cos\theta$	16	$\sqrt{12}(10r^5 - 12r^3 + 3r)\cos\theta$

In a C-Zernike polynomial, Z_1 corresponds to the piston term and is expressed as a constant 1. When the piston term is present, it is difficult to analyze the relationship between the optical wireless coherent-detection performance and the turbulence effect. A Shack–Hartmann wavefront sensor cannot detect the wavefront translation. Therefore, in the subsequent analysis, the piston item is ignored; Z_2 and Z_3 are tilt items; Z_4 corresponds to defocus items; Z_5 and Z_6 correspond to astigmatism items; Z_7 and Z_8 correspond to coma aberration items; and Z_{11} is spherical aberration items.

According to the analysis of Eqs. (6.43) and (6.44), the two main characteristics of C-Zernike polynomials [26] are as follows:

(1) Within the unit-circle domain, it is orthogonal:

$$\iint\limits_S Z_n^m(r,\theta)Z_{n'}^{m'}(r,\theta)d^2r = \frac{\pi}{n+1}\delta_{nn'}\delta_{mm'}. \tag{6.46}$$

(2) In the unit-circle domain, the root-mean-square value of the C-Zernike polynomial is 1:

$$\sqrt{\frac{\iint\limits_S \left(Z_n^m(r,\theta)\right)^2 rdrd\theta}{\iint\limits_S rdrd\theta}} = 1. \tag{6.47}$$

The first 15 wavefronts were modeled and simulated according to the C-Zernike polynomial expression, as shown in Fig. 6.7.

6.2.1.2 Chebyshev Polynomials

In research on adaptive-optics systems, the C-Zernike polynomial is the most common wavefront-description method. This is mainly owing to the characteristics of the C-Zernike polynomial itself. First, its unique orthogonality facilitates wavefront analysis and application in circular-aperture optical systems. Second, there is a linear relationship between it and the classic Seidel optical distortion. However, in an optical system with a square aperture, the C-Zernike polynomial is directly used on the square substrate, which loses its original orthogonality.

The Chebyshev and Legendre polynomials can be used to describe the wavefront distortion in a square aperture. To simplify the description of wavefront distortion in a square aperture, a Cartesian coordinate system is used. Considering the wavefront distortion described by the Chebyshev polynomial, the wavefront distortion generated by the signal beam after passing through atmospheric turbulence can be expressed as [27]

$$\Delta\varphi(x,y) = \sum_{i=1}^{I} b_i C_i(x,y), \tag{6.48}$$

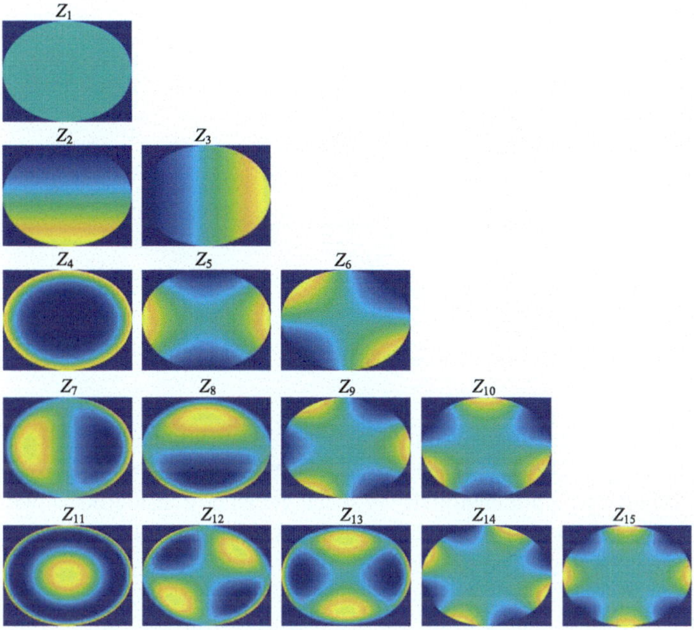

Fig. 6.7 First 15-order C-Zernike polynomial-surface diagrams

where b_i represents the ith Chebyshev polynomial coefficient, $C_i(x, y)$ represents the ith Chebyshev polynomial, I represents the number of Chebyshev polynomial items used, $x = r\cos\theta$ and $y = r\sin\theta$. The Chebyshev polynomials $C_i(x, y)$ over the orthonormal x and y variables in a square aperture are expressed as

$$C_i(x, y) = T_n(x)T_m(y), \tag{6.49}$$

where i is a polynomial-sort index starting with $i = 1$, and n and m are nonnegative integers. Let $T_0(x) = 1$ and $T_1(x) = x$. The orthonormal form $T_n(x)$ of the Chebyshev polynomial can be obtained recursively, as follows:

$$T_{n+1}(x) = 2xT_n(x) - T_{n-1}(x). \tag{6.50}$$

The product of any two Chebyshev polynomial functions in the interval $x \in [-1, 1]$ satisfies the following relationship:

$$\int_{-1}^{1} \frac{1}{\sqrt{1-x^2}} T_n(x)T_m(x)dx = \begin{cases} 0 & n \neq m \\ \pi & n = m = 1 \\ \frac{\pi}{2} & n = m \neq 1 \end{cases}. \tag{6.51}$$

Table 6.3 First 16-order Chebyshev polynomials

Number of items i	$C_i(x, y) = T_n(x)T_m(y)$	Number of items i	$C_i(x, y) = T_n(x)T_m(y)$
1	1	9	$2xy^2 - x$
2	x	10	$4y^3 - 3y$
3	y	11	$8x^4 - 8x^2 + 1$
4	$2x^2 - 1$	12	$4x^3y - 3xy$
5	xy	13	$(2x^2 - 1)(2y^2 - 1)$
6	$2y^2 - 1$	14	$4xy^3 - 3xy$
7	$4x^3 - 3x$	15	$8y^4 - 8y^2 + 1$
8	$2x^2y - y$	16	$16x^5 - 20x^3 + 5x$

In Eq. (6.51), after normalization, the function is orthogonal in the square area of $x \in [-1, 1]$ and $y \in [-1, 1]$. However, because the wavefront information of the beam is distributed two-dimensionally in space, it can be considered as two independent processes in the x and y directions. Thus, the orthogonality of the Chebyshev polynomials can be described as

$$\int_{-1}^{1}\int_{-1}^{1} C_i(x, y)C_{i'}(x, y)\frac{1}{\sqrt{1-x^2}\sqrt{1-y^2}}dxdy = \begin{cases} 0 \ i \neq i' \\ U \ i = i' \end{cases}, \tag{6.52}$$

where U is a constant normalized by the Chebyshev polynomial, expressed as

$$U = \begin{cases} \pi^2 \ n = m = 0 \\ \frac{\pi^2}{4} \ n = m \neq 0 \\ \frac{\pi^2}{2} \quad \text{else} \end{cases}. \tag{6.53}$$

The spatial distribution of the wavefront is represented by independent Chebyshev polynomials in the x and y directions. Table 6.3 gives the first 16-order normalized Chebyshev polynomials:

6.2.1.3 Legendre Polynomials

The wavefront distortion described by the Legendre polynomial is introduced below. The wavefront distortion generated by the signal beam in a square-aperture optical system after passing through atmospheric turbulence can be further expressed as

$$\Delta\varphi(x, y) = \sum_{i=1}^{L} e_i L_i(x, y), \tag{6.54}$$

where e_i represents the coefficient of the ith Legendre polynomial, $L_i(x, y)$ is the ith Legendre polynomial, and L is the number of Legendre polynomial terms used. The Legendre polynomials of the orthonormal x and y variables on the square aperture can be expressed as

$$L_i(x, y) = Q_l(x)Q_m(y), \tag{6.55}$$

where i is a polynomial-sort index starting with $i = 1$, and l and m are positive integers (including zero), where $n = l + m$. The orthonormal form $Q_n(x)$ of the Legendre polynomial yields

$$Q_n(x) = \sqrt{2n+1}P_n(x). \tag{6.56}$$

In Eq. (6.56), if $P_0(x) = 1$ and $P_1(x) = x$, the polynomial $P_n(x)$ can be obtained recursively, as follows:

$$P_{n+1}(x) = \frac{(2n+1)}{(n+1)}xP_n(x) - \frac{n}{(n+1)}P_{n-1}(x). \tag{6.57}$$

The product of any two Legendre polynomial functions in the interval $x \in [-1, 1]$ satisfies the following relationship [28]:

$$\int_{-1}^{1} P_n(x)P_m(x)dx = \frac{2}{2n+1}\delta_{nm}. \tag{6.58}$$

After normalization, the above formula is orthogonal in the square area of $x \in [-1, 1]$ and $y \in [-1, 1]$. However, because the wavefront information of the beam is distributed two-dimensionally in space, it can be regarded as two independent processes in the x and y directions. Thus, the orthogonality of Legendre polynomials can be described as

$$\frac{1}{4}\int_{-1}^{1}\int_{-1}^{1} L_i(x, y)L_{i'}(x, y)dxdy = \delta_{ii'}. \tag{6.59}$$

Two-dimensional Legendre polynomials, independent in the x and y directions, are used to represent the spatial distribution of the wavefront. Table 6.4 gives the first 16-order Legendre polynomials.

6.2.1.4 S-Zernike Polynomials

To be suitable for optical systems with square apertures, C-Zernike polynomials can be orthogonalized to obtain S-Zernike polynomials, which satisfy the square-domain

Table 6.4 Legendre polynomials of the first 16 orders

Number of items i	$L_i(x, y) = Q_l(x)\, Q_m(y)$	Number of items i	$L_i(x, y) = Q_l(x)\, Q_m(y)$
1	1	9	$\sqrt{15}/2(3xy^2 - x)$
2	$\sqrt{3}\, x$	10	$\sqrt{7}/2(5y^3 - 3y)$
3	$\sqrt{3}y$	11	$3/8(35x^4 - 30x^2 + 3)$
4	$\sqrt{5}/2(3x^2 - 1)$	12	$\sqrt{21}/2(5x^3y - 3xy)$
5	$3xy$	13	$5/4(3x^2 - 1)\,(3y^2 - 1)$
6	$\sqrt{5}/2(3y^2 - 1)$	14	$\sqrt{21}/2(5xy^3 - 3xy)$
7	$\sqrt{7}/2(5x^3 - 3x)$	15	$3/8(35y^4 - 30y^2 + 3)$
8	$\sqrt{15}/2(3x^2y - y)$	16	$\sqrt{11}/8(63x^5 - 70x^3 + 15x)$

orthogonality. Therefore, C-Zernike polynomials should first be orthogonalized in a square field. Using the Cartesian coordinate system, the M-term Zernike polynomial of a signal-beam wavefront can be expressed as

$$\Delta\varphi(x, y) = \sum_{i=1}^{M} a_i Z_i(x, y), \tag{6.60}$$

where $Z_i(x, y)$ is the ith Zernike polynomial, and a_i is the coefficient of the ith Zernike polynomial. To establish the orthonormal basis functions in the square area, a Gram–Schmidt orthogonal transformation is performed on the polynomial set $\{Z_i(x, y)\}$, and a new orthogonal-basis polynomial $\{S_i(x, y)\}$ is obtained.

$$S_1(x, y) = \frac{Z_1(x, y)}{\|Z_1(x, y)\|} \tag{6.61}$$

$$S_2(x, y) = \frac{Z_2(x, y) - \langle Z_2(x, y), S_1(x, y)\rangle S_1(x, y)}{\|Z_2(x, y) - \langle Z_2(x, y), S_1(x, y)\rangle S_1(x, y)\|} \tag{6.62}$$

$$S_m(x, y) = \frac{Z_m(x, y) - \sum\limits_{i=1}^{m-1} \langle Z_m(x, y), S_i(x, y)\rangle S_i(x, y)}{\left\| Z_m(x, y) - \sum\limits_{i=1}^{m-1} \langle Z_m(x, y), S_i(x, y)\rangle S_i(x, y) \right\|}, \quad m = 2, 3, \ldots, M \tag{6.63}$$

In Eqs. (6.61) to (6.63), $\langle\rangle$ represents the ensemble average, $\|\cdot\|$ is the two-norm of the vector, and the new orthogonal polynomial set $\{S_i(x, y)\}$ is obtained as the orthonormal base. Then, the wavefront on the square aperture is described by the S-Zernike polynomial as

$$\Delta\varphi(x, y) = \sum_{i=1}^{M} c_i S_i(x, y), \tag{6.64}$$

where c_i represents the coefficient of the ith S-Zernike polynomial, $S_i(x, y)$ is the ith S-Zernike polynomial, and M is the number of S-Zernike polynomial terms used. Using the Gram–Schmidt orthogonal transformation, the orthonormal S-Zernike polynomials on the x and y variables of the square aperture are defined. The ith S-Zernike polynomial can be expressed as [29]

$$S_i(x, y) = \sum_{j=1}^{i} M_{ij} Z_j(x, y), \tag{6.65}$$

where the M_{ij} function is an orthogonal-transformation matrix, $Z_j(x, y)$ is a Zernike polynomial, and the orthogonal transformation function can be expressed as

$$M_{ii}(x, y) = \frac{1}{\left[\frac{1}{Q}\iint\limits_{S}\left(\sum\limits_{i*=1}^{i-1} c_{i,i*}S_{i*} + Z_i\right)^2 dxdy\right]^{1/2}}, \tag{6.66}$$

where Q is the area of the square and i is a positive integer. The S-Zernike polynomial coefficients $c_{i,i*}$ are expressed as follows:

$$c_{i,i*} = -\frac{1}{Q}\iint\limits_{S} Z_i S_{i*} dxdy. \tag{6.67}$$

The lower-order wavefront distortion term after orthogonalization still has the basic characteristics of the C-Zernike-polynomial model method. Moreover, the Seidel optical aberration can be converted to its wavefront coefficient, which is used to analyze the wavefront distortion introduced in the atmospheric turbulent transmission. Table 6.5 gives the first 16-order normalized S-Zernike polynomials.

Considering the Chebyshev, Legendre, and S-Zernike polynomials describing the wavefront distortion in the square-aperture optical system, the first 15 wavefront types described by S-Zernike polynomials were simulated, as shown in Fig. 6.8.

The mode-method description can effectively decompose and reflect the wavefront information, and extract and analyze wavefront distortions separately after a mode expansion. More importantly, the orthogonalization process of the model method ensures that the distribution of the wavefront coefficients of each order does not change with the higher-order wavefront distortion; thus, it does not affect the calculation of the specific-order wavefront distortion.

Table 6.5 First 16 order S-Zernike polynomials

Number of items i	$S_i(x, y)$	Number of items i	$S_i(x, y)$
1	1	9	$\sqrt{5/31}(27x^2y - 35y^3 + 6y)$
2	$\sqrt{6}\,x$	10	$\sqrt{5/31}(35x^3 - 27xy^2 - 6x)$
3	$\sqrt{6}\,y$	11	$1/(2\sqrt{67})(315x^4 + 630x^2y^2 + 315y^4 - 240x^2 - 240y^2 + 31)$
4	$\sqrt{5/2}(3x^2 + 3y^2 - 1)$	12	$15/(2\sqrt{2})(x^2 - y^2)(7x^2 + 7y^2 - 3)$
5	$6xy$	13	$\sqrt{42}(5x^3y + 5xy^3 - 3xy)$
6	$3\sqrt{5/2}(x^2 - y^2)$	14	$3/(4\sqrt{134})(490x^4 - 360x^2y^2 + 490y^4 - 150x^2 - 150y^2 + 11)$
7	$\sqrt{21/31}(15x^2y + 15y^3 - 7y)$	15	$5\sqrt{42}(x^3y - xy^3)$
8	$\sqrt{21/31}(15x^3 + 15xy^2 - 7x)$	16	$\sqrt{55/1966}(315x^5 + 630x^3y^2 + 315xy^4 - 280x^3 - 324xy^2 + 57x)$

6.2.2 Beam-Field Distribution Model Under the Effect of Wavefront Distortion in a Circular Aperture

Because of the position and working nature of the transmitting/receiving antenna, a optical wireless coherent communication system will be affected by environmental factors such as atmospheric turbulence, which changes the shape and position of the received beam spot on the receiving end. Therefore, wavefront distortion reduces the imaging performance of the optical system, and the parameters involved in the optical design, such as the optical structure and aperture shape, affect the deviation of the optical system from predetermined imaging-quality requirements.

The beam-field distribution model of the distorted spot under different optical-system apertures was deduced, a simulation analysis was carried out, based on the theoretical model, and the influence of wavefront distortion on the beam-field distribution at the receiving-end surface of the optical wireless coherent communication system was studied.

To establish the beam-field distribution model of a distorted beam wavefront and simulate the transmission of the signal beam in atmospheric turbulence, the wave theory was adopted, and the beam-transmission process satisfied the Fresnel diffraction condition [30]. The complex amplitude distribution of the optical field at the receiving-end plane is

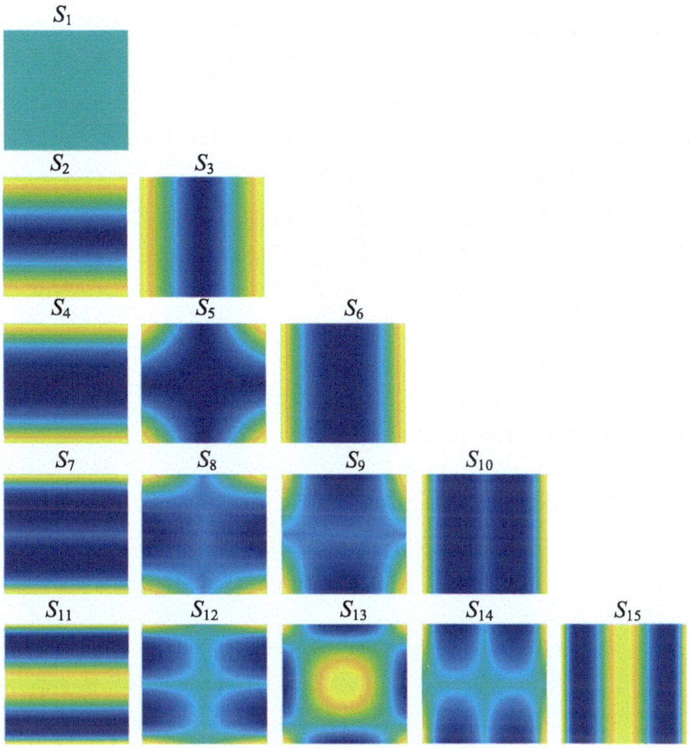

Fig. 6.8 First 15-order S-Zernike polynomial-surface diagrams

$$U(x, y, z) = -\frac{j}{2\lambda} \int \int_{\Sigma} U(x', y', 0) e^{jkl} \frac{(1 + \cos\theta')}{l} ds', \tag{6.68}$$

where $U(x', y', 0)$ is the complex amplitude of the initial beam field, θ' is the diffraction angle, and $k = 2\pi/\lambda$, where λ is the wavelength and l is the distance between the receiving and initial ends, which can be expressed as

$$l = \sqrt{(x - x_1)^2 + (y - y_1)^2 + z^2}, \tag{6.69}$$

where $(1 + \cos\theta')/2$ in Eq. (6.68) is used as the tilt factor. $\cos\theta'$ can be approximately expressed as

$$\cos\theta' \cong \frac{z}{l}. \tag{6.70}$$

When the signal beam is disturbed by environmental factors, such as atmospheric turbulence, the wavefront of the diffracted beam passing through the diffraction aperture is distorted. The influence of the phase wavefront distortion $\Delta\varphi$ on the transmission beam can be multiplied by a distortion phase-factor $e^{j\Delta\varphi}$ beam-field distribution function, which is expressed as

$$U(x, y, z) = -\frac{j}{2\lambda}\iint_{\Sigma} U(x', y', 0)e^{jk\Delta\varphi}e^{jkl}\frac{(1+\cos\theta')}{l}ds'. \tag{6.71}$$

After the distorted beam is diffracted and passes through the lens surface, the complex amplitude distribution of the beam field at the exit vertex of the lens is expressed as

$$U(x, y, z) = -\frac{j}{2\lambda}\iint_{\Sigma} U(x', y', 0)e^{jk\Delta\varphi}e^{jkR'}e^{jkl}\frac{(1+\cos\theta')}{l}ds'. \tag{6.72}$$

In Eq. (6.72), the radius of curvature R' of the lens can be expressed as

$$R' = (n-1)\sqrt{R_l^2 - (x^2 + y^2)} + d, \tag{6.73}$$

where n is the refractive index of the lens material, R_l is the curvature radius of the convex surface of the lens, and the central thickness of the lens is d. The beam-field distribution model under the influence of wavefront distortion is

$$I(x, y, z) = U(x, y, z) \cdot U^*(x, y, z) = |U(x, y, z)|^2. \tag{6.74}$$

The optical-field distribution on the receiving plane is an important factor that determines the performance of the receiving end of the optical system; it affects the mixing efficiency of the receiving-end system and directly affects the communication quality of the optical wireless coherent communication system. For simplicity, only the influence of a single wavefront distortion on the system transmission is discussed here.

Using the C-Zernike-polynomial mode method, the beam wavelength λ = 1550 nm, distance between the receiving end and the initial end $l = 1$ km, refractive index of the lens $n = 1.5062$, curvature radius of the convex surface of the lens R_l = 100 mm, and the lens-center thickness $d = 3$ mm. Figure 6.9 shows the simulation of the beam-field distribution under the influence of wavefront distortion in a circular-aperture optical system.

Using the C-Zernike polynomial to describe several primary wavefront distortions, the beam-field distribution at the receiving end passing through a circular-aperture optical system under the influence of a single primary wavefront distortion is shown in Fig. 6.9. Figure 6.9a shows the beam-field distribution of a signal beam received on the receiving-end surface, when the influence of wavefront distortion is not considered. Figure 6.9b–f reflect the optical-field distribution of the signal beam at the

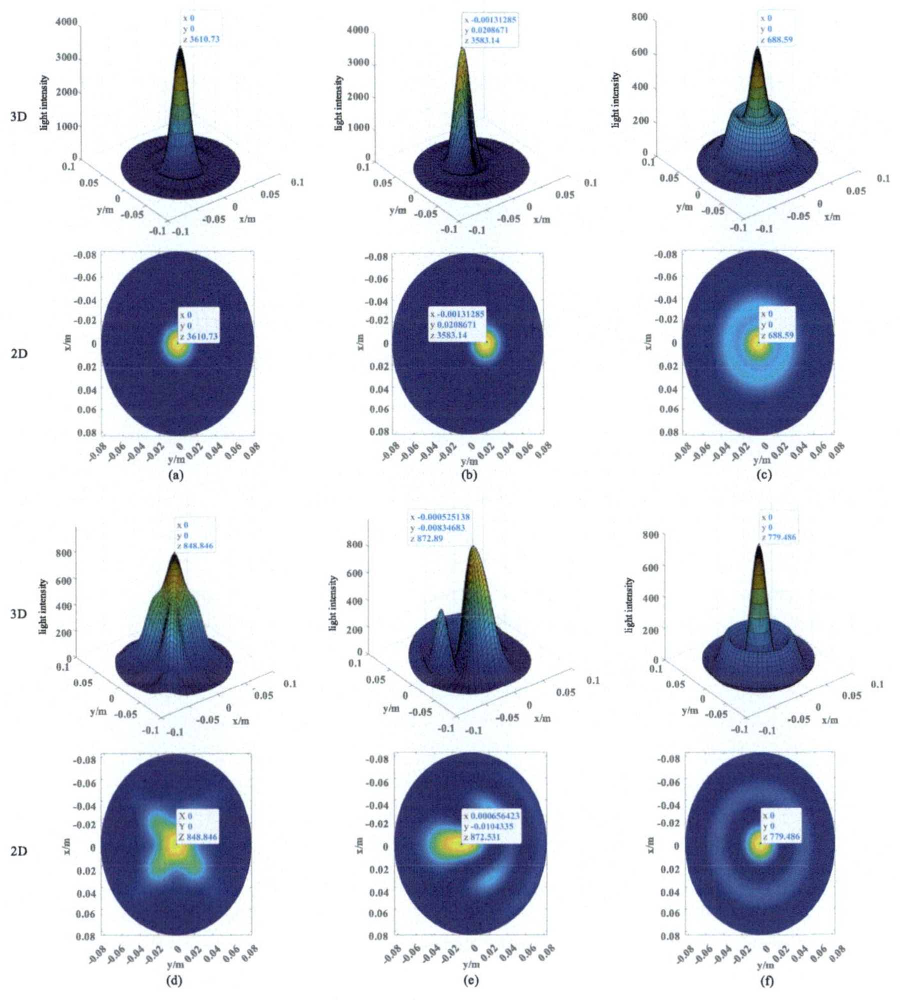

Fig. 6.9 Beam-field distribution model after introducing a single primary wavefront distortion into a circular aperture. **a** Beam-field spot influenced by ideal wavefront. Beam-field flare affected by **b** tilt term, **c** defocus term, **d** astigmatism term, **e** coma term, and **f** spherical aberration term

receiving end under the influence of wavefront distortions in the circular-aperture optical system.

It can be observed that, under the influence of a single primary wavefront distortion, the beam-field distribution of the signal beam at the receiving end changes. The introduction of a single primary wavefront distortion not only leads to an asymmetric beam-field distribution of the signal beam at the receiving end, but also affects the beam intensity to varying degrees. As shown in Fig. 6.9b, a tilt wavefront distortion causes the center position of the signal beam to shift at the receiving end, and the wavefront distortion of the coma term in Fig. 6.9e causes the beam-field spot to exhibit a nonuniform distribution.

It was concluded that the tilt and coma terms made the beam-field spot asymmetrical, and the defocus, astigmatism, and spherical-aberration terms caused the main-lobe energy of the spot to spread to the side lobe. Consequently, the matching degree of the optical field at the receiving end was reduced, which further affects the performance of the receiving-end system.

6.2.3 Beam-Field Distribution Model Under the Influence of Wavefront Distortion in a Square Aperture

Establishing an optical-field distribution model under the influence of wavefront distortion in a square area of an optical system requires further research. Some lasers use an optical system with a square aperture, and some optical components, such as lenses, diaphragms, and gratings, are also designed with a square aperture. Therefore, a study on the beam-transmission characteristics of wavefront distortion in a square-aperture optical system described by the S-Zernike polynomial mode method was conducted for the beam-field distribution model under the influence of wavefront distortion in a square aperture.

After the square-aperture laser sends out a signal beam, considering the diffraction of the square-aperture optical lens and atmospheric turbulence, the beam is received by the optical receiving antenna and reaches the focal plane of the receiving end. The beam-transmission structure in the square-aperture optical system is shown in Fig. 6.10. Σ is the diffraction aperture. The signal beam passes through the square aperture along the z-axis direction, and the incident beam passes through point (x', y') on the aperture Σ and arrives at point (x, y) along the z-axis direction.

The S-Zernike-polynomial mode method is used to describe the square-aperture optical system for fitting. The S-Zernike polynomial describes various primary wavefront distortions, and the simulation parameter settings are consistent with those in the circular-aperture optical system shown in Fig. 6.9.

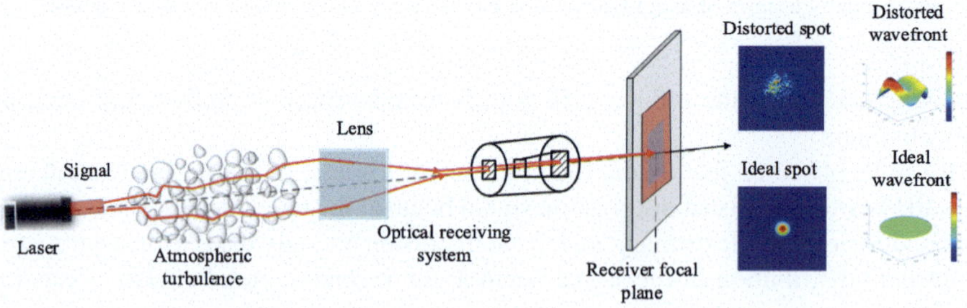

Fig. 6.10 Structure diagram of a beam transmission through a square aperture

Figure 6.11 shows the obtained beam-field distribution model under the influence of a single primary wavefront distortion. Figure 6.11 reflects the beam-field distribution of the distorted signal beam at the receiving end through the square aperture of the optical system. Figure 6.11a shows the ideal optical-field distribution of the signal beam detected at the receiving end.

As shown in Fig. 6.11b, the wavefront-distortion tilt term causes the position of the transmission-signal spot to shift as a whole, which means that the transmission direction of the signal beam has changed; however, the beam-field intensity obtained

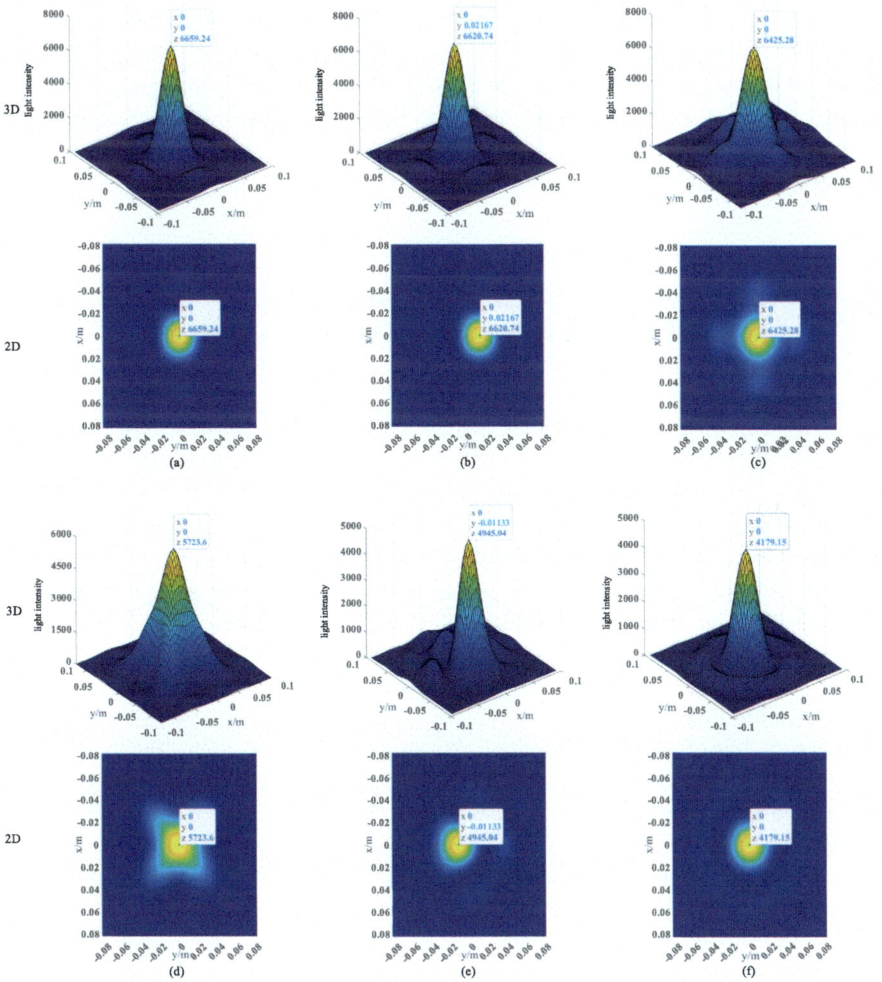

Fig. 6.11 Beam-field distribution model after introducing a single primary wavefront distortion into a square aperture. **a** Beam-field spot influenced by ideal wavefront. Beam-field flare affected by **b** tilt term, **c** defocus term, **d** astigmatism term, **e** coma term, and **f** spherical-aberration term

by the receiving end has no obvious attenuation. In Figs. 6.11c–f, both the defocus and astigmatism items lead to the symmetry degradation of the beam-field spot of the signal beam, the spherical-aberration item causes the signal beam spot to disperse at the receiving end, and the coma and spherical-aberration items cause a large attenuation of the beam-field intensity at the receiving end.

In summary, the tilt and coma wavefront distortions will lead to a non-uniform distribution of the beam-field spots of the signal beam. Although other wavefront distortions will affect the beam-field distribution of the signal beam, the redistributed beam field is still uniform. The coma and spherical aberrations greatly attenuate the signal-beam intensity at the receiving end.

Finally, under the same parameters, the beam-field intensity obtained by the optical system with a square aperture was greater than that obtained with a traditional circular aperture at the receiving end. This is because an optical system with a square aperture is more conducive to the synthesis layout of the signal beam and can capture more signal-beam energy, improving the performance of the optical wireless coherent-detection system.

6.3 Effect of Wavefront Distortion on the Performance of a Optical Wireless Coherent-Detection System

6.3.1 *Effects of Misalignment Caused by Wavefront Distortion*

In a optical wireless coherent communication system under the influence of atmospheric turbulence, the receiving end must have a strong ability to align and capture data, and the optical-field distribution must be uniform. However, environmental factors, such as atmospheric turbulence, lead to poor alignment of the signal beam reaching the receiving end. In severe cases, this will lead to a decrease in the stability of the optical wireless coherent communication link and even interrupt the communication link.

Therefore, it is of practical significance to study the impact of wavefront distortion on the alignment performance to effectively reduce the acquisition and tracking deviation in actual link construction, which is beneficial to the establishment and maintenance of optical wireless coherent communication links.

When a optical wireless coherent communication system is affected by atmospheric turbulence, the wavefront of the transmitted signal beam is distorted. When tilt and coma distortion components are present in the wavefront distortion, the position of the spot centroid shifts.

The wavefront distortion produced by an optical system can be quantitatively described by the optical-path difference between beam rays. The alignment deviation caused by wavefront distortion can be defined as $\theta\,(\theta_{\mathrm{Xmax}}, \theta_{\mathrm{Ymax}})$ of the emission-axis deflection, with or without wavefront distortion, taking the spot center as the origin.

As shown in Fig. 6.12, P1(x_1, y_1) is the peak beam-intensity position of the signal beam reaching the receiving end when there is no wavefront distortion. When the signal beam reaches the receiving end after passing through atmospheric turbulence, the peak beam intensity of the signal beam shifts to $P_2(x_2, y_2)$.

The average alignment deviation $\Delta\theta$ is defined as the optical-path difference between the peak signal-intensity positions, $P_1(x_1, y_1)$ and $P_2(x_2, y_2)$, at the receiving end. The average alignment deviation caused by the wavefront distortion can be expressed as

$$\Delta\theta=\sqrt{\theta_{\text{Xmax}}^2+\theta_{\text{Ymax}}^2}, \tag{6.75}$$

where θ_{Xmax} and $\theta_{\cdot\text{Ymax}}$ are the deviation components of the misalignment angle in the x and y directions, respectively, and can be expressed as

$$\theta_{\text{Xmax}}=\frac{x_2-x_1}{L} \tag{6.76}$$

$$\theta_{\text{Ymax}}=\frac{y_2-y_1}{L}, \tag{6.77}$$

where L is the transmission distance. The center position of the signal beam can be obtained from the centroid position of the first-order moment of the beam-intensity distribution on the x-axis and y-axis on the beam-receiving plane:

$$x_1=\frac{\int_{-\infty}^{\infty}\int_{-\infty}^{\infty}xI_0(x,y)dxdy}{\int_{-\infty}^{\infty}\int_{-\infty}^{\infty}I_0(x,y)dxdy} \tag{6.78}$$

$$x_2=\frac{\int_{-\infty}^{\infty}\int_{-\infty}^{\infty}xI_{re}(x,y)dxdy}{\int_{-\infty}^{\infty}\int_{-\infty}^{\infty}I_{re}(x,y)dxdy} \tag{6.79}$$

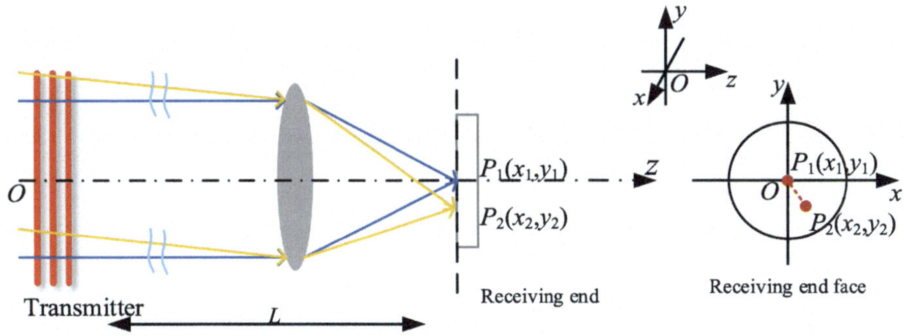

Fig. 6.12 Schematic diagram of alignment deviation

$$y_1 = \frac{\int_{-\infty}^{\infty}\int_{-\infty}^{\infty} yI_0(x, y)dxdy}{\int_{-\infty}^{\infty}\int_{-\infty}^{\infty} I_0(x, y)dxdy} \tag{6.80}$$

$$y_2 = \frac{\int_{-\infty}^{\infty}\int_{-\infty}^{\infty} yI_{re}(x, y)dxdy}{\int_{-\infty}^{\infty}\int_{-\infty}^{\infty} I_{re}(x, y)dxdy}. \tag{6.81}$$

In Eqs. (6.78)–(6.81), $I_0(x, y)$ is the beam-intensity distribution of the signal beam on the receiving-end surface under ideal conditions, and $I_{re}(x, y)$ is the beam-intensity distribution on the receiving-end surface when wavefront distortion occurs.

In a optical wireless coherent-detection system, the centroid algorithm is used to obtain the centroid-position coordinates of the collected signal beam affected by wavefront distortion. The position of the alignment deviation of the signal beam is calculated by comparison with the centroid-position coordinates of the ideal signal beam. The tracking algorithm is then used to adjust the signal beam for alignment so that the signal beam received by the optical wireless coherent-detection system has a beam-field distribution that is as uniform as possible and a high beam-intensity amplitude.

Simulation analyses were performed on the influence of the alignment deviation caused by each order of wavefront distortion on the system. Combining Eqs. (6.75)–(6.77), the main simulation parameters are set as follows: transmission distance $L = 1$ km, laser wavelength $\lambda = 1550$ nm, receiving-antenna aperture $D = 50$ mm, focal length of the optical system $f = 49.3876$ mm, refractive index of the lens $n = 1.5062$, and thickness of the center of the lens $d = 3$ mm. The x-direction, y-direction, and average alignment-deviation simulation results caused by each wavefront distortion are shown in Figs. 6.13–6.15.

Figure 6.13 shows the change curve of the alignment deviation in the x direction caused by lower-order wavefront distortion. The results show that the alignment

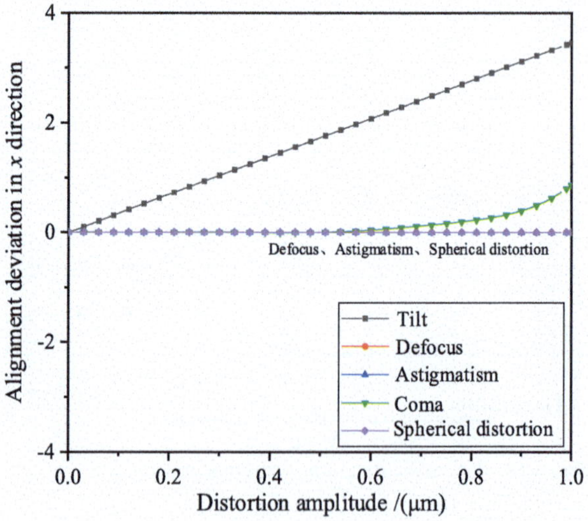

Fig. 6.13 Change curve of alignment deviation in the x direction caused by wavefront distortion

deviation in the x direction caused by the wavefront-distortion tilt term (Z_2) is more obvious. With an increase in the distortion amplitude, the alignment deviation in the x direction caused by the coma aberration term (Z_8) fluctuates slightly. The defocus, astigmatism, and spherical aberration terms (Z_4, Z_6, and Z_{11}) do not cause a misalignment in the x direction.

Figure 6.14 shows the change curve of the alignment deviation in the y direction caused by lower-order wavefront distortion. The simulation shows that as the distortion amplitude increases, the alignment deviation in the y direction caused by the wavefront-distortion tilt and coma items (Z_3, Z_7) becomes more obvious, whereas the wavefront-distortion defocus, astigmatism, and spherical-aberration items (Z_4, Z_5, and Z_{11}) do not cause misalignments in the y direction.

Figure 6.15 shows the change curve of the average alignment deviation caused by lower-order wavefront distortion. It can be observed from the simulation that with a gradual increase in the distortion amplitude, the alignment deviations caused by the wavefront-distortion tilt and coma also increase; the defocus, astigmatism, and spherical-aberration items do not cause alignment deviations.

Combined with the distorted beam-field distribution model, after the introduction of a single primary wavefront distortion in Sect. 6.2, it is concluded that the tilt and coma terms make the signal beam-field spot distribution asymmetric, and the defocus, astigmatism, and spherical-aberration items diffuse the energy of the main lobe of the spot to the side lobes, weakening the intensity of the central beam; however, the central position of the spot does not change.

In summary, the misalignment caused by symmetrically distributed wavefront distortions, such as defocus, astigmatism, and spherical aberration, is negligible. When there are asymmetrically distributed wavefront distortions, such as tilt and coma terms, the introduced alignment deviation is large. As the distortion amplitude increases, the alignment deviation gradually increases, and wavefront correction is

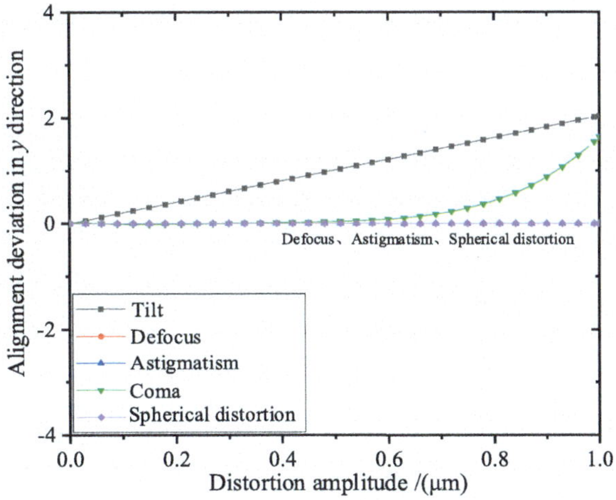

Fig. 6.14 Change curve of alignment deviation in the y direction caused by wavefront distortion

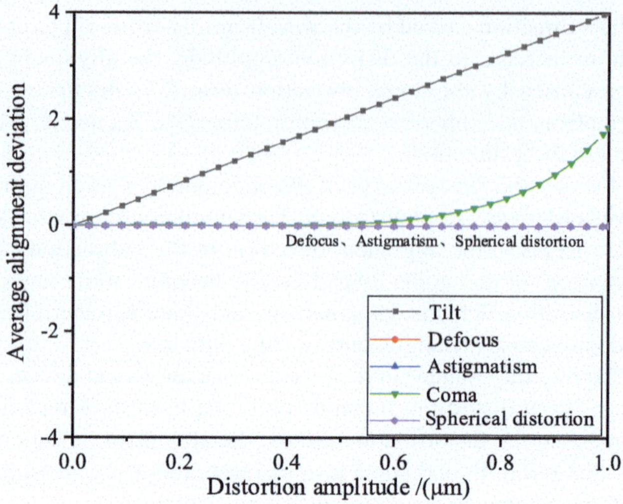

Fig. 6.15 Change curve of the average alignment deviation caused by wavefront distortion

required. Here, the alignment deviation caused by wavefront distortion can be eliminated by an algorithm, such that more-accurate signal-beam position information can be obtained.

6.3.2 Effect of Wavefront Distortion on the Mixing Efficiency

In a optical wireless coherent communication system, the existence of a turbulent channel distorts the wavefront of the signal beam carrying information, which causes a received signal beam-spot dispersion and wavefront phase mismatch. Furthermore, the mixing efficiency of the signal beam and local-oscillator beam in the optical wireless coherent-detection system at the receiving end is reduced, resulting in a decrease in the detection sensitivity of the system and an increase in the bit-error rate. Wavefront distortions caused by atmospheric turbulence and other environmental factors have become the main factors affecting the performance of optical wireless coherent-detection systems at the receiving ends.

After analyzing the effect of the alignment deviation at the receiving end caused by each wavefront distortion, the effect on the performance of the optical wireless coherent-detection system at the receiving end was analyzed. The main simulation conditions are set as follows: receiving aperture $D = 105$ mm, photodetector quantum efficiency $= \eta$ 0.8, Planck's constant $h = 6.63 \times 10^{-34}$ J·s, beam speed $c = 2.99 \times 10^{8}$ m/s, wavelength $= \lambda$ 1550 nm, signal optical power $P_s = 0.36$ μW, local-oscillator optical power $P_l = 0.36$ mW, effective noise bandwidth $\Delta f = 200$ MHz, and zenith angle $= \psi \pi/3$.

When the transmission distance L is 1 km, as the transmission distance increases, the relationship between the heterodyne-detection mixing efficiency and wavefront distortion in the x direction is as shown in Fig. 6.16. The relationship between the changes in the wavefront distortion in the y direction is as shown in Fig. 6.17.

As shown in Fig. 6.16, as the transmission distance L increases, the overall mixing efficiency of the system under the influence of wavefront distortions tends to decrease. Among them, the wavefront distortion of the tilt term in the x direction has the

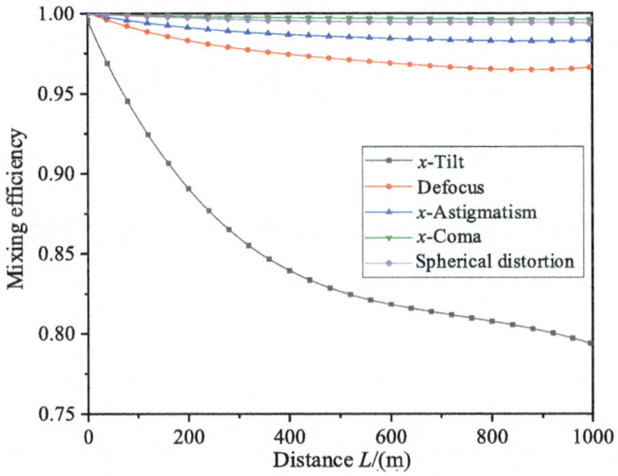

Fig. 6.16 Mixing efficiency variation with the transmission distance L under wavefront distortion in the x direction

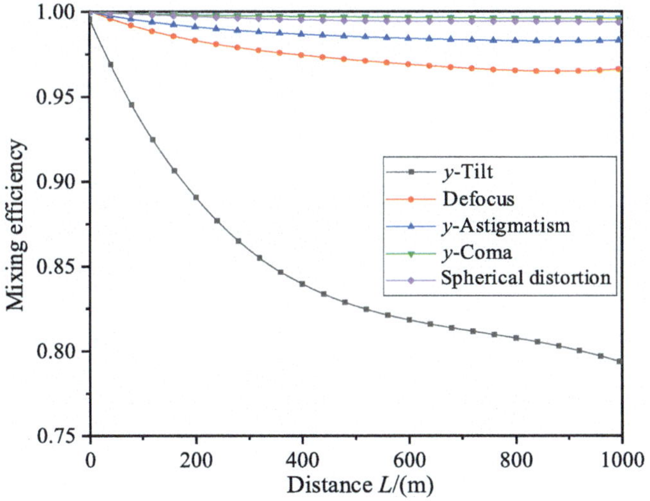

Fig. 6.17 Mixing-efficiency variation with the transmission distance L under wavefront distortion in the y direction

most obvious impact on the mixing efficiency of the system, because the overall tilt component accounts for 86.99% of the entire wavefront distortion. The wavefront distortion of the defocus term also has a serious impact on the mixing efficiency.

Under the influence of wavefront distortion in the y direction, the variation in the system mixing efficiency with the transmission distance L is shown in Fig. 6.17. Because only directional differences exist in the wavefront distortion in the x and y directions, the influence on the mixing efficiency of the optical wireless coherent-detection system at the receiving end is the same. When the transmission distance is 1 km, the mixing efficiency decreases by 20.6% under the influence of the y-direction tilt term. Therefore, it is necessary to separately correct and compensate for the tilt component of the wavefront distortion to improve the coherent-detection performance in actual wavefront correction.

6.3.3 Correction–Compensation Algorithm for Higher-Order Wavefront Distortion

The wavefront distortion caused by atmospheric turbulence is expanded using the Zernike polynomial model, in which the primary wavefront distortion mainly includes tilt (items 2 and 3), defocus (item 4), astigmatism (items 5 and 6), coma (items 7 and 8) and spherical aberration (item 11). The components of spherical aberration and above are called higher-order wavefront distortion. They will cause the signal beam spot to expand and produce an uneven distribution of the beam energy, which affects the mixing efficiency of the optical wireless coherent-detection system. Table 6.6 gives the primary wavefront distortions represented by the Zernike polynomials.

Table 6.6 Primary wavefront distortions represented by Zernike polynomials

Number of items i	$Z_i(r, \theta)$	Name
1	1	Piston
2	$2r\cos\theta$	x-tilt
3	$2r\sin\theta$	y-tilt
4	$\sqrt{3}(2r^2 - 1)$	Defocus
5	$\sqrt{6}\,r^2\sin 2\theta$	Oblique astigmatism
6	$\sqrt{6}\,r^2\cos 2\theta$	Vertical astigmatism
7	$\sqrt{8}(3r^3 - 2r)\sin\theta$	Vertical coma
8	$\sqrt{8}(3r^3 - 2r)\cos\theta$	Horizontal coma
9	$\sqrt{8}\,r^3\sin 3\theta$	Vertical trefoil
10	$\sqrt{8}\,r^3\cos 3\theta$	Oblique trefoil
11	$\sqrt{5}(6r^4 - 6r^2 + 1)$	Primary spherical

Equation (6.42) is the expansion of the wavefront distortion under a phase mismatch caused by atmospheric turbulence. The corresponding matrix form is expressed as

$$\Delta\varphi(r,\theta) = AZ^T, \tag{6.82}$$

where A is the coefficient vector of the Zernike polynomial, $Z = [Z_1(r, \theta), \ldots, Z_n(r, \theta)]$ is the polynomial vector formed by each Zernike polynomial, and Z^T is the transposition of the polynomial vector. According to the performance requirements of the optical wireless coherent-detection system, after the signal-beam wavefront distortion is corrected, wavefront $\Delta\varphi(r, \theta)$ is expressed as

$$\Delta\varphi(r,\theta)+\varphi_{dm}(r,\theta) = \varphi(r,\theta), \tag{6.83}$$

where $\varphi_{dm}(r, \theta)$ is the wavefront corrected and compensated for by a piezoelectric deformable mirror. According to the wavefront-correction technology of the adaptive-optics (AO) system, the closer the wavefront $\varphi(r, \theta)$ is to zero, the better the wavefront correction effect is; that is, the piezoelectric deformable mirror fully compensates for the signal-beam wavefront under ideal conditions:

$$\Delta\varphi(r,\theta) = -\varphi_{dm}(r,\theta). \tag{6.84}$$

To correct and compensate for the wavefront distortion, the piezoelectric deformable mirror adopts a proportional–integral–differential (PID) algorithm to control the voltage of the piezoelectric actuator that eliminates the hysteresis effect, thereby correcting the wavefront distortion of the actual optical system [31]. Wavefront $\varphi_{dm}(r, \theta)$ corrected by the piezoelectric deformable mirror is expressed as follows:

$$\varphi_{dm}(r,\theta) = \sum_{j=1}^{Y} V_j R_j(r,\theta), \tag{6.85}$$

where V_j is the voltage of the jth actuator, Y is a positive integer, and $R_j(r, \theta)$ is the wavefront influence function expanded as follows:

$$R_j(r,\theta) = \sum_{i=1}^{N} x_{ji} Z_i(r,\theta), \tag{6.86}$$

where x_{ji} is the Zernike-polynomial coefficient expanded using the wavefront-influence function of the piezoelectric deformable mirror. Substituting Eq. (6.86) into (6.85), the wavefront $\varphi_{dm}(r, \theta)$ corrected by the piezoelectric deformable mirror can be expressed as

$$\varphi_{dm}(r,\theta) = \sum_{i=1}^{N}\sum_{j=1}^{Y} V_j x_{ji} Z_i(r,\theta). \tag{6.87}$$

The corresponding matrix form is expressed as

$$\varphi_{dm}(r,\theta) = VXZ^T, \tag{6.88}$$

where $V = [V_1, \ldots, V_j]$ represents the control-voltage vector of the piezoelectric deformable mirror and X is the Zernike-polynomial coefficient matrix of the piezoelectric deformable-mirror influence function. The coefficient matrix X can be obtained by measurement, the voltage V_j is applied to the piezoelectric actuators individually, and the other actuator voltages are set to zero so that the coefficients of the influence function on a single actuator can be obtained. Then, according to Eqs. (6.88) and (6.82), we obtain

$$V = -X^+ A, \tag{6.89}$$

where X^+ is the generalized inverse matrix X of the influence-function coefficient matrix of the piezoelectric deformable mirror.

Here, the receiving end of the optical system converts the signal beam into an electrical signal; that is, the wavefront sensor outputs the wavefront error and drives the piezoelectric deformable mirror in the form of a voltage to generate the required wavefront. Then, the relationship between the wavefront error detected by the wavefront sensor and the piezoelectric deformable-mirror actuator voltage vector V and response-function vector R can be expressed as

$$S = R \cdot V. \tag{6.90}$$

The PID-algorithm correction principle of the piezoelectric deformable mirror is given below.

Figure 6.18 shows the PID-algorithm correction principle of the piezoelectric deformable mirror. The PID algorithm controls the signal error slope $e(t) = r(t) - y(t)$, where $r(t)$ is the expected slope and $y(t)$ is the detected slope. The proportional algorithm performs a simple and fast adjustment response to the current error slope $e(t)$ of the system; the integral algorithm eliminates the system's static error slope $\int e(t)dt$, thereby enabling the system to balance itself. The differential algorithm is mainly responsible for improving the dynamic characteristics of the error slope and controlling it according to the variation trend $de(t)/dt$ of the error slope.

The output is an iterative closed-loop feedback control with a corrected error slope $u(t)$, which is applied to the piezoelectric deformable mirror to eliminate wavefront distortion. The output control-signal error $u(t)$ loaded onto the piezoelectric deformable mirror can be expressed as

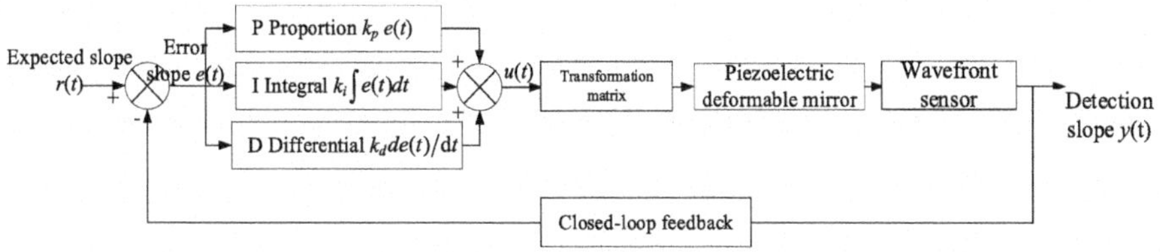

Fig. 6.18 PID-algorithm correction principle

$$u(t) = k_p\left[e(t) + \frac{1}{T_i}\int_0 e(t)dt + T_d\frac{de(t)}{dt}\right], \tag{6.91}$$

where k_p is the proportional coefficient, T_i is the integral time constant, and T_d is the differential time constant. The discrete form $u(w)$ of the wth sampling is expressed as

$$u(w) = k_p e(w) + k_i\sum_{i=0}^{w} e(i) + k_d[e(w) - e(w-1)], \tag{6.92}$$

where the integral k_i and differential k_d are the control coefficients of the PID algorithm, w is the sampling number, $e(w)$ is the discrete error corresponding to $e(t)$, and $e(w-1)$ is the amount of discrete error at sampling time $w-1$. The wavefront error detected by the wavefront sensor at the wth acquisition moment is converted into the driving voltage:

$$V_{dm}(w) = R^{+} \times S_{pid}(w), \tag{6.93}$$

where R^{+} is the generalized inverse matrix of the influence-function matrix R of the piezoelectric deformable mirror, $S_{pid}(w)$ is the corrected wavefront-slope error at the wth acquisition time, and $V_{dm}(w)$ is the corresponding driving voltage. The wavefront error corrected by the PID control algorithm can be expressed as [80]

$$S_{pid}(w) = k_p[S_e(w) - S_e(w-1)] + k_i S_e(w) + k_d[S_e(w) - S_e(w-1)], \tag{6.94}$$

where $S_e(w)$ is the wavefront error at time w. The piezoelectric deformable mirror is controlled by the PID algorithm to apply a positive voltage to each actuator, and an analog voltage input is used for feedback. When half of the maximum voltage is applied to each electrode, the piezoelectric deformable mirror has a flat surface and can be used for correcting the wavefront distortion. Local concave and convex states are generated when no voltage is applied or when the maximum voltage is applied.

By combining the wavefront-distortion correction technology of the adaptive-optics system with a optical wireless coherent-detection system, a piezoelectric deformable mirror is proposed as the receiving front-end wavefront corrector of the optical wireless coherent-detection system. The performance of the optical wireless coherent-detection system at the receiving end can be improved by correcting the higher-order wavefront distortion of the signal beam.

6.3.4 Simulation of Higher-Order Wavefront Distortion and Its Correction–Compensation Quantity

The signal beam corrects the front N-order wavefront distortion to further improve the performance of the optical wireless coherent-detection system in exchange for a small cost of system complexity. The main simulation settings are as follows: receiving aperture $D = 105$ mm, photodetector quantum efficiency $\eta = 0.8$, Planck's constant $h = 6.63 \times 10^{-34}$ J s, beam speed $c = 2.99 \times 10^8$ m/s, wavelength $\lambda = 1550$ nm, signal optical power $P_s = 0.36$ μW, local-oscillator optical power $P_l = 0.36$ mW, and effective noise bandwidth is $\Delta f = 200$ MHz.

The higher-order wavefront distortion of the signal beam is corrected and the relationship between the mixing efficiency of the heterodyne-detection system and the change in atmospheric-coherence length r_0 is obtained, as shown in Fig. 6.19.

It can be observed in Fig. 6.19 that with the increase of the wavefront correction order in heterodyne detection, the system mixing efficiency gradually increases. Compared with the system mixing efficiency when the wavefront distortion is not corrected, when the atmospheric-coherence length $r_0 = 0.18$ m, the first third-order wavefront distortion is corrected, the system mixing efficiency is increased from uncorrected 0.70 to 0.95, and a gain of 1.33 dB is obtained. When the atmospheric-coherence length $r_0 = 0.08$ m, after correcting the third-order wavefront distortion, the mixing efficiency of the system increases from 0.50 to 0.80, and the gain is 2.03 dB.

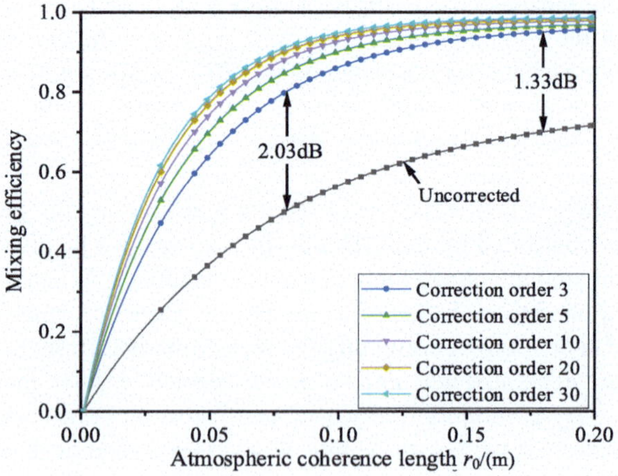

Fig. 6.19 Relationship between atmospheric-coherence length r_0 and mixing efficiency during heterodyne detection

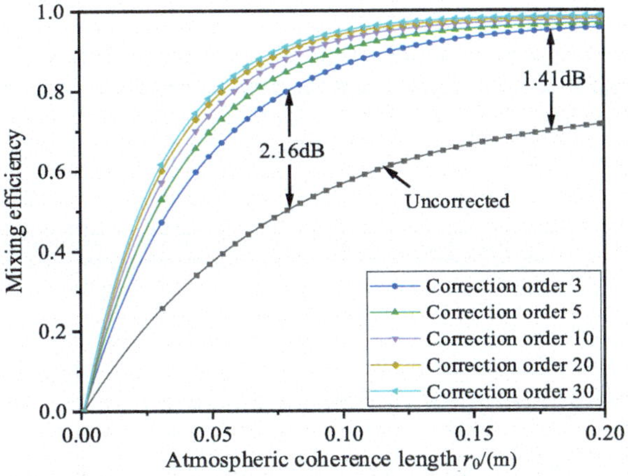

Fig. 6.20 Relationship between the atmospheric-coherence length r_0 and mixing efficiency in homodyne detection

Selecting the same simulation parameters as in Fig. 6.19, the higher-order wavefront distortion of the signal beam is corrected and the relationship between the mixing efficiency of the homodyne detection system and atmospheric-coherence length r_0 is obtained, as shown in Fig. 6.20.

As shown in Fig. 6.20, the influences of the order of the wavefront correction on the mixing efficiency of the system in homodyne and heterodyne detection are basically the same. Compared with the mixing efficiency of the homodyne-detection system when the wavefront distortion is not corrected, when the atmospheric-coherence length $r_0 = 0.18$ m, the mixing efficiency of the third-order wavefront distortion before correction is increased from the uncorrected 0.68 to 0.94, and the gain is 1.41 dB. After correcting the fifth-order wavefront distortion, a 1.55-dB gain is obtained, and a 1.60-dB gain is obtained after correcting the 10th-order wavefront distortion.

It was concluded that with an increase in the wavefront correction order, the mixing efficiency of the system at the receiving end was correspondingly improved, and the lower-order wavefront-distortion correction most obviously improved the mixing efficiency of the system.

6.3.5 Effect of Wavefront Distortion on the Performance Limit of a Coherent-Detection System

The maximum signal-to-noise ratio that a optical wireless coherent-detection system can achieve—that is, when the system mixing efficiency reaches the ideal extreme value—is generally called the detection limit of optical wireless coherent detection. It is also the theoretical limit of optical wireless coherent detection.

Combined with the performance index of the optical wireless coherent-detection system, based on the correction and compensation of the higher-order wavefront-distortion Zernike-polynomial mode method, over- or under-compensation of the wavefront distortion leads to a loss of system performance at the receiving end, and overcompensation increases the system complexity and algorithm redundancy.

This section discusses the simulation and analysis of the compensation using the optimal correction order of wavefront distortion on the atmospheric-turbulence effect, which further improves the performance limit of the optical wireless coherent-detection system at the receiving end.

As the intensity of atmospheric turbulence increases, we consider a larger phase-compensation mode to reduce the wavefront distortion, thereby achieving the best compensation for the mixing efficiency of the coherent-detection system at the receiver.

The simulation conditions were set as follows: receiving aperture $D = 105$ mm, corresponding to D/r_0 values of 0.2, 2, 5, and 8; quantum efficiency of the photodetector $\eta = 0.8$; Planck's constant $h = 6.63 \times 10^{-34}$ J s; beam speed $c = 2.99 \times 10^8$ m/s; wavelength $\lambda = 1550$ nm; signal optical power $P_s = 0.36$ μW, local-oscillation optical power $P_l = 0.36$ mW, and effective noise bandwidth $\Delta f = 200$ MHz. When the wavefront correction order N changes, the extreme values of the heterodyne-detection mixing efficiency and the BER, calculated are obtained using numerical calculations. The simulation results are shown in Fig. 6.21.

Figures 6.21a, b show that the mixing efficiency significantly improves for a certain number of wavefront correction orders. The mixing efficiency converges to a stable value when the number of correction orders increases. Among them, the influence of the lower-order wavefront correction on the mixing efficiency and BER is evident. The corresponding mixing efficiencies increase to 97%, 51%, 12%, and 4% when D/r_0 is 0.2, 2, 5, and 8, respectively; only the tilt term is corrected. Following the correction of the defocus term, the mixing efficiency increases to 98%, 55%, 16%,

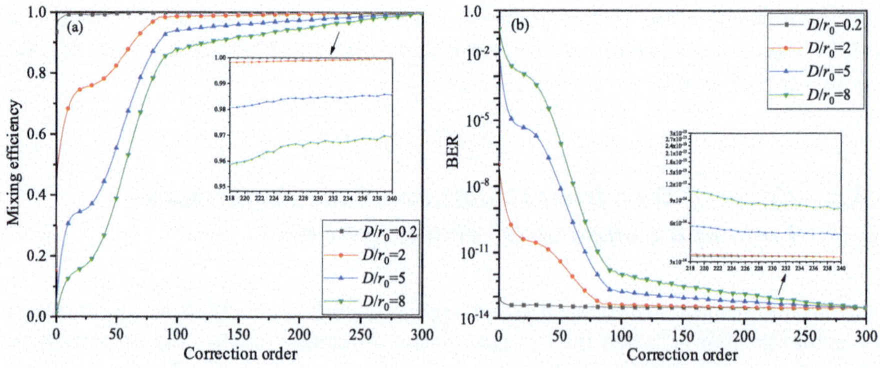

Fig. 6.21 Relationship between the wavefront correction and **a** mixing efficiency and **b** BER in heterodyne detection

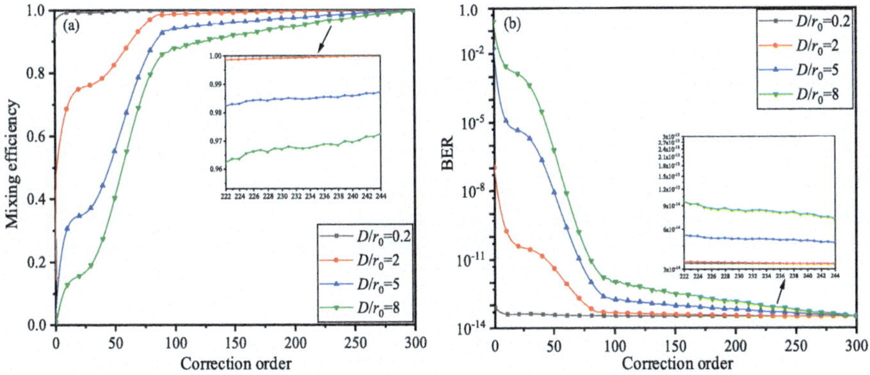

Fig. 6.22 Relationship between the wavefront correction and **a** mixing efficiency and **b** BER in homodyne detection

and 6%, and following the correction of the astigmatism term, the mixing efficiency increases to 99%, 60%, 22%, and 8%, respectively.

When D/r_0 is 0.2 in heterodyne detection, only the fifth-order wavefront correction is completed, the mixing efficiency reaches the extreme value, and the BER is 5.03×10^{-14}. When D/r_0 is 2, the 85th-order wavefront-correction mixing efficiency reaches its extreme value, and the BER is 5.01×10^{-14}. When D/r_0 is 5, the mixing efficiency below 231 orders of wavefront correction reaches an extreme value, and the BER is 5.05×10^{-14}. When D/r_0 is 8, it is necessary to correct the mixing efficiency below 264 orders to reach the extreme value, and the BER is 5.04×10^{-14}.

Selecting the same simulation parameters as shown in Fig. 6.21, Fig. 6.22 shows the relationship curves of the mixing efficiency and BER with the change in the wavefront correction order when D/r_0 takes different values.

Figures 6.22a and 6.22b show that the effect of the lower-order wavefront correction in homodyne detection on the mixing efficiency and BER at the receiver side of the system is the same as that in heterodyne detection. When D/r_0 is 0.2, the wavefront correction is the same as that under heterodyne detection; only the correction of the fifth-order mixing efficiency converges to 1, and the BER is 5.03×10^{-14}.

When D/r_0 is 2, a correction of the 87th-order mixing efficiency is required to reach an extreme value, and the BER is 5.0×10^{-14}. When D/r_0 is 5, the correction of the mixing efficiency below order 234 reaches an extreme value, and the BER is 5.07×10^{-14}. When D/r_0 is 8, the mixing efficiency must be corrected below order 269 to converge to 1, and the BER is 5.0×10^{-14}.

Comparing Figs. 6.21 and 6.22, we find that under the influence of atmospheric turbulence, the wavefront correction order required for homodyne detection to reach the wavefront-distortion correction limit is higher than that required for heterodyne detection, owing to the higher requirement for spatial phase matching.

According to the trend of wavefront-distortion correction with the mixing efficiency, when the atmospheric-coherence length r_0 is certain and the correction of the wavefront distortion is of a specific order, there exists a theoretically optimal receiving aperture D that maximizes the correction gain.

When the wavelength $\lambda = 1550$ nm, signal optical power $P_s = 0.36$ μW, local-oscillation optical power of $P_l = 0.36$ mW, and atmospheric-coherence length $r_0 = 0.05$ m. When the wavefront aberration correction order N varies, the numerical simulation of the relationship between the receiving aperture D and the correction gain variation. Figure 6.23 shows the simulation results of the curve-correction gain varying with the receiving aperture D for the different detection methods, and the optimal correction-gain results are shown in Fig. 6.24.

Figure 6.23 shows that as the wavefront correction order increases, the correction gain increases gradually, and as the receiving aperture D increases, the correction gain reaches a maximum and then shows a downward trend. Comparing Fig. 6.23a, b,

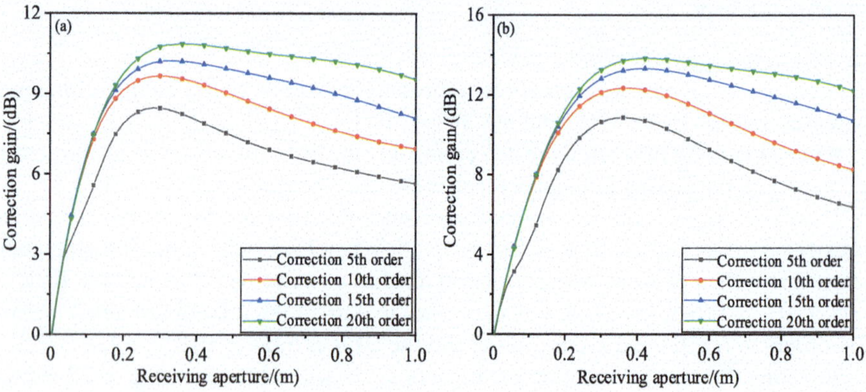

Fig. 6.23 When the wavefront correction order is different, the correction gain varies with the receiving aperture. **a** Heterodyne detection. **b** Homodyne detection

Fig. 6.24 Variation of the wavefront correction order and optimal correction gain at $r_0 = 0.05$ m

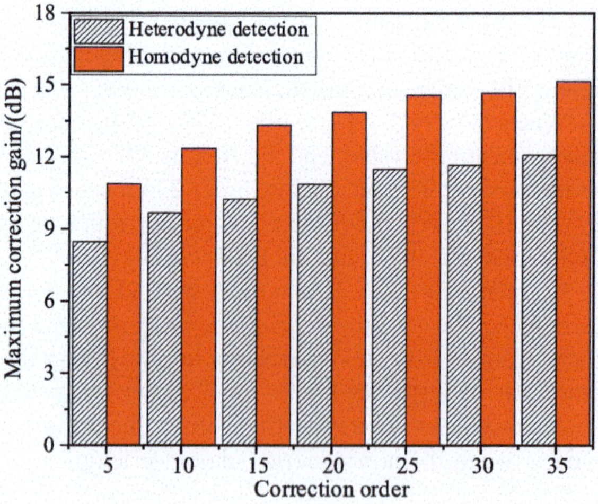

the correction gain obtained by homodyne detection is greater than that obtained by heterodyne detection because the detection sensitivity of homodyne detection itself is better than that of heterodyne detection.

Observing Fig. 6.24, we find that when the D/r_0 ratio is 7.6, the best heterodyne-detection correction gain is 8.5 dB, and the best homodyne-detection correction gain is 11.2 dB, when correcting the wavefront distortion of the first five orders. The best heterodyne-detection correction gain is 10.3 dB and the best homodyne-detection correction gain is 13.3 dB, when correcting the wavefront distortion of the first 15 orders. The correction gain increases with the wavefront correction order.

When the wavefront-distortion correction is below the 35th order, the maximum heterodyne-detection correction gain rises to 12.1 dB, and the maximum homodyne-detection correction gain rises to 15.1 dB. The homodyne-detection correction gain is approximately 3 dB higher than the heterodyne-detection correction gain.

Considering different detection methods, the same simulation parameters as those in Fig. 6.23 were selected. Figure 6.25 shows the curves of the correction gain changing with the receiving aperture D in heterodyne detection and homodyne detection when the atmospheric-coherence length $r_0 = 0.01$ m. The optimal correction-gain results are shown in Fig. 6.26.

Figure 6.25 shows that as the wavefront correction order increases, the correction gain increases gradually, and as the receiving aperture D increases, the correction gain reaches a maximum. And then as the receiving aperture D continues to increases, the correction gain showed oscillation decrease. The oscillation mainly comes from the fact that with the increase of turbulence intensity, the energy of the main lobe of the wavefront distortion spreads to the side lobes, which leads to a decrease in the matching degree of the optical field at the receiving end, and thus, the oscillation phenomenon.

Figure 6.26 shows that the correction gain increases as the wavefront correction order gradually increases. When correcting the first five orders of wavefront

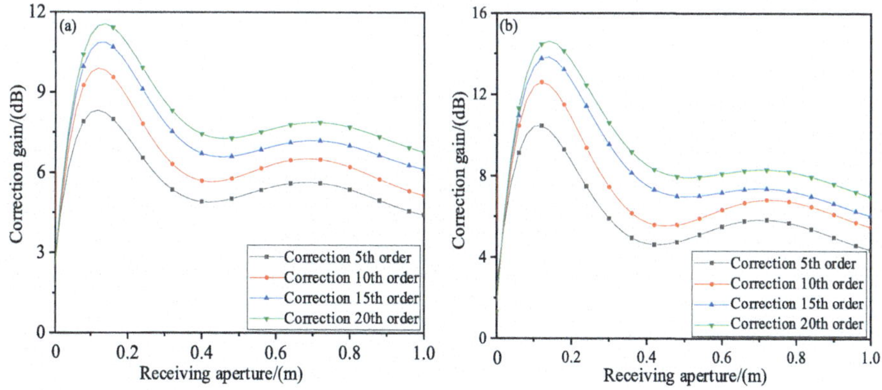

Fig. 6.25 When the wavefront correction order differs, the correction gain varies with the receiving aperture. **a** Heterodyne detection. **b** Homodyne detection

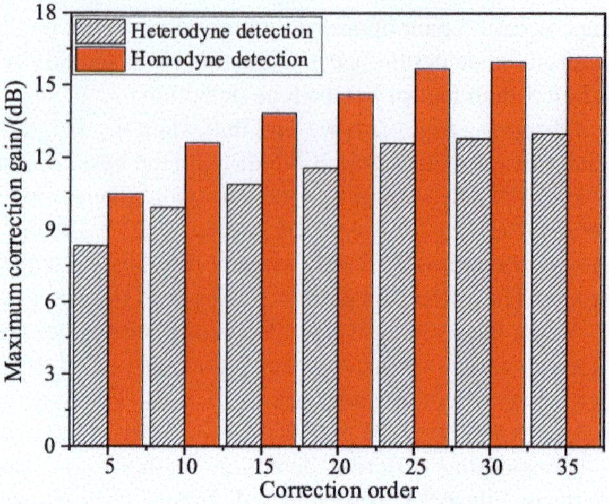

Fig. 6.26 Variation of the wavefront correction order and optimal correction gain at $r_0 = 0.01$ m

distortion, the best heterodyne-detection correction gain was 8.3 dB, and the best homodyne-detection correction gain was 11.1 dB. When the wavefront distortion was corrected to below 35 orders, the maximum heterodyne-detection correction gain increased to 12.5 dB, and the maximum homodyne-detection correction gain rose to 15.5 dB. The correction gains for heterodyne and homodyne detection were optimal when the D/r_0 ratio was 11.5. We found that the optimal correction gain for homodyne detection was generally 3 dB higher than that for heterodyne detection.

Comparing the simulation results of receiving aperture D and the correction gain under different turbulence conditions, it is concluded that as the turbulence intensity increases, the correction gain is more sensitive to the selection of the receiving aperture. Under a specific turbulence intensity, different orders of wavefront-distortion correction have suitable receiving apertures to maximize the correction gain, which provides a quantitative reference for the selection of an optical system in practice.

6.4 Experimental Research

Based on the theoretical analysis of wavefront distortion in optical wireless coherent-detection systems, an experimental optical path was constructed to measure and analyze the mixing efficiency and bit-error rate of the optical wireless coherent-detection system at the receiving end. Figure 6.27 shows the experimental principle of wavefront correction in a optical wireless coherent-detection system.

The wavefront-correction principle of the optical wireless coherent communication system is illustrated in Fig. 6.27. The Koheras BasiK module laser emits a 1550-nm signal beam through the transmitting antenna, which enters the atmospheric-turbulence channel. It receives the signal beam through the receiving

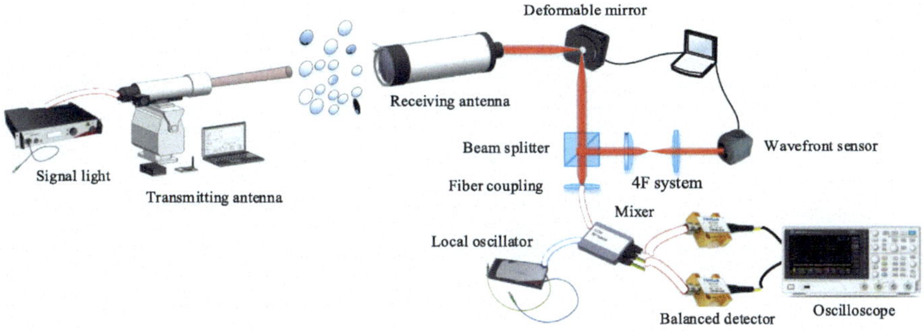

Fig. 6.27 Schematic diagram of a wavefront-correction optical wireless coherent communication system

antenna, and enters the coherent-detection system at the receiving end with adaptive wavefront-distortion correction.

The signal beam received by the receiving antenna is divided into two beams using a splitter prism, after passing through 40 piezoelectric deformable mirrors in a parallel-exit direction. One beam passes through the 4F collimating and shrinking system composed of lenses L1 and L2, detects the wavefront through the HASO4-NIR wavefront sensor, and transmits the detected wavefront information to the computer. This drives the 40-element piezoelectric deformable mirror to correct the wavefront distortion, and constitutes an adaptive-optics closed-loop correction system.

The other beam is coupled into the fiber after the closed-loop correction of the adaptive-optics system, and then mixed with the local oscillator in the COH24 mixer. The photoelectric conversion is realized through a balanced detector and the IF signal is displayed on the oscilloscope.

6.4.1 Indoor Wavefront-Distortion Correction Experiment

In the experiment, the D/r_0 value was converted to a random corresponding Zernike coefficient of the deformable mirror, which corresponds to the driving voltage of the deformable mirror. The atmospheric-turbulence intensity $D/r_0 = 2$, and the corresponding Zernike coefficients were obtained. The experimental parameters were set as follows: receiving aperture $D = 105$ mm and wavelength $\lambda = 1550$ nm. Integral-gain control was used with an integral control coefficient $k_i = 0.05$, and 2000 closed-loop calibration experiments were conducted. The tilt and defocus terms of the wavefront distortion were corrected, and the wavefront Zernike coefficients were obtained, as shown in Fig. 6.28.

Figure 6.28 shows the distribution diagram of the uncorrected Zernike coefficients and after correcting the tilt and defocus items in the indoor experiment. It can be seen

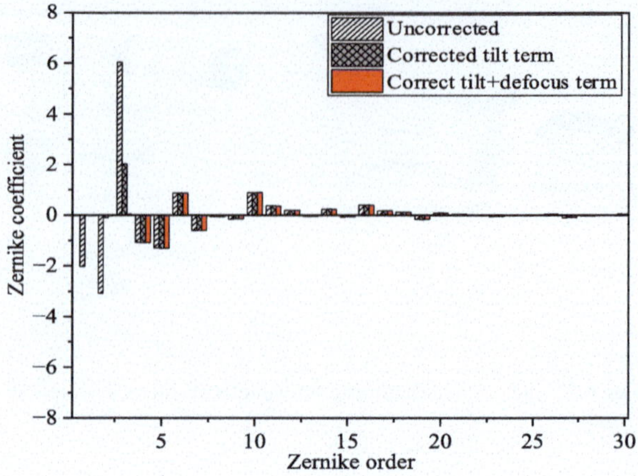

Fig. 6.28 Zernike coefficients after wavefront correction

that when the wavefront is not corrected, the wavefront coefficients of the tilt and defocus items are larger. The tilt and defocus items account for most of the wavefront distortion.

After using the fully corrected closed-loop control algorithm to correct the tilt and defocus items, it can be seen that the corresponding lower-order corrected wavefront-distortion coefficient is almost zero, and the wavefront-distortion correction is good. Moreover, the wavefront distortion of the beam phase is independent and orthogonal in space. When the wavefront-distortion tilt and defocus terms are corrected, the remaining orders are unaffected.

Figure 6.29 shows the peak-to-valley (PV) curve after the closed-loop wavefront correction when $D/r_0 = 2$. As shown in Fig. 6.29, the PV value of the system at the open-loop time is 32.85 μm. After approximately 500 iterations, the PV value of the system decreases to a stable state. The PV value after correcting the tilt term is stable at 13.2 μm. The PV value after correcting the defocus term is stable at 12.6 μm. The PV value after correcting the astigmatism is stable at 10.1 μm. The analysis shows that with an increase in the wavefront correction order, the PV value gradually decreases, and the wavefront correction effect is good.

When the detector sensitivity $R = 0.6$ A/W, beam speed $c = 2.99 \times 10^8$ m/s, quantum efficiency $\eta = 0.8$, Planck's constant $h = 6.63 \times 10^{-34}$ J·s and detector bandwidth $B = 40$ GHz, after tilt, defocus and astigmatism wavefront distortions are corrected, Fig. 6.30 shows the mixing efficiency and bit-error rate curve of the receiver system.

Figure 6.30 shows the relationship between the mixing efficiency and bit-error rate of the optical wireless coherent-detection system after different orders of wavefront corrections and the number of iterations, which is 2000. It can be seen that when the number of iterations is approximately 500, the mixing efficiency and bit-error rate of the system tend to be stable. After correcting the tilt term, the mixing efficiency is maintained at 36%, and the bit-error rate is maintained at the order of 10^{-5}. After correcting the defocus term, the mixing efficiency is maintained at 47%, and the

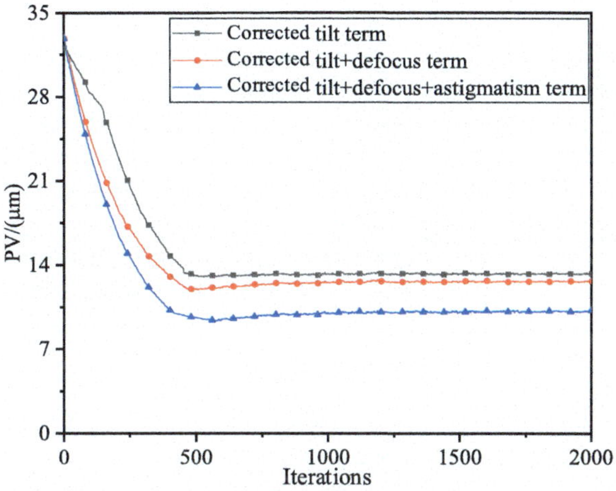

Fig. 6.29 PV value after wavefront correction when $D/r_0 = 2$

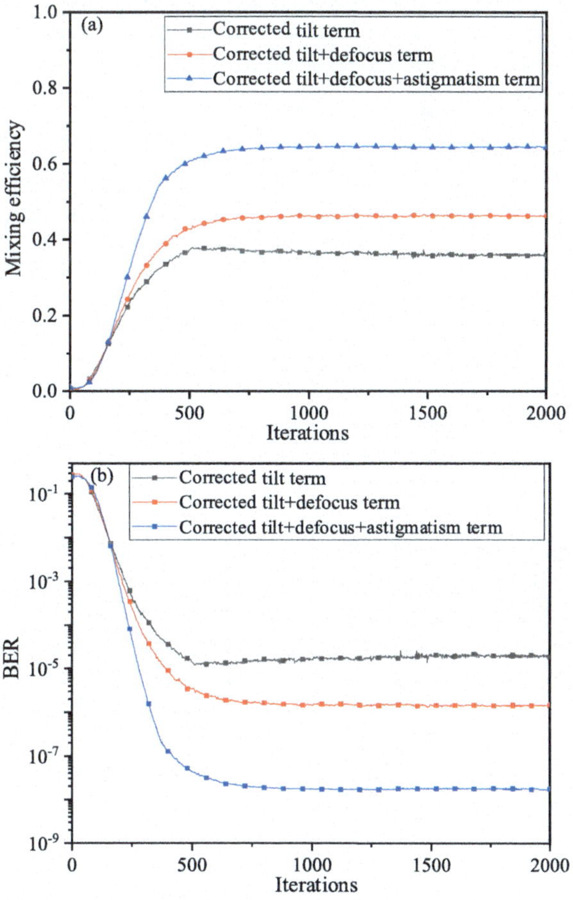

Fig. 6.30 Relationship between the wavefront-distortion correction and **a** mixing efficiency and **b** bit-error rate when $D/r_0 = 2$

bit-error rate is maintained at the order of 10^{-6}. After correction to astigmatism, the mixing efficiency is maintained at 62%, and the bit-error rate is maintained at the order of 10^{-8}.

When the experimental environment and parameters are identical to those shown in Fig. 6.30, IF signals with different correction orders were collected. The amplitude changes of the IF signals before and after wavefront correction are shown in Fig. 6.31.

Figure 6.31a shows the amplitude-variation waveform of the IF signal at the receiving end without wavefront correction. Figures 6.31b–d show the amplitude-variation waveform of the IF signal after wavefront correction to tilt, defocus, and astigmatism, respectively. Figure 6.31a shows that the measured amplitude waveform of the IF signal without wavefront correction is disordered, and its peak value is only 385.7 mV.

Figure 6.31b shows that the amplitude waveform of the IF signal will fluctuate slightly after only correcting the tilt term, and its peak value is 744.2 mV. In Fig. 6.31c, after the amplitude of the IF signal is corrected for the defocus term, the waveform changes smoothly, and the peak-to-peak value of the IF signal is 872 mV. In Fig. 6.31d, the amplitude of the IF signal is smooth after astigmatism correction, and its peak value is 1.104 V, which is conducive to demodulation signal processing at the back end.

6.4.2 *Wavefront-Distortion Correction Experiment in an External Field*

To more accurately and generally analyze the impact of wavefront distortion on the optical wireless coherent-detection system, a complete correction algorithm was analyzed for the 1.2-km data collected in the wavefront-distortion correction experiment, according to the experimental link described in this section.

The experimental parameters were set as follows: communication distance $L =$ 1.2 km, optical aperture $D = 105$ mm, signal wavelength $\lambda = 1550$ nm, detector sensitivity $R = 0.6$ A/W, carrier optical frequency $v = 1.94 \times 10^{14}$ Hz, electronic charge $e = 1.6 \times 10^{-19}$ C, quantum efficiency $\eta = 0.8$, Planck's constant $h = 6.63 \times 10^{-34}$ J·s, detector bandwidth $B = 40$ GHz, and atmospheric-coherence length $r_0 = 0.055$ m. The integral-gain control coefficient $k_{\mathrm{i}} = 0.05$. Figure 6.32 shows the change in the Zernike coefficient before and after wavefront correction at different orders in the closed-loop state, after 2000 iterations.

Figure 6.32 shows the Zernike-coefficient distribution of the wavefront measured before and after correction to the tilt and defocus terms at a transmission distance of 1.2 km. It can be seen that when the wavefront is not corrected, the wavefront coefficients of the tilt, defocus, and spherical aberration terms are larger; that is, the overall distortion component is more serious.

Because the tilt and defocus terms account for the majority of the wavefront distortion, after correcting the tilt and defocus terms, respectively, it can be seen

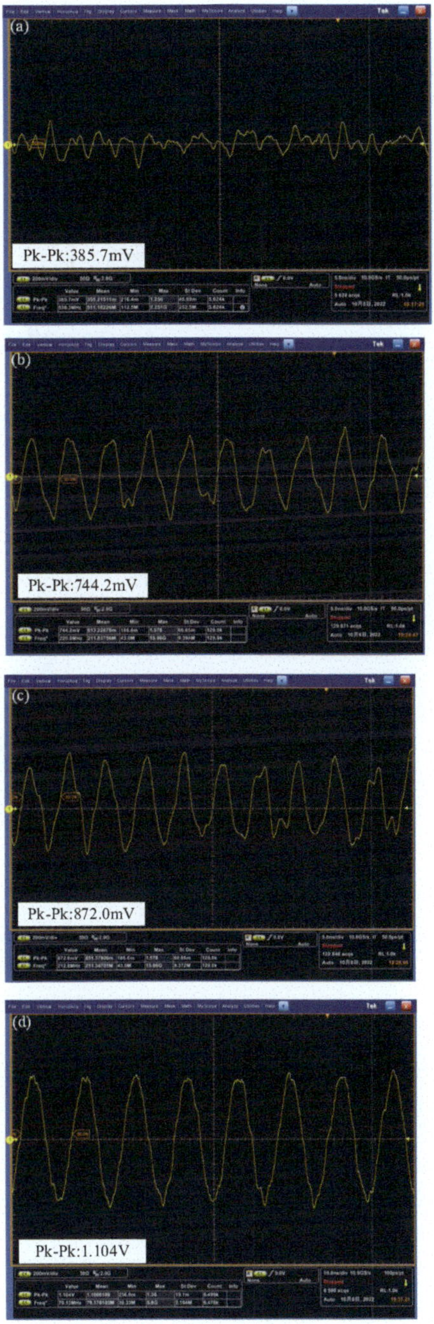

Fig. 6.31 Amplitude change of IF signal before and after wavefront correction. **a** Uncorrected; **b** corrected tilt term; **c** corrected defocus term; **d** corrected astigmatism term

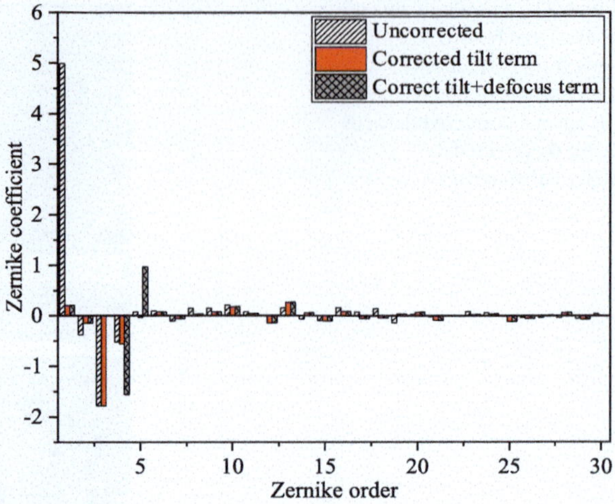

Fig. 6.32 Zernike coefficient before and after wavefront correction for a transmission distance of 1.2 km

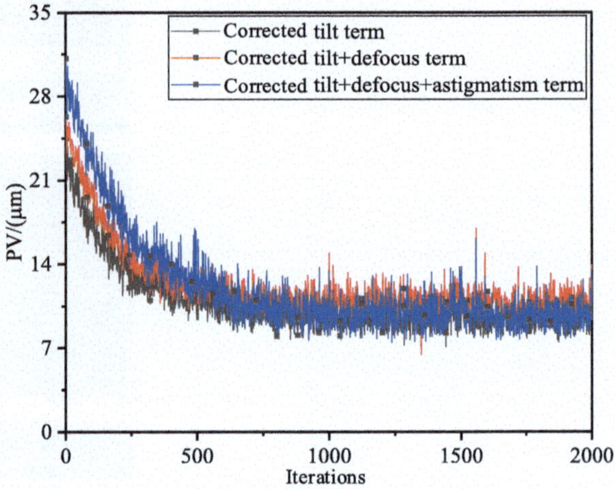

Fig. 6.33 PV curve of the system after wavefront correction with a transmission distance of 1.2 km

that the wavefront-distortion coefficient of the corresponding lower-order wavefront correction is significantly attenuated. However, the correction effect differs from that of the indoor and 600-m field experiments, and the wavefront coefficient, corrected by unstable factors such as the transmission distance, still fluctuates. Because the wavefront tilt and defocus components account for most of the wavefront-distortion components, the corrected wavefront-distortion components are mainly reflected in the wavefront tilt and defocus components.

When the experimental environment and parameters are identical to those shown in Fig. 6.32, the wavefront coefficients under different correction orders are obtained. The PV curve of the system after wavefront correction is shown in Fig. 6.33, and the mixing-efficiency change and bit-error rate curves are shown in Fig. 6.34.

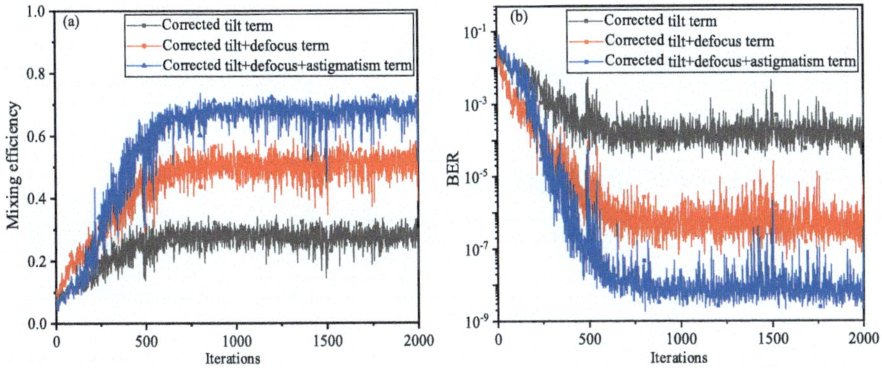

Fig. 6.34 Relationship between different wavefront corrections and **a** the mixing efficiency and **b** the bit-error rate at a transmission distance of 1.2 km

Figure 6.33 describes the PV value after closed-loop wavefront correction for a transmission distance of 1.2 km. From Fig. 6.33, it is observed that the PV value after wavefront-distortion correction of different orders decreases significantly. After approximately 500 iterations, the PV value of the system approaches a stable state. The analysis shows that with an increase in the wavefront correction order, the PV value gradually decreases, and the wavefront-correction effect improves.

According to Fig. 6.34a, in the field experiment with a transmission distance of 1.2 km, when adaptive-optics closed-loop correction is used to correct the tilt term, the mixing efficiency tends to stabilize at 500 sampling points and remains at approximately 36%. When the wavefront distortion is corrected for the defocus term, the mixing efficiency tends to be approximately 52%; and when the astigmatism term by wavefront distortion is corrected, the mixing efficiency tends to be approximately 70%.

Figure 6.34b shows that after correcting only the tilt term, the BER of the optical wireless coherent-detection system decreases from 10^{-1} to 10^{-4} after correction. After correction for the defocus term, the BER tends to 10^{-7}. When the astigmatism term is corrected, the BER decreases to 10^{-8}. After approximately 500 iterations, the mixing efficiency and bit-error rate of the optical wireless coherent-detection system stabilizes.

6.4.3 1.3-km Multi-beam Propagation Experiment

To further verify the correctness of the influence of multi-beam-propagation wavefront-distortion correction on the mixing efficiency and bit-error rate of the receiving system in the theoretical analysis, a wavefront-correction experiment of a optical wireless coherent communication system was conducted.

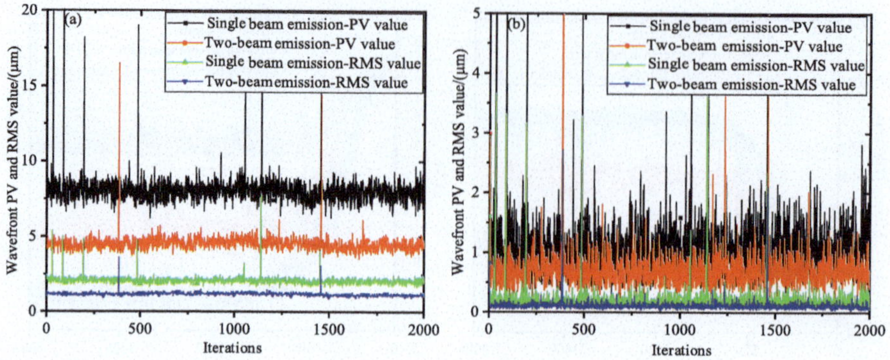

Fig. 6.35 Wavefront peak–valley and root-mean-square values of single-beam and two-beam propagation. **a** Uncorrected. **b** Corrected

The experimental link was from the sixth teaching building of Xi'an University of Technology to Kaisen Fujing Yayuan. The communication distance was 1.3 km. The experimental period was December 3, 2022. The weather was overcast, and the northerly wind was grade 2. When the communication distance $L = 1.3$ km, the receiving aperture $D = 105$ mm, the signal optical power was approximately 1 mW, and the local-oscillator optical power was approximately 10 mW.

The wavefront peak–valley and root-mean-square values before and after correction were measured for single-beam and two-beam propagation, as shown in Fig. 6.35. The self-noise inside the wavefront sensor leads to synaptic data in the measurement of the wavefront slope and phase.

Figure 6.35 shows that without wavefront-distortion correction, the mean values of the wavefront peak-to-valley and root-mean-square values for a single-beam transmission are 8.06 μm and 2.13 μm, with variances of 1.59 μm^2 and 0.06 μm^2, respectively. The mean values of the wavefront peak-to-valley and root-mean-square values for a two-beam transmission are 4.56 μm and 1.19 μm, with variances of 0.06 μm^2 and 0.01 μm^2, respectively. This further indicates that the wavefront-smoothing effect and jitter-suppression effect under a multi-beam transmission are better than those under a single-beam transmission, and the suppression effect becomes more obvious as the number of beam transmissions increases.

After using a 40-unit piezoelectric deformable mirror for wavefront correction, the mean peak–valley and root-mean-square values of the wavefront during a single-beam transmission at the transmitting end are 1.15 μm and 0.22 μm, with variances of 1.68 μm^2 and 0.05 μm^2, respectively. The mean values of the peak–valley and root-mean-square values of the wavefront during a two-beam transmission are 0.73 μm and 0.19 μm, with variances of 0.23 μm^2 and 0.01 μm^2, respectively. It can be concluded that after using the two-beam transmission, the corrected wavefront convergence at the receiving end is smoother.

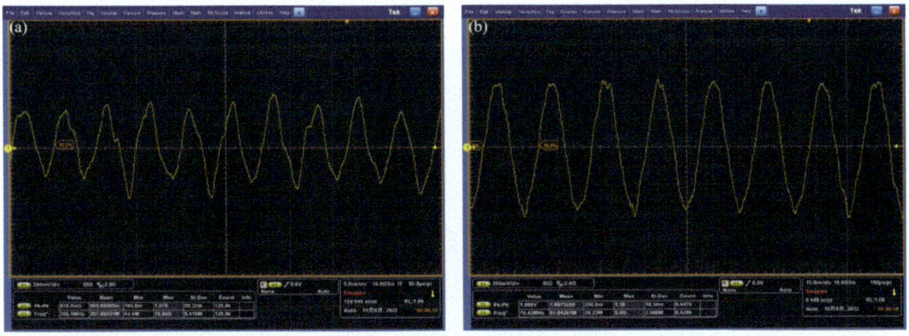

(a) Single-beam propagation at transverse scanning speed 5 ns/div (b) Two-beam propagation at transverse scanning speed 5 ns/div

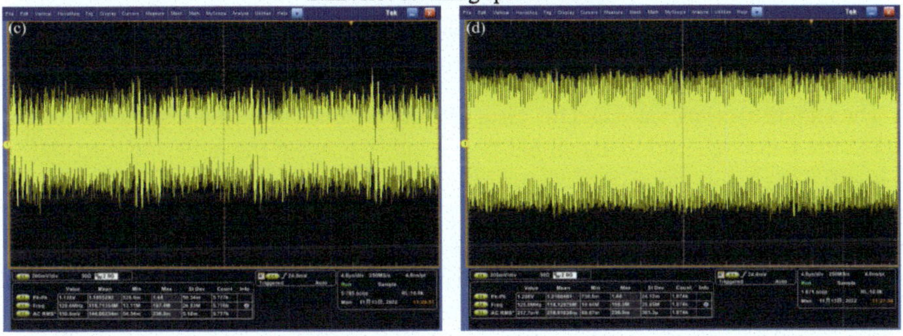

(c) Single-beam propagation at a lateral scanning speed of 4 μs/div (d) Two-beam propagation at a lateral scanning speed of 4 μs/div

Fig. 6.36 Intermediate-frequency signal after wavefront correction

When the experimental environment and parameters are the same as those shown in Fig. 6.35, the intermediate-frequency signal was collected after wavefront correction for single-beam and two-beam transmissions. The amplitude changes of the IF signal are shown in Fig. 6.36.

Figure 6.36 shows the intermediate-frequency signal waveform output by the detector after wavefront correction at the receiving end for single-beam and two-beam transmissions. It can be seen that when a single beam is transmitted, the transverse scanning speed is 5 ns/div. After correction, the sine-wave shape of the intermediate-frequency signal is relatively irregular. When the transverse scanning speed is 4 μs/div, the amplitude fluctuation of the intermediate-frequency signal is more obvious.

After using two beams for transmission, the sine waveform of the intermediate-frequency signal is regular at a transverse scanning speed of 5 ns/div. At a transverse scanning speed of 4 μs/div, the amplitude of the intermediate-frequency signal is relatively stable. Therefore, the gain brought about by a multi-beam transmission can be directly reflected in the amplitude stability of the intermediate-frequency signal, which is beneficial for subsequent demodulation processing.

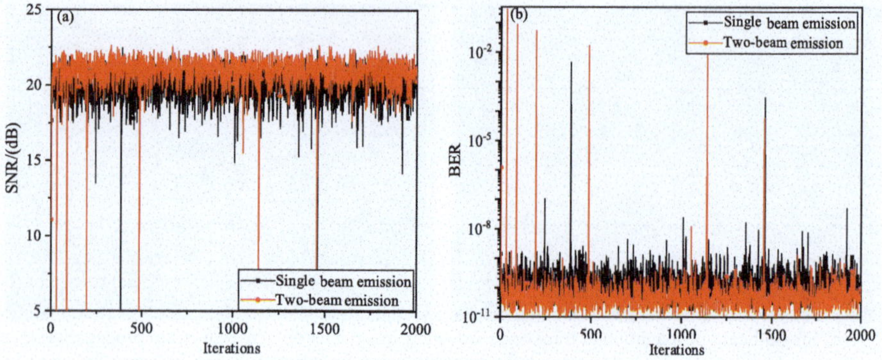

Fig. 6.37 After wavefront correction for single-beam and two-beam transmissions. **a** SNR. **b** BER

The signal-to-noise ratio and bit-error rate curves obtained after wavefront correction at the receiving end using single-beam and two-beam transmissions are shown in Fig. 6.37.

From Fig. 6.37, it can be seen that using two-beam transmissions has advantages over single-beam transmissions in terms of the signal-to-noise ratio and bit-error rate and has less fluctuation. When transmitting a single beam, the error rate remains stable on the order of 10^{-9}. When transmitting multiple beams, the error rate remains stable on the order of 10^{-10}. Furthermore, it has been proven that multi-beam transmissions are more effective for improving the gain of optical wireless coherent-detection systems, compared to single-beam transmissions.

References

1. Wu JL, Ke XZ, Yang SJ, Ding DQ (2022) Effect of multi-beam propagation on free-space coherent optical communications in a slant atmospheric turbulence. J Opt 24(7):075601
2. Hyde MW, Tyler GA (2016) Temporal coherence effects on target-based phasing of laser arrays. J Opt Soc Am A 33(10):1931–1937
3. Ke XZ, Wu JL (2019) Principles and applications of wireless optical coherent communication. China Science Publishing & Media Ltd., Beijing
4. Liu C, Chen S, Li X et al (2014) Performance evaluation of adaptive optics for atmospheric coherent laser communications. Opt Express 22(13):15554–15563
5. Andrews LC, Phillips RL (2005) Laser beam propagation through random media. SPIE Press, Washington
6. Zhai C (2021) A novel idea: calculating anisotropic turbulence only by Kolmogorov structure constant C_n^2 and power law α. Results Phys 19:103483-1–103483-4.
7. Li N, Qiao C, Zhang P et al (2019) Research of laser propagation and its phase compensation in non-Kolmogorov turbulence atmosphere. Chin J Quant Electron 36(06)745–751
8. Li Z, Liao Z, Liang J et al (2022) Research and prospects of atmospheric turbulence model and atmospheric channel model. Opt Commun Technol 1(00):1–14
9. Kong Y, Ke X, Yang Y (2015) Influence research of atmospheric turbulence on space coherent optical communications. Laser Optoelectron Progress 52(8):080601-1–080601-7

10. Liu H, Ji Y, Liu L (2012) Effect of aberration on performance of the bit error rate in an inter-satellite coherent optical communication receiving system. Acta Optica Sinica 32(01):46–51
11. Geng C, Li F, Zuo J et al (2020) Fiber laser transceiving and wavefront aberration mitigation with adaptive distributed aperture array for free-space optical communications. Opt Lett 45(7):1906–1909
12. Konwar S, Boruah BR (2020) Leveraging the orthogonality of Zernike modes for robust free-space optical communication. Commun Phys 3(1):1–9
13. Noll RJ (1976) Zernike polynomials and atmospheric turbulence. J Opt Soc Am 66(3):207–211.
14. Hu PH, Stone J, Stanley T (1989) Application of Zernike polynomials to atmospheric propagation problems. J Opt Soc Am A 6(10):1595–1608
15. Erol B, Altiner B, Adali E et al (2019) H-infinity Suboptimal controller design for adaptive optic systems. Trans Inst Meas Control 41(8):1–14
16. Paneva-Konovska J (2014) A family of hyper-Bessel functions and convergent series in them. Fractional Calculus Appl Anal 17(4):1001–1015
17. Soto-Quiros P, Torokhti A (2017) Optimal transforms of random vectors: the case of successive optimizations. Signal Process 132(16):183–196
18. Wang F (2020) Research on pre-compensation of beam phase distortions of ground-to-satellite uplink laser communication. Harbin Institute of Technology, Harbin
19. Chang HR, Wen HJ, Ning L (2001) Adaptive-optics compensation by distributed beacons for non-Kolmogorov turbulence. Appl Opt 40(21):3441–3449
20. Cheng MJ, Guo LX, Li JT (2017) Effect of aberration correction on beam wander of electromagnetic multi-Gaussian shell-model beams in anisotropic turbulence. Optik 144:1–14
21. Tang H, Guo PZ (2014) Phase compensation in non-Kolmogorov atmospheric turbulence. Optik 125:1227–1230
22. Liu HL, Lu YF, Xia J et al (2016) Radial phased-locked partially coherent flat-topped vortex beam array in non-Kolmogorov medium. Opt Express 24(17):19695–19712
23. Yalçın A, Yahya B, Gökçe MC (2019) Average channel capacity in anisotropic atmospheric non-Kolmogorov turbulent medium. Opt Commun 451:129–135
24. Area I, Dimitrov DK, Godoy E (2017) Recursive computation of generalised Zernike polynomials. J Comput Appl Math 312(2):58–64
25. Wang Q, Tong S, Xu Y (2013) On simulation and verification of the atmospheric turbulent phase screen with Zernike polynomials. 42(7):1907–1911.
26. Wu Y (2021) Predictive control and experimental research on wavefront distortion in wireless optical coherent communication. Xi'an University of Technology, Xi'an
27. Aftab M, Graves LR, Burge JH et al (2019) Rectangular domain curl polynomial set for optical vector data processing and analysis. Opt Eng 58(9):095105-1–095105-10
28. Xiang R (2015) Research on active beam quality control technology of high energy solid flat noodles laser. Sichuan: China Academy of Engineering Physics
29. Mahajan VN, Dai GM (2007) Orthonormal polynomials in wavefront analysis: analytical solution. J Opt Soc Am A 24(9):2994–3016
30. Díaz JA (2019) Comment on "Computer simulation of Fresnel diffraction from rectangular apertures and obstacles using the Fresnel integrals approach." Opt Laser Technol 121(105819):1–3
31. Fan Z, Dai Y, Tie G et al (2018) Hysteresis characteristics and closed loop correction of unimorph deformable mirror. Infrared Laser Eng 47(10):1020001-1–1020001-6

Chapter 7
Difference of Wavefront Distortion of Beams with Different Wavelengths Passing Through Atmospheric Turbulence

Beam transmission through atmospheric turbulence distorts the wavefront, and the dispersion effect of the transmission medium leads to differences in the wavefront distortion of beams with different wavelengths. In a dual-wavelength adaptive-optics system, the distortion difference is passed into the closed-loop correction system and finally produces a correction residual. Therefore, it is important to improve the correction accuracy by studying the transmission of beams of different wavelengths in the atmosphere and the change characteristics of the distortion difference.

7.1 Transmission of Beams of Different Wavelengths in Atmospheric Turbulence

An important corollary of Maxwell's system of equations is that a beam is an electromagnetic wave and that the speed of light is the fastest speed [1]. Therefore, when analyzing the propagation of a beam in a turbulent atmosphere, it is important to analyze the beam transmission according to its fluctuating equations, starting from the fluctuating nature of the beam.

The main component of a beam when it propagates in a turbulent atmosphere is its electrical component, which corresponds to the set of Maxwell's equations [2]:

$$\nabla^2 E + k^2 n^2 E + 2\nabla[E \cdot \nabla \ln n] = 0. \tag{7.1}$$

This equation is also known as d'Alembert's equation, where E is the electric component of the electromagnetic vector of the beam, $k = 2\pi/\lambda$ is the wavenumber of the beam, λ is the wavelength, n is the refractive index of the turbulent atmosphere, and ∇^2 is the Laplace operator. The third term in the equation is related to the beam polarization. When the beam propagates in a turbulent atmosphere, ignoring the

X. Ke and C. Ke, *Performance-Improvement Limits in Optical Wireless Coherent Communication*, Optical Wireless Communication Theory and Technology,
https://doi.org/10.1007/978-981-96-2000-5_7

effect of polarization, the third term can be removed. The electric component can be expressed as [3]

$$\nabla^2 E + k^2 n^2 E = 0. \tag{7.2}$$

Four general methods are used to solve d'Alembert's equation for beam transport in atmospheric turbulence: the geometric-optics approximation method, the perturbation method, the statistical-optics method, and the generalized Huygens–Fresnel principle.

7.1.1 *Transmission Analysis Method Laser in Atmospheric-Turbulence*

7.1.1.1 Geometric-Optics Approximation Method

The geometric-optics method is a beam-transport analysis method that assumes that when a laser beam propagates in atmosphere, it can be approximated without considering the beam-diffraction phenomenon, and only considering the refraction and reflection of the beam in the transport medium. The geometric-optics approximation is based on the condition $\sqrt{\lambda L} \ll l$, where $\sqrt{\lambda L}$ is the Fresnel scale, and l is the intra-turbulent scale. This method can be used to solve problems, such as spot drift and angle-of-arrival undulations, during beam transmissions.

In the geometric-optics approximation, the equation for the scalar fluctuations of the optical field is [4]

$$\nabla^2 \ln(E) + [\nabla \ln(E)]^2 + k^2 n^2 = 0. \tag{7.3}$$

By substituting amplitude A and phase S of the beam into Eq. (7.3), the d'Alembert equations for amplitude and phase are obtained as follows, respectively:

$$\nabla^2 A/A - (\nabla S)^2 + k^2 n^2 = 0 \tag{7.4}$$

$$\nabla^2 S + 2\nabla \ln A \cdot \nabla S = 0. \tag{7.5}$$

Substituting the refractive index n, phase S, and perturbation of the logarithmic amplitude A into the above equations, assuming that the beam wave is a plane wave propagating along the z-direction with amplitude A_0, a set of geometric-optics equations is obtained [5]:

$$\begin{cases} \nabla^2 S_1 + 2k\frac{\partial \chi}{\partial z} = 0 \\ \frac{\partial S_1}{\partial z} = kn_1 \end{cases}, \tag{7.6}$$

where S_1 denotes the phase-undulation variation and χ denotes the amplitude-undulation variation. When the beam wave propagates from $z = 0$ to $z = L$, the solution is

$$\left\{\begin{array}{c} S_1(x, y, L) = k \int\limits_0^L n_1(x, y, z)dz \\ \chi(x, y, L) = \frac{1}{2}\left\{\left[n_1(x, y, L) - n_1(x, y, 0)\right] - \int\limits_0^L \int\limits_0^z \nabla_{\perp}^2 n_1(x, y, \xi)d\xi dz\right\} \end{array}\right., \tag{7.7}$$

where x and y are the horizontal and vertical coordinates of the points on the receiving cross section, respectively, L is the beam-transmission distance, and n_1 is the refractive-index distribution on the transmission path.

7.1.1.2 Generalized Huygens–Fresnel Principle Approach

The Huygens–Fresnel principle states that the beam field at any point on a receiving plane is the superposition of beam fields at different points throughout the source [6]. Therefore, under the paraxial approximation condition, the generalized Huygens–Fresnel principle states that the beam field at the receiving end is [7]

$$\begin{array}{c} U(x, y, z) = \frac{ik}{2\pi z} \exp(ikz) \iint U_0(x_0, y_0) \\ \exp\left[-\frac{ik}{2z}\left[(x - x_0)^2 + (y - y_0)^2\right]\right]dx_0 dy_0 \end{array}, \tag{7.8}$$

where $U(x, y, z)$ is the beam-field distribution at the receiver end, $U_0(x_0, y_0)$ is the beam field at point (x_0, y_0) on the transmitter end, and a double integration is the integration of the point sources over the entire transmitter beam-field cross section to obtain the beam field at the receiver end after the point-source superposition. In addition, the phase-change $\exp(ikz)$ of the beam field at the receiving end is introduced, after the beam-transmission distance L.

The same principle as the distance introduces the phase change. The perturbation of the transmission path caused by the turbulent atmosphere also causes a phase change in the optical field at the receiving end. According to the generalized Huygens–Fresnel principle, the phase-perturbation term of the beam wave after passing through the turbulent atmosphere is

$$\exp(\chi + iS_1). \tag{7.9}$$

By substituting Eq. (7.9) into Eq. (7.8), the expression for the beam field at the receiving end, after a beam wave is transmitted over distance L in a turbulent atmosphere, is obtained as

$$\begin{array}{c} U(x, y, z) = \frac{ik}{2\pi z} \exp(ikz) \int U_0(x_0, y_0) \\ \exp\left[-\frac{ik}{2z}\left[(x - x_0)^2 + (y - y_0)^2\right]\right] \exp(x + iS_1)dx_0 dy_0 \end{array}. \tag{7.10}$$

The generalized Huygens–Fresnel principle is extensible, and factors other than the transmission distance and turbulent-atmospheric disturbances can be considered in the form of exponential terms for the entire transmission process.

7.1.1.3 Micro-perturbation Methods

Two early perturbation theories that received considerable attention were the Born approximation and the Rytov approximation. The Born approximation was applied to the integral equation for scattering, which can be derived directly from the Schrödinger fluctuation equation corresponding to a quantum system. The Rytov approximation was first used to develop the transmission of acoustic waves. It was later considered by Russian scientists as a smooth perturbation method and was extended to migrate to the optical-approximation method; both methods are restricted to the weak-turbulence region, while the parabolic-equation method is applicable to multiple atmospheric conditions [8].

Born approximated the solution of the fluctuation Eq. (7.2) by considering the perturbation term to be additive. To facilitate the solution, the refractive-index undulation is first expressed [9]:

$$n^2 = [n_0 + n_1]^2 \simeq 1 + 2n_1, \ |n_1| \ll 1, \tag{7.11}$$

where n_0 denotes the mean value of the random fluctuation of the refractive index, the changing refractive index centers around the mean value with a small random fluctuation, and n_0 is the amount of change, using the Taylor expansion to solve for the primary function of the amount of change.

Assuming that the refractive-index fluctuations satisfy a Gaussian distribution with a mean value of 0 and that the beam propagates in the positive direction along the z-axis, the beam field at $z = L$ can be expressed as

$$U = U_0 + U_1 + U_2 + \cdots, \tag{7.12}$$

where U_0 denotes the ideal optical field without turbulence and $U_i(i > 1)$ denotes the ith-order scattering and turbulence-disturbance term. The higher the order of the disturbance term, the smaller its component.

The advantage of the Born approximation is that the fluctuation equation with random variables and spatially dependent coefficients is transformed into a chi-square equation and a set of linear non-chi-square equations. Each non-chi-square equation in the set of linear non-chi-square equations can be solved by Green's function.

The Rytov approximation method transforms the electromagnetic-wave equation under weak-turbulence conditions into

$$U(\mathbf{r}, L) = U_0(\mathbf{r}, L) \exp[\psi(\mathbf{r}, L)], \tag{7.13}$$

where $\psi(\mathbf{r}, L)$ is the complex phase perturbation due to atmospheric turbulence. The Rytov approximation can be used to directly solve the wave-conduction equation, obtaining the first- and second-order solutions.

Most of the early work in Rytov's theory used only first-order perturbation terms, which are proportional to the first-order Born approximation, also known as the single-scattering approximation. First-order perturbation is sufficient to compute statistics of interest, such as the logarithmic-amplitude variance, phase variance, intensity and phase correlation functions, and wave-structure functions.

7.1.1.4 Statistical-Optics Methods

The geometric-optics and Rytov microturbulence methods are limited, and are more accurate and reasonable for the analysis of beam transport in weakly turbulent regions with isotropic homogeneity. However, for a more complex strong-turbulence region, the limitations of these two analysis methods are evident. Therefore, for the beam-transmission problem in strongly turbulent regions, the solution of the correlation coefficient and correlation statistical moment is more accurate and consistent with the experimental-data verification.

The first-order statistical moment represents the variation in the mean value of the amplitude, assuming that the expression of the optical field after transmission through the atmospheric channel is $U(x, y, z)$ and the expression of the optical field at the transmitting end is $U_0(x_0, y_0, 0)$. Then, the first-order statistical moment of the amplitude of the optical field after transmission is [10]

$$U(x, y, z) = U_o(x_0, y_0, 0) \exp\left[-\int_0^t \alpha(z^{\cdot})dz^{\cdot}\right], \tag{7.14}$$

where $\alpha(z^{\cdot})$ indicates the attenuation and beam-coherence degradation caused by the attenuation, scattering, and reflection of the beam field by atmospheric turbulence. It depends mainly on the characteristics of the atmospheric turbulence itself, such as turbulence intensity, inner turbulence scale, outer turbulence scale, and other parameters. The single-point second-order statistical moment represents the average value of the beam intensity, which is based on the relation that the beam intensity is the square of the amplitude, and the second-order statistical moment of the beam intensity is [10]

$$I(x, y, z) = I_0(x, y, 0) \exp\left[-2\int_0^z \alpha(z^{\cdot})dz^{\cdot}\right]. \tag{7.15}$$

The second-order statistical relationship between the two points represents the coherence of the optical field after the beam transmission, and the larger the value of the second-order statistical matrix of the beam intensity at this point, the better the coherence of the optical field after transmission. The fourth-order matrix usually represents the correlation between multiple points on the beam field, comparing two, and can be used to characterize the spatial-correlation properties of the beam field. The fourth-order statistical moment of two points is [11]

$$\begin{aligned} M_4 &= \langle u(\rho_1, z)u(\rho_2, z)u^*(\rho_1, z)u^*(\rho_2, z)\rangle \\ &= \langle I(\rho_1, z), I(\rho_2, z)\rangle. \end{aligned} \tag{7.16}$$

7.1.2 Plane-Wave Phase-Undulation Difference Analysis

Two plane waves with wavelengths λ_1 and λ_2, transmitted in an atmospheric channel, are received at the receiving end by an antenna with aperture D. The wavefront phases of the two beams measured in the receiving-aperture plane are ϕ_1 and ϕ_2, respectively, and the phases vary irregularly with time. The wavefront phase undulation at moment t can be defined as [12]

$$\theta_i(x, y) = \phi_i(x, y) - \frac{1}{A}\iint dxdy\phi_i(x, y)W(x, y), \quad i = 1, 2, \tag{7.17}$$

where $\theta_i(x, y)$ is the wavefront-phase undulation of a plane wave in the plane of the receiving aperture, A is the area of the receiving-aperture region, and $W(x, y)$ defines the receiving aperture with diameter D:

$$W(y) = \begin{cases} 1 & |y| < D/2 \\ 0 & |y| > D/2 \end{cases}. \tag{7.18}$$

The mean-square value of the phase-undulation variance of two beams with different wavelengths is defined as follows:

$$\overline{\Delta\theta^2} = \frac{1}{A}\iint [\theta_1(x, y) - \theta_2(x, y)]^2 W(x, y)dxdy, \tag{7.19}$$

where $\overline{\Delta\theta^2}$ is the mean-square value of the phase difference between the wavefront-phase undulations at the same point on the receiving aperture for plane waves of two different wavelengths:

$$\begin{aligned} \sigma_\theta^2 &= \overline{\langle\Delta\theta^2\rangle} \\ &= \frac{1}{A^3}\iiint dxdydtW(x, y)W(t) \times \langle[\phi_1(x, y) - \phi_2(x, y)] \\ &\quad \times[\phi_1(x, y) - \phi_1(t) - \phi_2(x, y) + \phi_2(t)]\rangle, \end{aligned} \tag{7.20}$$

where $\langle[\phi_1(x, y) - \phi_2(x, y)]\rangle$ can be defined as the mutual-covariance function of the difference in phase undulation between the two wavelengths:

$$C_\phi(x - y; \lambda_1, \lambda_2) = \langle\phi_1(x)\phi_2(y)\rangle. \tag{7.21}$$

Based on the mutual-covariance function, we can obtain the structural function of the difference in phase undulation for plane waves of different wavelengths at the same point on the receiving aperture [13]:

$$D_\phi(x - y; \lambda_1, \lambda_2) = 2[C_\phi(0; \lambda_1, \lambda_2) - C_\phi(x - y; \lambda_1, \lambda_2)]. \tag{7.22}$$

By substituting Eq. (7.21) into Eq. (7.20), calculating, and organizing it, we obtain

$$\sigma_\theta^2 = \frac{1}{A^2}\int\int \mathrm{d}x\mathrm{d}yW(x)W(y)$$
$$\left[\frac{1}{2}D_\phi(x - y; \lambda_1, \lambda_1) + \frac{1}{2}D_\phi(x - y; \lambda_2, \lambda_2) - D_\phi(x - y; \lambda_1, \lambda_2)\right] \tag{7.23}$$

According to the results obtained in [11], the phase-structure function is a wavelength-dependent function; therefore, the phase-structure function corresponding to the mutual covariance between the two wavelengths is [11]

$$\begin{aligned} D_\phi(x - y, \lambda_1, \lambda_2) &= \frac{1}{k_1 k_2} D_\Omega(x - y, \lambda_1, \lambda_2) \\ &= \frac{8.16k^2(1-\nu^2)}{2\pi} \int_0^L dzC_n^2(Z) \\ &\quad \times \int_0^\infty \mathrm{d}KK^{-8/3}[1 - J_0(K\rho)] \\ &\quad \times \left[\cos\left(\frac{K^2}{k}\frac{\nu}{1-\nu^2}L - Z\right) + \cos\left(\frac{K^2}{k}\frac{L-Z}{1-\nu^2}\right)\right], \end{aligned} \tag{7.24}$$

where $k = (k_1 + k_2)/2$, $k_1 = 2\pi/\lambda_1$, and $k_2 = 2\pi/\lambda_2$. $\upsilon = |k_1 - k_2|/(k_1 - k_2)$, K is the spatial wavenumber, Z is the distance of the optical-signal transmission, and L is the total transmission distance. $J_0(K\rho)$ is the Bessel function with respect to the spatial wavenumber and the plane vector, where $\rho = x - y$. $C_n^2(z)$ is the atmospheric refractive-index structure constant on the transmission path used to characterize the atmospheric-turbulence intensity. Substituting $\upsilon = |k_1 - k_2|/(k_1 - k_2)$ into Eq. (7.24), it can be expanded to obtain the following expressions for beams of different wavelengths [14]:

$$
\begin{aligned}
D_\phi(x-y,\lambda_1,\lambda_2) &= \frac{1}{k_1k_2}D_\Omega(x-y,\lambda_1,\lambda_2)\\
&= \tfrac{8.16k^2(1-\nu^2)}{2\pi}\int_0^L \mathrm{d}ZC_n^2(Z)\\
&\quad\times\int_0^\infty \mathrm{d}KK^{-8/3}[1-J_0(K\rho)]\\
&\quad\times\left\{\cos\left[\frac{(L-Z)K^2}{2}\frac{k_1-k_2}{k_1k_2}\right]+\cos\left[\frac{(L-Z)K^2}{2}\frac{k_1+k_2}{k_1k_2}\right]\right\}
\end{aligned}
\tag{7.25}
$$

Substituting Eq. (7.25) into Eq. (7.23) yields

$$
\begin{aligned}
\sigma_\theta^2 &= \frac{1}{A}\int d\overline{\rho}\left[\frac{1}{2}D_\phi(\rho;\lambda_1)+\frac{1}{2}D_\phi(\rho;\lambda_2)-D_\phi(\rho;\lambda_1,\lambda_2)\right]\\
&\quad\times\frac{2}{\pi}\left\{\cos^{-1}\left(\frac{\rho}{D}\right)-\frac{\rho}{D}\left[1-\left(\frac{\rho}{D}\right)^2\right]^{1/2}\right\},
\end{aligned}
\tag{7.26}
$$

where ρ is the distance from a point on the receiving section to the center of the circle and D is the radius of the receiving section. Using the rotational symmetry of the structural function of the circular receiving aperture, Eq. (7.27) can be obtained after further simplification [15]:

$$
\begin{aligned}
\sigma_\theta^2 &= \frac{16}{\pi D^2}\int_0^D \rho\mathrm{d}\rho\times\left[\frac{1}{2}D_\phi(\rho;\lambda_1)+\frac{1}{2}D_\phi(\rho;\lambda_2)-D_\phi(\rho;\lambda_1,\lambda_2)\right]\\
&\quad\times\left\{\cos^{-1}\left(\frac{\rho}{D}\right)-\frac{\rho}{D}\left[1-\left(\frac{\rho}{D}\right)^2\right]^{1/2}\right\}.
\end{aligned}
\tag{7.27}
$$

After substituting Eq. (7.25) into Eq. (7.27), using trigonometric symmetry to simplify and combine like terms, and integrating over the entire structural aperture, we obtain

$$
\begin{aligned}
\sigma_\theta^2 &= \frac{4.08}{\pi}\int_0^L C_n^2(Z)\mathrm{d}Z\int_0^\infty K^{-8/3}\mathrm{d}K\\
&\quad\times\left\{1-\left(\frac{4}{KD}\right)^2\left[J_1\left(\frac{KD}{2}\right)\right]^2\right\}\\
&\quad\times\left\{\cos\left[\frac{(L-Z)K^2}{2k_1}\right]-\cos\left[\frac{(L-Z)K^2}{2k_2}\right]\right\},
\end{aligned}
\tag{7.28}
$$

where D is the diameter of the receiving aperture, L is the beam-transmission distance, and K is the spatial wavenumber. C_n^2 is the atmospheric refractive-index structure constant, k_1 and k_2 are the wavenumbers corresponding to wavelengths λ_1 and λ_1, and $J_1\left(\frac{KD}{2}\right)$ is a class of Bessel functions, respectively [16]:

$$J_1(x) = \sum_{m=0}^{+\infty} \frac{(-1)^m}{m!\Gamma(m+2)} \left(\frac{x}{2}\right)^{2m+1} \tag{7.29}$$

$$J_1(KD) = 1 - \Gamma_0(KD)\frac{z}{L}. \tag{7.30}$$

The cosine function in the last term of the formula is calculated from the sum-and-difference product formula and Taylor expansion:

$$\cos\left[\frac{(L-Z)K^2}{2k_1}\right] - \cos\left[\frac{(L-Z)K^2}{2k_2}\right] = -\frac{K^4(L-z)^2(k_1^2-k_2^2)}{8k_1^2k_2^2}. \tag{7.31}$$

After obtaining the turbulent power spectrum, the expression for the difference between the wavefront aberrations of plane waves of different wavelengths can be calculated.

When two plane beams of different wavelengths are transmitted along a horizontal transmission path, wavefront phase distortion and undulations are produced by the influence of atmospheric turbulence.

In Eq. (7.20), the difference between the wavefront-phase aberrations of two wavelength beams is taken as the mean-square value to measure the magnitude of the difference in phase aberrations between the two beams, afterwards, it is calculated to obtain Eq. (7.28). It can be seen that the difference between the phase undulations is mainly influenced by the wavelength, but also by the transmission distance, the atmospheric refractive index structure constant, the size of the receiving aperture, and the number of spatial waves.

7.1.3 *Spherical-Wave Phase-Undulation Difference Analysis*

The dual-wavelength adaptive-optics correction system is a correction method that obtains wavefront-phase information via indirect measurements. In practice, it is more common to use spherical-wave beams as signal beams and beacon beams; therefore, we analyzed the phase-undulation error between spherical waves of different wavelengths, based on the previous plane-wave analysis.

As shown in Fig. 7.1, in a horizontal atmospheric-channel link with a transmission distance L, ρ is the marker point at the transmitter side on the phase plane of the two waves, ρ_1 is the marker point at the receiver side on the phase plane of the two waves, and there is a certain distance between the two marker points. The difference between the phase aberrations of the two wavelength beams is calculated at the same point,

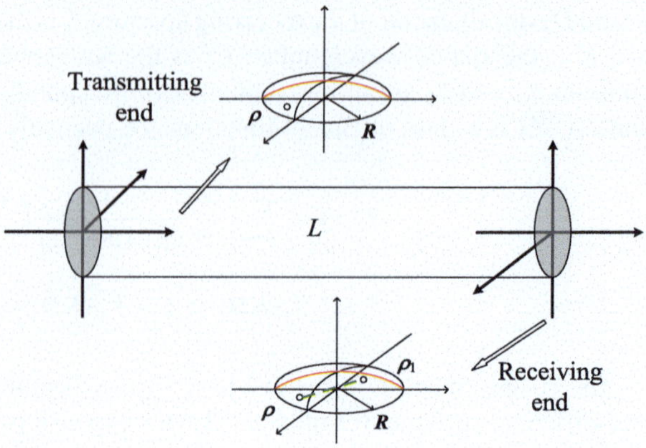

Fig. 7.1 Wavefront phase diagram with transmitter and receiver

and the structural function of the phase-distortion difference between spherical waves of different wavelengths on the receiving aperture is then calculated by comparing the two points.

The optical field at the receiving end, i.e., the wavefront function, is [17]

$$U(\rho) = \frac{k}{2\pi i L} \iint_{-\infty}^{+\infty} \mathrm{d}^2\rho_1 U_0(\rho_1) \times \exp\left[i\frac{k}{2L}(\rho - \rho_1)^2 + \chi_L(k, \rho, \rho_1) + iS_L(k, \rho, \rho_1)\right]. \tag{7.32}$$

The phase function of a spherical wave of $\lambda = 2\pi/k$ from the transmitter plane ρ_1 to the receiver plane ρ is the complex amplitude expression of the wavefront phase function $\left[\chi_L(k, \rho, \rho_1) + iS_L(k, \rho, \rho_1)\right]$, where $U_0(\rho_1)$ is the original phase function of the transmitter. A spherical-wave beacon beam of wavelength $\lambda_2 = 2\pi/k_2$ is transmitted with the signal beam in the common optical path. Under the same transmission conditions, the phase function of the beacon beam at the receiving end is

$$\varphi_S(\rho_1) = (k_2/2L)\rho_1^2 + S_L(k_2, 0, \rho_1). \tag{7.33}$$

The basic principle of adaptive-optics correction is to superimpose a complex conjugate phase that generates an aberrated phase on the aberrated phase. After superimposing the complex conjugate phase of the signal beam on the signal beam, we obtain the corrected optical-field phase of the signal beam. The expression of the complex-conjugate phase superimposition process is

$$\Delta S(\rho) = S_L(k, 0, \rho) - S_{-L}(k_2, 0, \rho), \tag{7.34}$$

where $S_{-L}(k_2, 0, \rho)$ denotes the complex-conjugate phase of the wavefront phase obtained from the beacon-beam measurement, and $S_L(k, 0, \rho)$ denotes the signal-beam wavefront aberration phase. Then, $\Delta S(\rho)$ is the phase deviation due to the difference in wavelength, i.e., the phase residual after correction by the adaptive-optics system. We use a coefficient-averaged mean-squared value of the aberrations $\Delta S(\rho)$ to facilitate the observation of their variation under turbulent time-varying characteristics; the expression is

$$d_{\Delta S}(\rho_1, \rho_2) = \langle [\Delta S(\rho_1) - \Delta S(\rho_2)^2] \rangle, \tag{7.35}$$

where ρ_1 and ρ_2 are the points at which the wavefront-undulation phase difference values are obtained at different moments.

In each homogeneous turbulent flow, the mean-square expression for the phase difference of the phase undulation owing to different wavelengths is given according to the Rytov-variance perturbation approximation theory [6]:

$$\begin{aligned} d_{\Delta S}(\rho_1, \rho_2) = 4\pi^2 \int_0^L \mathrm{d}x \int_0^\infty \mathrm{d}\kappa \kappa \phi_n(\kappa) \\ \times \left[1 - \Gamma_0(\kappa|\rho_1 - \rho_2|)\frac{x}{L}\right] \\ \times \left[k \cos\frac{\kappa^2(L-x)x}{2kL} - k_2 \cos\frac{\kappa^2(L-x)x}{2k_2 L}\right]^2, \end{aligned} \tag{7.36}$$

where Γ_0 is the gamma function, L is the communication-link optical-signal transmission distance, κ is the spatial wavenumber, $k = 2\pi/\lambda$ is the signal-beam wavenumber, and $k_2 = 2\pi/\lambda_2$ is the beacon-beam wavenumber. The Kolmogorov turbulence power spectrum is

$$\Phi_n(k) = 0.033 C_n^2 \kappa^{-11/3}, \quad 2\pi/L_0 < \kappa < 2\pi/l_0, \tag{7.37}$$

where C_n^2 is the atmospheric refractive-index structure constant, and L_0 and l_0 are the outer and inner turbulent scales, respectively. Substituting the Kolmogorov-turbulence power spectrum into Eq. 7.36 [18] yields the following [19]:

$$d_{\Delta S}(\rho) \cong \begin{cases} a_1 C_n^2 (k-k_2)^2 \kappa_m 7/3 L^3 \rho^2, & \kappa_m \rho \ll 1, \\ a_2 C_n^2 (k-k_2)^2 L \rho^{5/3}, & \kappa_m \rho \ll 1, \end{cases} \tag{7.38}$$

where $\rho = \rho_1 - \rho_2$. $\rho_1 = \rho_2$ denotes the value of the phase undulation of the signal and beacon beams at the same point on the receiving plane; C_n^2 is the atmospheric refractive-index structure constant, $\kappa_m = 5.92/l_0$ where l_0 denotes the internal turbulent scale, a_1 and a_2 are the numerical-calculation constants under the two expressions, respectively, and denotes the equivalent width of the collimated beam.

The mean-square value of the difference between the wavelength of the signal beam and the phase of the beacon-beam wavefront aberration received by the system in a circular receiving aperture at a certain radius is used as a measure of the magnitude of the wavefront-distortion difference after beams of different wavelengths pass through atmospheric turbulence. It is also a correction residual after the adaptive-optics system corrects the signal beam according to the beacon-beam wavefront-phase information.

7.2 Adaptive-Optics System Correction-Residual Analysis

Adaptive-optical systems have played a great role in applications in various disciplines, such as astronomy, medicine, and communication; however, their measurement errors and correction errors have limited their correction accuracy and they have become one of the bottlenecks limiting the system performance [20]. Therefore, an analysis of the calibration error of adaptive-optics systems and a description of the sources of errors and their effects are necessary to guide the selection of appropriate design parameters, according to their specific applications in the fields of medicine, laser communication, and imaging [21].

7.2.1 AO-System Error Model

Several major sources of errors in adaptive-optics (AO) systems are the system bandwidth error, beam-cone effect, unequal-halo error, and noise error [22]. A dual-wavelength spatial laser communication system mainly includes a signal beam and beacon beam combined with beam-transmitting and receiving optical systems, wavefront sensors, deformation mirrors, and control systems. The signal beam and the beacon beam are co-emitted or not, the measurement error of the wavefront sensor, the fitting-surface error of the deformation mirror, and the unequal-halo error caused by the atmospheric dispersion of the signal and beacon beams are the sources of the system-correction residuals. The magnitude is measured by the mean-squared difference of the wavefront-phase undulations produced by these errors. Assuming that the errors are independent of each other, the total error can be calculated as follows:

$$\sigma_{sum} = \sqrt{\sigma_{Noise}^2 + \sigma_{fit}^2 + \sigma_{calib}^2 + \sigma_{iso}^2 + \sigma_{cone}^2}, \tag{7.39}$$

where σ_{sum} denotes the total system correction residual and σ_{cone} denotes the non-common optical path error. σ_{iso} denotes the unequal-halo error, which includes both the dispersion unequal-halo error and the unequal-halo error caused by the zenith-angle refraction. σ_{fit} denotes the deformation mirror-type fitting error and σ_{Noise}

denotes the electronic noise and errors caused by the mechanical vibration of the system. The system fitting, measurement, and frequency-matching errors are the main errors of adaptive-optics systems, and are introduced below. In addition, the measurement error caused by different wavelengths is introduced in detail.

7.2.2 Phase-Fitting Errors of the Deformed Mirror

The wavefront corrector generates the inverse-conjugate phase of the aberrated phase and is the central component of the adaptive-optics system. The wavefront corrector generates a corrected front pattern or changes the refractive index of the transmission medium under the control of the system, which in turn changes the optical range of the optical wavefront for phase correction [23].

To ensure the accuracy of the aberration-wavefront correction by the adaptive-optics system, the wavefront corrector must have sufficient spatial degrees of freedom to well fit the aberration to be corrected. Moreover, whether the response speed far exceeds the temporal frequency of the perturbed wavefront and whether the spatial frequency of the mirror resolution satisfies the Greenwood frequency [24] determine the correction accuracy of the deformation mirror. The deformation mirror can be considered as a filter, and the part of the aberration power spectrum with a spatial frequency higher than the cutoff frequency determined by the Nyquist sampling criterion will not be corrected [25].

From the literature [26], it is possible to obtain the following equation for the phase-shape fitting error:

$$\sigma_{fit}^2 = 0.274(D/r_0)^{5/3}N_a^{-5/6}\,, \tag{7.40}$$

where D is the physical diameter of the receiving antenna at the receiver end; r_0 is the atmospheric-turbulence coherence constant; and N_a is the number of submirrors in the deformation mirror. Considering that the shape of the deformation mirror can be square, prolate, etc., and has been commonly used for the middle-masking ratio water-draining problem of the Mak–Cass antenna, the equation is simplified to [27]

$$\sigma_{fit}^2 = \mu(d/r_0)^{5/3}, \tag{7.41}$$

where μ is a constant associated with the phase shape; it is 0.335 for square mirrors, 0.339 for hexagonal-splicing mirrors, and 0.221 for continuously actuated deformation mirrors, according to the literature [28]. The error offset is obtained by taking the mean value of the closed-loop system test offset, and the corrected phase-shape matrix is obtained, based on the reconstruction matrix and the system-response matrix operation. The error offset is obtained after the difference between the mean value and the response phase-shape mean variance. The errors caused by phase fitting can be determined by drawing the erroneous phase shape and then obtaining the root-mean-square (RMS) of the error phase shape according to the offset [29].

7.2.3 Non-common Optical-Path Errors

To detect faint targets, reference stars are required to indirectly detect the disturbance of the wavefront phase caused by atmospheric turbulence. To facilitate the targeting and tracking of acquisition, pointing, and tracking (APT) systems, space laser communication deliberately selects beacon beams with better transmittance under complex atmospheric conditions to indirectly obtain wavefront-phase distortion information.

Commonly used beacon beams include the Rayleigh beacon, sodium guide star, and other types; however, the routes of the beacon beam and target source beam through atmospheric turbulence are not the same, and the wavefront-sensor measurement of the beacon-wavefront phase does not match the signal-beam phase. In space laser communication, the non-common optical-path emission of the signal and beacon beams leads to the same problem. There is a certain angular error between the guide star and the target, which means that the path through atmospheric turbulence during transmission is not exactly the same. The error angle between the target beam and the guide star can be expressed as [30]

$$\theta_0 = \left[2.91k^2 \sec^{8/3} \zeta \int C_n^2(h) h^{5/3} dh\right]^{-3/5}, \tag{7.42}$$

where θ_0 is the error angle, ζ is the azimuthal angle, k is the spatial wavenumber, h is the transmission-path length, and C_n^2 is the atmospheric refractive-index structure constant. Then, the corresponding unequal-halo error can be calculated according to the following equation [25]:

$$\sigma_{iso}^2 = a(\theta/\theta_0)^{5/3}, \tag{7.43}$$

where $a = kD/2$ and D is the diameter of the receiving antenna. A more detailed expression for Eq. (7.43) was obtained using the Mellin transform [31]:

$$\sigma_{iso}^2 = 2.606k_0^2 \int_0^\infty dh C_n^2(h) \int dk k^{-8/3}(1 - J_0[k\theta h]), \tag{7.44}$$

where J_0 is the zero-order Bessel function and the other parameters are the same as before.

The above error is caused by the non-common optical path between the guide star and the target beam source. However, the unequal-halo error also includes the dispersion unequal-halo error; that is, the wavelengths between the guide star and the target-beam source are different, and there is a certain error in the wavefront aberration between the two beams, owing to atmospheric dispersion.

7.2.4 Wavefront-Sensor Measurement Errors

The wavefront sensor converts the wavefront-phase undulation into the output of the offset between the spot center of mass in the measurement plane, relative to the calibration point, and generates a wavefront-reconstruction matrix based on the offset. The sources of its measurement error are the limited number of sub-apertures and low spatial resolution. The detector itself is also an electronic component, and its internal electronic noise and dark current also affect the number of received photons, the noise of the camera output process, and other factors. The measurement offset is internally passed through the wavefront-correction system, which eventually leads to correction errors.

The lens subarray in the wavefront sensor images a spot on the focal plane, forming a subspot array. Readout noise, including preamplifier noise and A/D circuit-conversion noise occurs during the sensor reading of the slope information of the spot array. The pixel-sampling noise can be expressed in the form of a variance [32]:

$$\sigma_{xcr}^2 = \frac{\sigma_A^2}{V} + \frac{\sigma_r^2}{V^2} ML\left(\frac{L^2 - 1}{12} + X_c^2\right), \tag{7.45}$$

where σ_{xcr}^2 denotes the variance of the center-detection error due to readout noise and photon-undulation noise, σ_A^2 is the equivalent Gaussian width of the spot, and σ_r^2 is the readout noise variance. M and L denote the sampling-window areas in the focal plane for each sublens, and V is the total number of photon events within the window.

The output signal of the wavefront sensor contains a certain dark-level signal. Its effect on the detection accuracy is comparable to that of random noise and affects both the systematic and random errors of the center-of-mass detection. The systematic error of the center-of-mass detection can be expressed as

$$\Delta_{xc}^2 = (\overline{x_c} - \overline{x_p})^2, \tag{7.46}$$

where Δ_{xc}^2 denotes the systematic error of the center-of-mass detection, $\overline{x_c}$ denotes the detection position, $\overline{x_p}$ denotes the actual position of the spot center of mass, and Δ_{xc}^2 denotes the square of the error. The variance of the center-of-mass position is calculated using the following formula:

$$\sigma_{xc}^2 = \left(\frac{SNR}{1 + SNR}\right)^2 \sigma_{xp}^2 + \left(\frac{1}{1 + SNR}\right)^2 \sigma_{xn}^2, \tag{7.47}$$

where σ_{xp}^2 denotes the variance of its own system-detection error, σ_{xn}^2 denotes the variance of the background-level center-of-mass position, and SNR is the system signal-to-noise ratio. Hence, the above equation can accurately describe the wavefront-sensor center-of-mass detection-system error under the influence of the background level. The random error generated by the plasmonic detection can then be expressed as

$$\sigma_{xc}^2 = \left(\frac{SNR}{1+SNR}\right)^2 \frac{K\sigma_A^2}{V_p} + \left(\frac{1}{1+SNR}\right)^2 \frac{\sigma_r^2}{V_n^2}\left[\frac{LM\,(L^2-1)}{12} + LMx_n^2\right], \tag{7.48}$$

where σ_{xc}^2 is the random detection-error variance and K is the number of analog-to-digital conversion units caused by a photon. By combining the two, the total center-of-mass detection error σ_x^2 under the influence of background-level noise is obtained as

$$\begin{aligned}\sigma_x^2 &= \sigma_{xc}^2 + \Delta_{xc}^2 \\ &= \left(\frac{1}{1+SNR}\right)^2 \left\{(SNR)^2 \frac{K\sigma_A^2}{V_p} + \frac{\sigma_r^2}{V_n^2}\left[\frac{LM\,(L^2-1)}{12} + LMx_n^2\right] + (\overline{x_c} - \overline{x_p})^2\right\}.\end{aligned} \tag{7.49}$$

In addition, the larger the effective photosensitive area of the wavefront sensor, the larger the truncation-threshold selection, which will also affect the detection accuracy of the sensor. On the premise that the system noise follows a Gaussian distribution, the larger the effective area, the higher the signal-to-noise ratio of the system, which will reduce the detection error. If the detection-threshold selection is too large or too small, it will increase the detection error; hence, the parameters must be selected to measure the performance of multiple aspects, compromising to obtain the most value [33]. In summary, the wavefront-sensor error group has four major categories: truncation errors, sampling errors, photon-noise and readout noise-errors, and dark-level noise errors [34].

7.2.5 Optical-Component Polarization Chromatic-Aberration Errors

Adaptive-optical systems have a large number of optical components and surface-coated elements. The coating can increase the beam-passage efficiency and reduce the reflected beam. Examples include the coated glass of optical antennas, coated reflection mirrors, and polarization beam-splitting prisms in the optical path. Beams incident at an angle to the surface of the coated elements produce a certain amount of polarization effect, resulting in a two-way delay and beam-energy attenuation, producing additional wavefront aberrations in the receiving focal plane.

As shown in Fig. 7.2, three-dimensional coordinates are established in the direction of the perpendicular paper surface and the parallel straight surface, and the incident-beam electric vector is divided into the s-component and p-component. The vibration direction of the s-component is always perpendicular to the paper surface, and the vibration direction of most electric vectors in the p-component

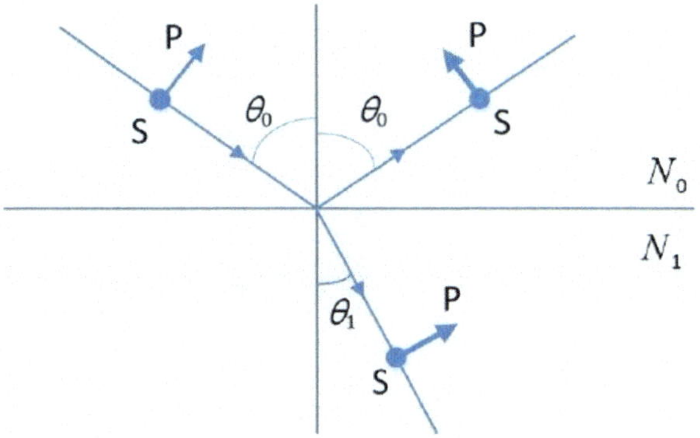

Fig. 7.2 Reflection and refraction of the beam on a dielectric interface [20]

is parallel to the paper surface; however, *p* contains part of the *s*-component. The medium has different refractive indices for the *s* and *p* components, resulting in a wavefront-distortion difference between the two polarization components, which is called polarization chromatic aberration [25, 35].

Corresponding to the example of the reflected beam in Fig. 7.2, the reflection coefficients of the *s*- and *p*-components can be calculated as

$$r_s = \frac{-\sin(\theta_0 - \theta_1)}{\sin(\theta_0 + \theta_1)} = |r_s|\exp(i\phi_s) \tag{7.50}$$

$$r_p = \frac{\tan(\theta_0 - \theta_1)}{\tan(\theta_0 + \theta_1)} = |r_p|\exp\left(i\phi_p\right), \tag{7.51}$$

where r_p and r_s are the reflectances of the *p*- and *s*-components of the beam on the boundary surface, θ_1 and θ_0 are the angle of reflection and angle of incidence, respectively, and N_0 and N_1 are the refractive indices of the medium on both sides of the incident boundary surface, respectively. ϕ_p and ϕ_s are the wavefront phases of the reflected *s*- and *p*-polarized beams, respectively.

According to Eqs. (7.50) and (7.51), when the beam is incident at a certain angle, the reflected phases of the *s* and *p* beams are as shown in Fig. 7.3, from which we can see that the phase difference between the *p* and *s* beams increases with changes in the wavelength and angle of incidence.

The Cassegrain optical antenna is used as an example for the polarization chromatic-aberration analysis. In space laser communication systems, the Maksutov–Cassegrain (Mak–Cass) antenna or Cassegrain optical antenna are often used to transmit optical signals. The Cassegrain optical-antenna reflector-type parabolic beam is mostly incident to the mirror at a certain angle, to increase the reflectivity. It is also coated on the parabolic-reflector surface to increase the reflectivity. Both of these cause polarization aberrations.

As shown in Fig. 7.4, the parallel beam incident to the primary mirror converges to the secondary mirror, the primary mirror from the edge to the center of the angle gradually increases, and the angle between the incident beam and the normal, which

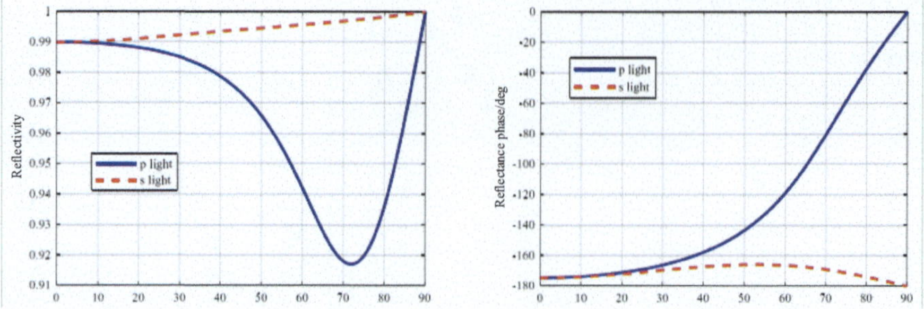

Fig. 7.3 Curves of the reflectivity and reflection phase of the *s* beam and *p* beam with the incidence angle [20]

is the angle of incidence, gradually increases. According to the analysis of the optical-antenna structure shown, with the increase in the angle of incidence, the phase difference between the *p* and *s* components increases.

Polarization aberration is a function of wavelength; hence, there are differences in the polarization chromatic aberration produced by the system under different wavelengths. Figure 7.5 shows the wavefront-polarization chromatic aberrations of 589-nm and 1300-nm beams, respectively. The RMS of the 589-nm wavefront-polarization chromatic aberration is 0.0180 rad and that of the 1300-nm wavelength beam is 0.0189 rad. The polarization chromatic aberration of the short wavelength is smaller than that of the long wavelength. The out-of-focus chromatic aberration between the two wavelengths reaches 1.144 rad, which is the main component of the correction residuals.

In addition to the aforementioned errors, mixing errors, beam-intensity flickering, and detector-calibration errors cause calibration errors in AO systems.

Fig. 7.4 Cassegrain beam-expansion system [20]

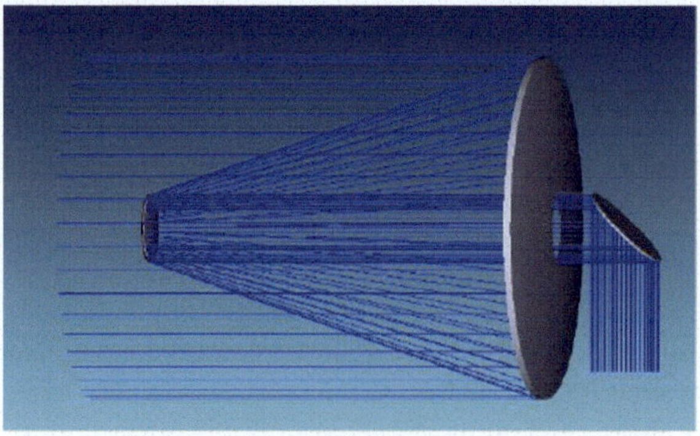

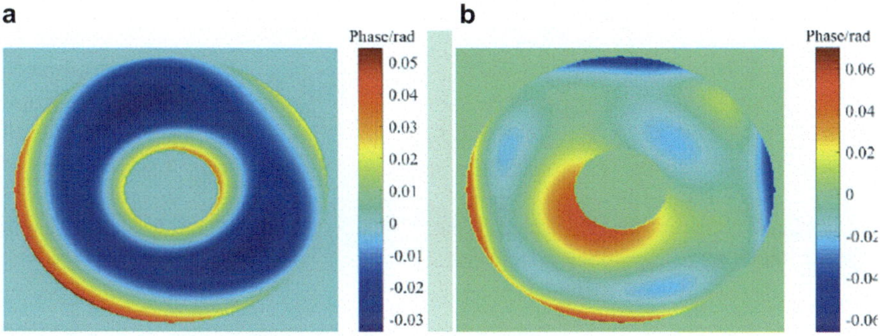

Fig. 7.5 Images of wavefront variations [20]. **a** 1319 nm. **b** 589 nm

7.3 Analysis of Wavefront Distortion Difference of Different-Wavelength Beams in Weak-Turbulence Regions

7.3.1 Angle-of-Arrival Undulation

The effect of atmospheric turbulence on a laser beam is related to the beam waveform, beam diameter, turbulence intensity, internal and external scales, and other factors. The essence of this effect is to degrade the beam coherence, which leads to a series of macroscopic physical manifestations, including beam-intensity scintillation, beam drift, spot expansion, and arrival-angle undulation [36].

When the turbulence is weak, there is more large-scale turbulence within the turbulence, and it primarily causes random deflections of the beam-propagation direction, along with diffraction and arrival-angle undulations in the received cross section [37]. As shown in Fig. 7.6, the transmitted signal beam passes through the receiving optical antenna after transmission through the atmospheric channel. The spot on the receiving focal plane is affected by its random jitter, which, in turn, affects the received-beam energy and fiber-coupling efficiency.

In arrival-angle undulation, the angle between the normal of the wavefront-envelope phase plane of the beam under the influence of turbulence and the normal of the laser wavefront-envelope phase plane unaffected by turbulence undulates with time around the mean value, as shown in Fig. 7.7. The beam vector produces a random polarization and arrives at the receiving cross section in the direction of its propagation with a deviation from the original propagation direction [38].

The phase difference between the center of the circle and any point ρ on the circumference of the circle is ΔS; thus, the angle of arrival α for a point at the center of the circle is

$$\alpha = \Delta S/(k\rho). \tag{7.52}$$

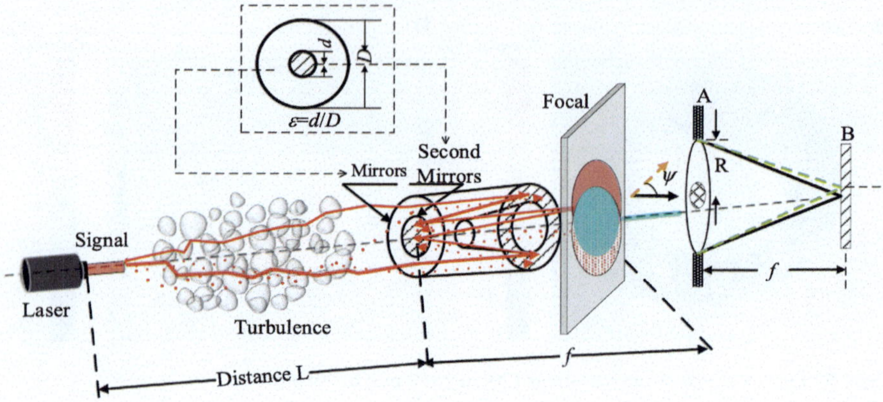

Fig. 7.6 Spot jitter caused by angle-of-arrival fluctuation

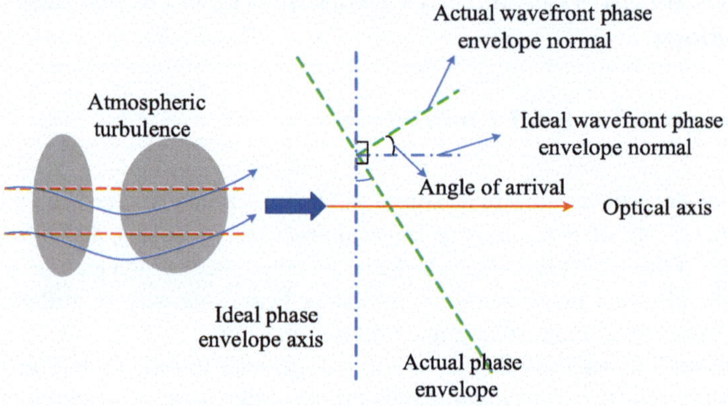

Fig. 7.7 Transmitted beam propagates in the atmosphere to produce the angle of arrival

When the radius of the circular region ρ is sufficiently small, the numerical magnitude of the angle of arrival at the center of the circle is obtained; thus, the limit of the defining equation for the angle of arrival is taken as

$$\alpha = \lim_{\rho \to 0} \Delta S/(k\rho). \tag{7.53}$$

The actual wavefront phase surface is an irregular and randomly undulating state; therefore, the arrival angles at different location points are random and arbitrary. In a practical case, the arrival angle can be calculated from the wavefront phase slope as follows:

$$\alpha = \lim_{\rho \to 0_k} \frac{1}{k} \cdot \frac{\partial D_S(\rho)}{\partial \rho}, \tag{7.54}$$

where $D_S(\rho)$ is the wavefront-phase structure function, which differs for plane waves, spherical waves, and Gaussian beams [39], as well as different atmospheric-turbulence power-spectrum conditions [40]. Therefore, we can obtain a generalized expression for the variance in the angle-of-arrival undulation [41]:

$$\begin{aligned}\sigma_{AF}^2 &= \lim_{\rho \to 0} \frac{1}{k^2} \cdot \frac{\partial^2 D_S(\rho)}{\partial \rho^2} \\ &= \pi^2 L \int_0^1 \mathrm{d}\xi \int_0^{+\infty} \mathrm{d}\kappa^3 \Phi_n(\kappa) h(\kappa, \xi)\kappa,\end{aligned} \tag{7.55}$$

where σ_{AF}^2 is the angular variance of arrival, L is the optical-signal transmission distance, $\Phi_n(\kappa)$ represents the turbulent power spectrum, and $h(\kappa, \xi)$ is a function of the different waveform weights representing the plane, spherical, or Gaussian beams of interest. The specific expression is [42]

$$\begin{cases} h_p(\kappa, \xi) = \left(1 + \frac{k}{\kappa^2 L} \sin\left(\frac{\kappa^2 L}{k}\right)\right) \mathrm{A}\left(\frac{D_G \kappa}{2}\right) \\ h_s(\kappa, \xi) = \left(1 + cos\left(\frac{\kappa^2 \xi (1-\xi) L}{k}\right)\right) \xi^2 \mathrm{A}\left(\frac{D_G \kappa \xi}{2}\right) \end{cases}, \tag{7.56}$$

where p represents the plane wave, s represents the spherical wave, and A is the aperture function corresponding to the circular receiving aperture. The Gaussian beam corresponds to a specific form that can be expressed as [42]

$$\begin{cases} h_p(\kappa, \xi) \approx 2\mathrm{A}\left(\frac{D_G \kappa}{2}\right) \\ h_s(\kappa, \xi) \approx 2\xi^2 \mathrm{A}\left(\frac{D_G \kappa \xi}{2}\right) \end{cases}. \tag{7.57}$$

Arrival-angle undulation is a low-order aberration that occupies 70% to 85% of the wavefront aberrations in space laser communication systems. It causes random spot jitter on the detector surface and affects the tracking and aiming of the APT system [43]. It also affects the power of the received optical signal in the finite receiving aperture [44] and even the fiber-coupling efficiency in the tail-signal fiber-coupling process [45]. For these reasons, researchers have conducted extensive research on arrival-angle undulation as a turbulent optical effect.

In 2003, Consortini et al. developed a mathematical model for the simple relationship between the internal and external turbulent scales and the angle-of-arrival scintillation, and measured the internal and external turbulent scales in the laboratory using the angle-of-arrival scintillation and beam-intensity scintillation [46].

In 2007, Italo Toselli and Larry. C. Andrews pointed out the phase structure constants as well as the angle-of-arrival undulation spectra under the weakly turbulent spectral Kolmogorov spectral line, and established universally applicable turbulence spectra as well as corresponding phase structure functions, and corresponding angle-of-arrival scintillation spectra [47].

In 2008, Conan, Borgnino, Ziad, and Martin et al. derived the spatio-temporal correlation function of the arrival angle based on the covariance function, and analyzed the spatio-temporal correlation of the arrival-angle undulation [48].

In 2017, Safari et al. simulated the performance degradation of free-space optical communication systems caused by angular fluctuations, such as beam distortion and angle-of-arrival (AOA) fluctuations, in the presence of atmospheric turbulence and transceiver vibrations [49]. Expressions for the interruption probability of coherent and direct detection in the scattering and thermal-noise limits were derived.

In 2018, Voelz, Wijerathna, and Muschinski et al. analyzed the angle-of-arrival generation in regions of strong and weak turbulence by means of ray tracing and fluctuating optics [50]. They showed the aperture-filtering effect on the angle of arrival, and justified the conclusions using numerical computer simulations and analyses.

7.3.2 Gaussian-Beam Arrival-Angle Undulation Correlation Function at Different Wavelengths

The correlation between Gaussian beams with wavelengths of λ_1 and λ_2 can be described by the mutual-correlation function [51] after the beams are transmitted through the same atmospheric channel with arrival-angle undulations of α_{λ_1} and α_{λ_2}, respectively, at the receiving end:

$$\left\langle \alpha_{\lambda_1}\alpha_{\lambda_2}\right\rangle = \frac{F^2}{4k_1k_2w_0^2}\int d^4\rho \exp\left(-\frac{2\rho^2}{w_0^2}\right) \times \left\langle \frac{\partial s_1(F,\rho_1,\lambda_1)}{\partial \rho_1}\cdot\frac{\partial s_2(F,\rho_2,\lambda_2)}{\partial \rho_2}\right\rangle T(\vec{\rho}) \tag{7.58}$$

$$T(\vec{\rho}) = \begin{cases} 1 & \rho \le D \\ 0 & \rho > D \end{cases}, \tag{7.59}$$

where $\left\langle \alpha_{\lambda_1}\alpha_{\lambda_2}\right\rangle$ denotes the reciprocal covariance between the corresponding angles of arrival undulations of the two wavelength beams, $F = \frac{z}{1+(z/f)^2}$ is the radius of curvature of the beam-wavefront surface, z is the beam-transmission distance, f is the beam parameter, $k_1 = \frac{2\pi}{\lambda_1}$ and $k_2 = \frac{2\pi}{\lambda_2}$ are the wave vectors of beams of different wavelengths, and ω_0 is the beam waist radius. D is the pupil diameter of the receiving optical system, and $\rho = \rho_1 - \rho_2$ denotes the distance between ρ_1 and ρ_2, two points on the wavefront surface; when $\rho = 0$, the two points coincide.

In Eq. (7.58) $\langle \ldots\ldots \rangle$ denotes the average system synthesis; and its specific expression is [52]

$$\left\langle \frac{\partial s_1(F,\rho_1,\lambda_1)}{\partial \rho_1}\cdot\frac{\partial s_2(F,\rho_2,\lambda_2)}{\partial \rho_2}\right\rangle = D_S''(\rho,\lambda_1,\lambda_2)\frac{\eta^2}{\rho^2} + D_S'(\rho,\lambda_1,\lambda_2)\left(1-\frac{\eta^2}{\rho^2}\right) \tag{7.60}$$

$$D_S = (L, \rho, \lambda_1, \lambda_2) = \frac{1}{1-h^2}\left[D_S(L, \rho, \lambda) - \frac{1}{h}D_x(Lh, \rho, \lambda)\right], \tag{7.61}$$

where $D_S(\rho, \lambda_1, \lambda_2)$ is a dual-frequency phase-structure function. $h = \frac{k_1-k_2}{k_1+k_2}$, $\lambda = \frac{2\pi}{k}$, $k = \frac{2k_1k_2}{k_1+k_2}$, and D_x are logarithmic-amplitude structure functions, and $D_S(L, \rho, \lambda)$ is a phase structure function. Substituting Eqs. (7.60) and (7.61) into Eq. (7.58), the correlation functions of the angular-arrival undulations of Gaussian beams at different wavelengths are obtained as follows:

$$G_\alpha(Z, \lambda_1, \lambda_2) = \sigma_\alpha^2(\lambda_1, \lambda_2) = \frac{0.101\pi^2\Gamma(1/6)}{7.963\sqrt{2}}C_n^2F^3(zR)^{-1/3} \times \frac{16}{17} \times \mathrm{Re}$$
$$\left[\frac{F\left(1, 1/6, 23/6; \frac{1}{1+ig_1}\right)}{(1+ig_1)^{1/6}} + \frac{F\left(1, 1/6, 23/6; \frac{1}{1+ig_2}\right)}{(1+ig_2)^{1/6}}\right], \tag{7.62}$$

where $G_\alpha(Z, \lambda_1, \lambda_2)$ is the number of interrelationships between the angular arrivals of Gaussian beams of different wavelengths; $g_1 = \frac{F(k_1+k_2)}{k_1k_2\alpha_0^2}$, $g_2 = \frac{F(k_1-k_2)}{k_1k_2\alpha_0^2}$, and $F(a, b, c, z)$ are Gaussian hypergeometric functions; R is the wavefront equivalent radius of curvature; and C_n^2 is the atmospheric refractive-index structure constant.

7.3.3 Numerical Relationship of the Gaussian-Beam Wavefront Phase at Different Wavelengths

The analysis of the wavefront-phase aberrations generated after the beam passes through the atmospheric channel using the geometric-optics method requires certain conditions to be met:

1. There are large- and small-scale vortices inside the atmospheric turbulence; the larger turbulent vortices deflect the beam passing through them, whereas the small vortices are diffracted by the beam emission. When the turbulence intensity is weak, the number of large vortices inside is high, and the phase aberrations produced by the Gaussian beam can be analyzed using geometric-optics methods [53].
2. The analysis of the wavefront-phase aberrations and the calculation of statistical-data characteristics are the results of long-term observations and satisfy the steady-state process criterion [54].
3. In moderate or weak turbulence regions, the turbulent power spectrum satisfies the Kolmogorov spectral line, and the isotropic condition is satisfied inside the medium.

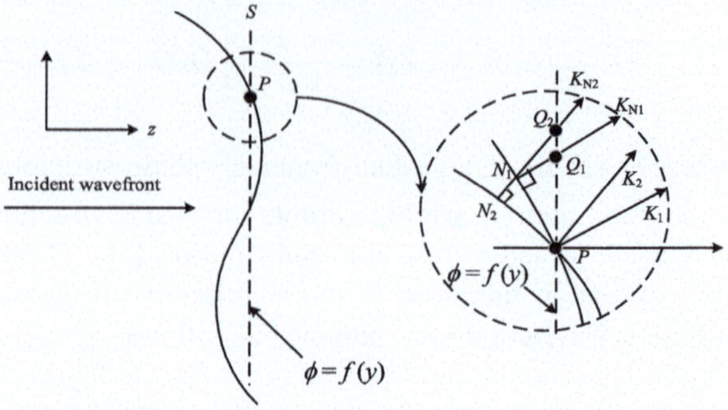

Fig. 7.8 Incident-wavefront section

Therefore, we analyzed the differences between the wavefront-phase aberrations in the paraxial region [55] for Gaussian beams of different wavelengths using a geometric-optics approach.

As shown in Fig. 7.8, the incident beam is incident along the *z*-axis in the forward direction, and the entire wavefront is distributed within the XOY plane. The wavefront-measurement surface *s* is perpendicular to the *z*-axis, and the phase-distribution function of the incident beam is at the measurement-line position. For wave $\varphi = f(y)$, *P* is a point on the measurement wavefront; owing to the wavefront-phase distortion, there is a phase difference between point *P* on the wavefront and the surrounding adjacent positions.

Assume that the signal-beam wavelength is λ_1 and the beacon-beam wavelength is λ_2. In the vicinity of point *P*, surface points N_1 and N_2 are selected along the two beams of equal-phase wavefronts, The distance between two points PN_1 and PN_2 is ρ, and the beam-transmission distance is *L*.

The propagation direction of the wavefront at points N_1 and N_2 are K_{λ_1} and K_{λ_2}, respectively, and the two propagation vectors intersect the measurement surface *S* at points Q_1 and Q_2. When ρ is very small and tends to 0, triangles N_1Q_1P and N_2Q_2P can be considered as right triangles. Then, the phase difference between the two points Q_1P and Q_2P is the corresponding wavefront phase difference between the beacon beam and the signal beam. The corresponding signal-beam wavefront phase difference is φ_{λ_1}, and the beacon-beam wavefront phase difference is φ_{λ_2}. The signal-beam and beacon-beam wavefront aberration phases under the above conditions can be expressed as

$$\varphi_{\lambda_2} = \varphi_p - \varphi_Q = \frac{2\pi}{\lambda_2} N_2 Q_2 = \frac{2\pi}{\lambda_2}(n_p - n_Q)L \tag{7.63}$$

$$\varphi_{\lambda_1} = \varphi_p - \varphi_{Q1} = \frac{2\pi}{\lambda_1} N_1 Q_1 = \frac{2\pi}{\lambda_1}(n_p - n_{Q1})L. \tag{7.64}$$

As the signal and beacon beams are transmitted along the same path, the air refractive index changes by the same amount between the corresponding point positions, such that

$$n_p - n_{Q1} = n_p - n_{Q2}. \tag{7.65}$$

When the signal beam and beacon beam are transmitted in the same atmospheric channel, the random changes in the refractive index in the atmospheric turbulence have different modulation effects on the Gaussian beams of different wavelengths. There is a certain difference between the wavefront undulating aberrations, i.e., the phase difference between points PQ_2 and PQ_1 is not the same, corresponding to the wavefront-aberration phase difference:

$$\begin{aligned}\Delta &= \left|\varphi_{\lambda_1} - \varphi_{\lambda_2}\right| \\ &= (\varphi_p - \varphi_Q) - (\varphi_p - \varphi_{Q1}) \\ &= 2\pi(n_p - n_Q)L\left|\frac{\lambda_1 - \lambda_2}{\lambda_1\lambda_2}\right|.\end{aligned} \tag{7.66}$$

The relationship between the signal-beam wavefront phase φ_{λ_1} at point P and the fiducial-beam wavefront phase φ_{λ_2} is

$$\varphi_{\lambda_1} = \frac{\frac{2\pi}{\lambda_2}(n_p - n_Q)L \pm \Delta}{\frac{2\pi}{\lambda_2}(n_p - n_Q)L} \cdot \varphi_{\lambda_2}. \tag{7.67}$$

Substituting Eqs. (7.65) and (7.66) into Eq. (7.67), the relationship between the aberration phases of the Gaussian beams at different wavelengths is obtained as follows:

$$\varphi_{\lambda_1} = \left(\frac{\lambda_2}{\lambda_1} \pm \left|\frac{\lambda_1 - \lambda_2}{\lambda_1\lambda_2}\right|\right)\varphi_{\lambda_2}. \tag{7.68}$$

According to this relationship, the wavefront-aberration phase difference of beams of different wavelengths can be calculated, based on the known wavelengths of the signal and beacon beams, beam-transmission distance L, and refractive-index change. The wavefront-aberration measurement of the signal beam can also be calculated based on that of the beacon beam.

7.3.4 Numerical Simulations

7.3.4.1 Transmission Characteristics of Gaussian Beams of Different Wavelengths in the Atmospheric Channel

Under atmospheric conditions with outer turbulent scale $L_0 = 5$ m, inner turbulent scale $l_0 = 0.1$ m, and atmospheric refractive-index structure constant $C_n^2 = 1 \times 10^{-20}$ $m^{-2/3}$ [56], Gaussian beams with wavelengths of 1550 nm, 850 nm, 632.8

nm, and 530 nm are transmitted in the same atmospheric channel with transmission distance $L = 10$ km. The simulation results of the wavefront phase distortion afterwards are shown in Fig. 7.9. From Fig. 7.9, it can be seen that under the same transmission conditions, the phase distortion of the Gaussian beam varies significantly with the wavelength, which indicates that atmospheric turbulence modulates the Gaussian beam differently at different wavelengths.

The undulating variance of the Gaussian-beam wavefront-phase aberration versus the wavelength is shown in Fig. 7.10. From Fig. 7.10, it can be seen that the wavefront-phase aberration variance decreases with increasing wavelength, indicating that atmospheric turbulence perturbs long waves less than short waves; i.e., the short-wave phase aberration is higher than that of the long wave.

If a long-wave beacon beam is used to correct a short-wave signal beam, the wavefront will be undercorrected. The correction residual increases with the increase of the wavelength difference between the two, and finally the compensation effect tends to disappear. If a short-wave beacon beam is used to correct a long-wave signal beam, the correction amount is larger than the actual phase distortion, and the correction residual increases rapidly with the increase of the wavelength. Finally, when the wavelength difference reaches a certain value, the correction residual will reach the phase-distortion error when there is no correction.

Figure 7.11 shows the wavelength difference of the phase aberrations of different Gaussian beams, where the signal-beam wavelength is 1550 nm. From the figure,

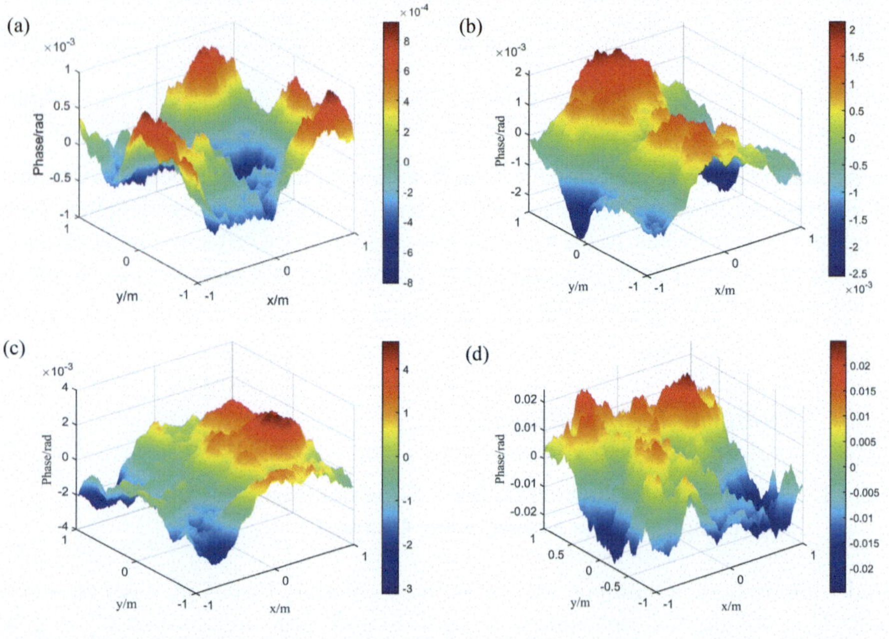

Fig. 7.9 Wavefront phase distortion of Gaussian beams with different wavelengths in the same atmospheric turbulence. **a** 1550 nm; **b** 850 nm; **c** 632.8 nm; **d** 530 nm

Fig. 7.10 Relationship between the mean-square value of the distorted wavefront and wavelength

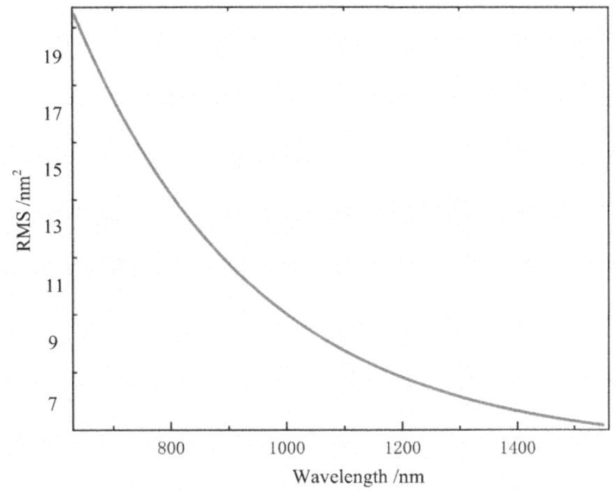

it can be seen that the smaller the wavelength difference between the signal beam and the beacon beam, the smaller the aberration-phase difference, which means that only when the signal-beam and beacon-beam wavelengths are very close to each other can the beacon-beam wavefront aberration be approximated as the signal-beam wavefront aberration.

In addition, it can also be concluded that when the signal-beam wavelength is larger than the beacon-beam wavelength, the phase relationship between the two wavelength beams should take the '-' sign, and when the signal-beam wavelength is smaller than the beacon-beam wavelength, the relationship should take the '+' sign.

Fig. 7.11 Change of the distortion-phase difference of different wavelength beams with the wavelength difference

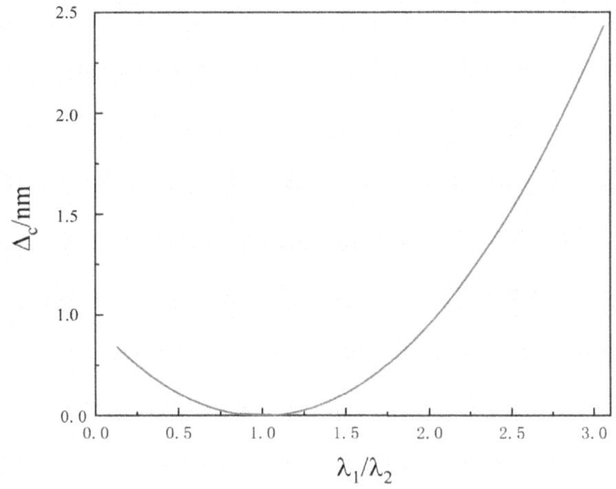

7.3.4.2 Angle-of-Arrival Undulation Correlation for Gaussian Beams of Different Wavelengths

The angle-of-arrival undulation correlation is related to multiple factors, such as the transmission distance, atmospheric refractive-index structure constant, beam waist radius, wavelength, and distance between two points in the wavefront space. In this section, the influence of the transmission distance, beam-wavelength difference, and turbulence intensity on the correlation is analyzed. The following wavelengths are used: 632.8, 750, 950, and 1280 nm, and the atmospheric refractive-index structure constant $C_n^2 = 1 \times 10^{-20}\ \text{m}^{-2/3} - C_n^2 = 1 \times 10^{-18}\ \text{m}^{-2/3}$.

When the bundle radius is $\omega_0 = 0.01$ m and the atmospheric refractive-index structure constant is $C_n^2 = 1 \times 10^{-20}\ \text{m}^{-2/3}$ [56], Fig. 7.12 shows the variation of the mutual relation number between different wavelength arrival-angular undulations with the transmission distance. As shown in the figure, the number of interrelationships between different wavelength arrival-angle undulations gradually increases with increasing transmission distance. This phenomenon arises because of the statistical correlation between the wavefront aberrations of different wavelengths at the same turbulent perturbation. The smaller the difference in wavelengths between the two beams at the same distance, the larger is the value of the correlation number between the corresponding arrival-angular undulations.

The variation curves of the interrelationships between the arrival-angle undulations of Gaussian beams of different wavelengths with the intensity of atmospheric turbulence are shown in Fig. 7.13. From Fig. 7.13, it can be observed that the interrelationship number between the arrival-angle undulations of Gaussian beams of different wavelengths gradually increases as the atmospheric refractive-index structure constant decreases. The smaller the wavelength difference between the beacon and signal beams, the larger the value of the correlation number between the corresponding arrival-angle undulations of beams of different wavelengths. This, in turn,

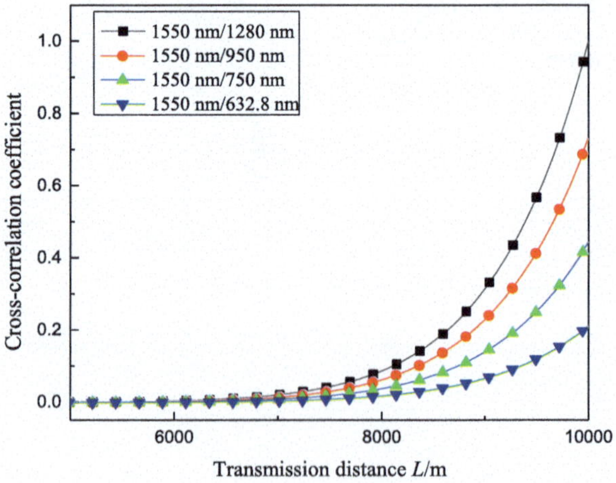

Fig. 7.12 Variation of the correlation coefficient between angle-of-arrival fluctuations of different wavelengths and transmission distances

indicates a higher consistency of the aberration information between the arrival-angle undulations of different wavelengths.

Figure 7.14 shows the effect of the beam waist radius on the correlation between the angular arrivals of Gaussian beams of different wavelengths; the vertical axis is the number of correlations after taking log10. The figure shows that the number of correlations between the arrival-angle undulation of different wavelengths of Gaussian beams gradually decreases as the beam waist radius increases, whereas the smaller the difference in wavelengths between the two beams, the higher the number of correlations.

From the above analysis, it can be seen that the correlation between the arrival-angle fluctuations of Gaussian beams of different wavelengths decreases with the increase of the atmospheric refractive-index structure constant, and under the

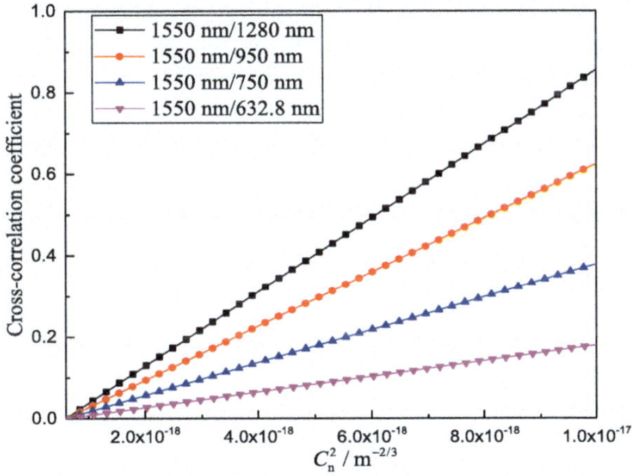

Fig. 7.13 Variation of the correlation coefficient between angle-of-arrival fluctuations of different wavelengths and the refractive-index structure constant

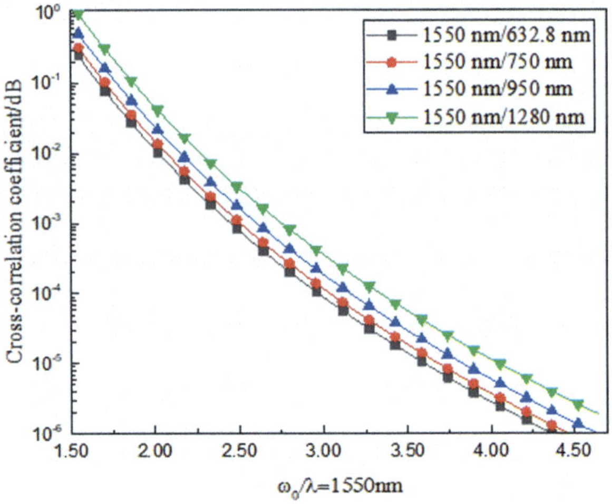

Fig. 7.14 Variation of the correlation coefficient between angle-of-arrival fluctuations of different wavelengths and the waist radius

same transmission distance and atmospheric refractive-index structure constant, the arrival-angle fluctuations of short waves are higher than those of long waves. In addition, under the same wavelength, the sensitivity of the arrival-angle fluctuations to the transmission distance is much higher than that to the atmospheric refractive-index structure constant.

7.4 Analysis of Wavefront Distortion Difference of Different-Wavelengths in Strong-Turbulence Regions

The essence of the optical effect of turbulence is the random distribution of the refractive index caused by the instability of the air temperature and pressure, which are continuously distributed and redistributed in different spaces over time. The energy carried by vortices inside the turbulence is continuously dispersed as the vortex size decreases until it is finally depleted. Therefore, turbulence is also classified as strong, medium-intensity, or weak turbulence [57].

The correlations and phase relations between Gaussian beams in the weak-turbulence region were analyzed in Sect. 7.3. In this section, the correlations and phase differences between Gaussian beams of different wavelengths in the strong-turbulence region are analyzed.

7.4.1 *Strong Turbulence*

In turbulence theory, the refractive-index structure function is used to describe the spatial distribution of the refractive index within atmospheric turbulence [58]:

$$D_n(r) = C_n^2 r^{2/3}, \tag{7.69}$$

where r represents the two points in the internal turbulence space. C_n^2 is the atmospheric refractive-index structure constant, which is related to environmental factors, such as altitude, temperature, atmospheric pressure, and wind speed and can measure the turbulence intensity. According to previous studies, the atmospheric turbulence is weakly turbulent when $C_n^2 < 10^{-17}$ $m^{-2/3}$; it is strongly turbulent or higher when $C_n^2 \geq 10^{-13}$ $m^{-2/3}$ [58]. The magnitude of the values does not vary significantly for a constant horizontal altitude; however, the effect of changes in altitude on the structural constants for slope paths, and uplink and downlink breaks is non-negligible and requires a specific analysis [59].

When the turbulence is located in a strong-turbulence region, the number of vortices with smaller scales inside the turbulence is higher, and the rate of time variation and spatial variation of the dissipation zone is faster. When the turbulence produces stronger diffraction effects, both the inner and outer turbulence scales must be considered. The Kolmogorov turbulence spectrum was used to describe turbulence in Sect. 7.3 [60]:

$$\Phi_n(\kappa) = 0.033C_n^2\kappa^{-11/3}, \quad 1/L_0 \ll \kappa \ll 1/l_0, \tag{7.70}$$

where $\Phi_n(\kappa)$ is the turbulence power spectrum, κ is the spatial wavenumber, and l_0 and L_0 are the inner and outer turbulence scales, respectively.

The turbulence power spectrum considers the spatial wavenumber in the isotropic region of the medium distribution, which is a more accurate description of the inertial region. However, it is not used completely for strongly turbulent regions, and the description error is larger. When the influence of the inner and outer turbulent scales cannot be ignored in the recalculation, and the dissipation range is further extended to noninertial subregions, Tatarskii introduced an *e*-exponential factor based on the Kolmogorov spectrum [61]:

$$\Phi_n(\kappa) = 0.033C_n^2\kappa^{-113}\exp\left(-\frac{\kappa^2}{\kappa_m^2}\right), \quad \kappa \gg 1/l_0; \kappa_m = 5.92/l_0. \tag{7.71}$$

However, this spectrum has singularities in the input region when taking extreme values, and is anisotropic in each; therefore, the Tatarskii spectrum is again improved for practical use so that it can maintain all homogeneous properties, even in strongly turbulent regions. The improved model is [62]

$$\Phi_n(\kappa) = \begin{cases} \frac{0.033C_n^2}{\left(\kappa^2+\kappa_0^2\right)^{1/6}}, & 0 \le \kappa \ll 1/l_0 \\ 0.033C_n^2\frac{\exp(-\kappa^2/\kappa_m^2)}{\left(\kappa^2+\kappa_0^2\right)^{1/6}}, & 0 \le \kappa < \infty \end{cases}, \tag{7.72}$$

where $\kappa_0 = 2\pi/L_0$. The first spectral model in Eq. 7.72 contains only the outer turbulent scale, and the second spectral model contains both the inner and outer turbulent scales. The exponential form of the second equation containing the outer turbulent scale is

$$\Phi(k) = 0.033C_n^2\kappa^{-11/3}[1 - \exp(-\kappa^2/\kappa_0^2)], \quad 0 \ll \kappa \ll 1/l_0, \tag{7.73}$$

where κ_0^2 is the parameter related to the outer turbulence-scale parameter; $\kappa_0 = c_0/L_0$, where c_0 is the scale parameter; the value is related to the actual application size, and its value is generally in the strong-turbulence region. The integration of this spectral line after any value in the input region is still convergent and satisfies the isotropic property. Moreover, the statistical properties and isotropy are the same within a reasonable range of wavenumbers; therefore, this spectral line is also used for analyzing the difference in phase distortion of the Gaussian-beam wavefront in the strong-turbulence region in this section.

According to the above discussion of the strong-turbulence power spectrum, we can summarize some characteristics of the strong-turbulence region compared to those of the weak-turbulence region.

1. The wavefront shape of the signal beam through strong turbulence is sharper and more complex and no longer satisfies the conditions for smooth integration regions or integration curves and geometric-optics analysis.
2. In addition to the phase distortion in the strongly turbulent region, intensity scintillation occurs again at the receiving end, and the spot is torn into a series of small scattered fragments [63].
3. The wavefront phase is no longer continuous, and many wavefront phase zeros appear, making it impossible to recover a continuous and smooth wavefront phase.

Based on the above characteristics of the strongly turbulent region, the error between the phase aberrations of the wavefronts of beams of different wavelengths cannot be analyzed using the geometric-optics analysis method at this time; instead, the phase-structure function method is used. It calculates the correlation function and the phase-structure function using the turbulent power spectrum applicable to the strongly turbulent region, and analyzes the influence of the internal and external turbulent scales, the spatial wavenumber, and other factors on the difference values.

7.4.2 Error Structure Function of Wavefront Distortion Difference of Different-Wavelengths Beams in Strong-Turbulence Regions

In the strong-turbulence region, the signal beam of wavelength λ_1 and the beacon beam of wavelength λ_2 are transmitted in a common optical path. The spatial optical communication system uses a Mak–Cass antenna to transmit the beam. After passing through the antenna, the Gaussian beam can be approximated as a uniform distribution with equal intensity in all parts in the receiving plane. The phase in the strong-turbulence region is torn into a discontinuous scattered distribution in the receiving cross section. For easy understanding and calculation, according to the Martov approximation,

$$\langle n(R_1)n(R_2)\rangle = B_n(R_1 - R_2) = \delta(z)A(r_1 - r_2), \tag{7.74}$$

where $n(\cdot)$ is the refractive index of air at a point on the transmission-path cross section, B_n is the mutual-correlation function between different points, and $A(r_1 - r_2)$ is the proportionality coefficient. To ensure the premise of uniform homogeneity and strong turbulence in the region, its value is provisionally set to 1. We can receive the signal beam at the receiving end, and the beacon-beam wavefront phase function can be approximated as a response function of the phase modulated by turbulence, as follows:

$$\phi_{\lambda_1} = k_1 \int_0^L \delta_n(\vec{r}, z)dz \tag{7.75}$$

$$\phi_{\lambda_2} = k_2 \int_0^L \delta_n(\vec{r}, z)dz, \tag{7.76}$$

where ϕ_{λ_1} and ϕ_{λ_2} are the wavefront-phase functions at wavelengths λ_1 and λ_2, L is the transmission-link length, r is the position vector at the receiving cross section, and $\delta(\vec{r}, z)$ is the shock-response function. k_1 and k_2 are the wavenumbers corresponding to wavelengths λ_1 and λ_2, respectively. From the above two expressions of the wavefront phase, it can be seen that the response of the phase function to atmospheric turbulence at a point on the receiving cross-section is composed of the $\delta(\vec{r}, z)$ shock-response function at the back and the wavenumber $k = 2\pi/\lambda$ at the front; thus, the wavefront phase distortion is different for different wavelengths after passing through turbulence.

The dispersion unequal-halo error $\triangle\ \phi$ between the phases of the beams at different wavelengths can be obtained from the difference in the corresponding wavefront phases, as follows:

$$\begin{aligned}\triangle\ \phi(\vec{r}) &= \phi_{\lambda_1} - \phi_{\lambda_2}\\ &= k_1 \int_0^L \delta_n(\vec{r}, z)dz - k_2 \int_0^L \delta_n(\vec{r}, z)dz.\end{aligned} \tag{7.77}$$

From the above equation, it can be observed that the unequal-halo error between the wavefront-phase aberrations of beams of different wavelengths is affected by the wavelength difference, transmission distance, distance between two points on the receiving cross section, and atmospheric turbulence. In the strong-turbulence region, the wavefront phase changes rapidly, which, in turn, leads to the aberration difference between the two wavefronts also changing rapidly and randomly. It does not satisfy the condition of a strictly smooth stochastic process, whose average value can only remain constant for a fairly short period of time.

To solve this problem, we do not study the difference in $\triangle\ \phi(\vec{r})$ itself, but the amount of change in the difference $\triangle\ \phi(\vec{r}) - \triangle\ \phi(\vec{r}, \triangle\ t)$, which can be considered as the relative change that satisfies the smooth stochastic process. This value with a slowly varying process can be described by the phase-difference structure function, as follows:

$$D_{\triangle\phi}(\vec{r}) = \left\langle [\triangle\ \phi(\vec{\rho} + \vec{r}) - \triangle\ \phi(\vec{r})]^2 \right\rangle. \tag{7.78}$$

The total amount of the phase-structure function consists of stable quantities and a small fraction of small variations, whose spatial distribution on the receiving cross section is expressed as

$$\begin{aligned} D_{\Delta\phi}(\vec{r}) &= \left\langle [\Delta\,\phi(\vec{\rho}+\vec{r}) - \Delta\,\phi(\vec{r})]^2 \right\rangle \\ &= 2[B_{\Delta\phi}(\vec{\rho}) - B_{\Delta\phi}(\vec{r})], \end{aligned} \tag{7.79}$$

where B is the correlation function because the phase-response function is defined as the shock response uniformly distributed on the receiving surface; therefore, the effect of atmospheric turbulence on the optical-phase surface is equivalent to the distribution of the refractive index. Thus, the relationship between the phase-difference structure function and the random-distribution structure function of the refractive index can be determined as follows:

$$\begin{aligned} D_{\Delta\phi}(\vec{r}) &= \left\langle \left[\left(k_1 \int_0^L \delta_n(\vec{r_1}, z)dz \right) - \left(k_2 \int_0^L \delta_n(\vec{r_2}, z)dz \right) \right]^2 \right\rangle \\ &= (k_1 - k_2)^2 \left\langle \left[\left(\int_0^L \delta_n(\vec{r_1}, z)dz \right) - \left(\int_0^L \delta_n(\vec{r_2}, z)dz \right) \right]^2 \right\rangle \\ &= (k_1 - k_2)^2 D_n(\vec{r}), \end{aligned} \tag{7.80}$$

where $D_n(\vec{r})$ is the phase-structure function of atmospheric turbulence. According to the relationship between the turbulent power spectrum and the phase-structure function, its expression is

$$\begin{aligned} D_n(\vec{\rho}) &= 2[B_n(\vec{r}+\vec{\rho}) - B_n(\vec{r})] \\ &= 8\pi \int_0^\infty \kappa^2 \Phi(\kappa) \left[1 - \frac{\sin(k\rho)}{k\rho} \right] d\pi, \end{aligned} \tag{7.81}$$

where $B_n(\vec{r})$ is the number of interrelationships between two points on the spot, $\vec{\rho}$ is the position vector between two points on the receiving cross section, κ is the spatial wavenumber, and $\Phi(\kappa)$ is the spatial turbulence power spectrum. The spatial power spectrum of the turbulent flow is the modified Kolmogorov power spectrum according to the discussion in Sect. 7.5.1; the specific form of the expression is shown in Eq. (7.72). The McLaughlin-series expansion was applied to the following equation: $\left[1 - \frac{\sin(k\rho)}{k\rho}\right]$. The McLaughlin-series expression is as follows:

$$f(x) = \sum_{n=0}^{\infty} \frac{f^{(n)}(0)}{n!} x^n. \tag{7.82}$$

The even terms of the result after the expansion converge on the real number region, so for all the first n terms in the convergence domain, and the sum result is:

$$\left[1 - \frac{\sin(k\rho)}{k\rho}\right] = \sum_{n=1}^{\infty} \frac{(-1)^{n-1}}{(2n+1)!} (k\rho)^{2n}. \tag{7.83}$$

Substituting the result into the refractive-index structure–function expression and calculating the integral, we obtain

$$\begin{aligned} D_n(\vec{\rho}) &= 8\pi \int_0^{\infty} \kappa^2 \Phi(\kappa) \left[1 - \frac{\sin(k\rho)}{k\rho}\right] d\pi \\ &= 1.685 C_n^2 \kappa_l^{-2/3} \left\{ {}_1F_1\left[-\frac{1}{3}; \frac{3}{2}; \frac{\kappa_l^2 \rho^2}{4}\right] - 1 + 2.470 \left[1 - {}_1F_1\left[\frac{1}{6}; \frac{2}{3}; -\frac{\kappa_l^2 \rho^2}{4}\right]\right] \right. \\ &\quad \left. -0.071 \left[1 - {}_1F_1\left[\frac{1}{4}; \frac{2}{3}; -\frac{\kappa_l^2 \rho^2}{4}\right]\right] \right\}, \end{aligned} \tag{7.84}$$

where $\kappa_l = 3.3/l_0$ is the parameter of the internal turbulent scale, C_n^2 is the atmospheric refractive-index structure constant, ${}_1F_1(a; b; x)$ is the first class of merging hypergeometric functions, and its calculation is approximated by the relative numerical magnitude of the variables to determine its approximate expression. When the distance ρ between two points on the receiving section satisfies the condition $l_0 \ll \rho \ll L_0$, the large parameter approximation is used:

$${}_1F_1(a; c; x) \sim \frac{\Gamma(c)}{\Gamma(c-a)} x^{-a}, \tag{7.85}$$

where $\Gamma(\cdot)$ is the gamma function. Applying the approximation formula to Eq. (7.84), the approximation result is obtained as

$$\begin{aligned} D_n(\vec{\rho}) &= 1.685 C_n^2 \kappa_l^{-2/3} \left\{ \left[0.9420 \left(\frac{\kappa_l^2 \rho^2}{4}\right)^{\frac{1}{3}} - 1\right] + \left[2.470 \left(1 - 0.7639 \left(-\frac{\kappa_l^2 \rho^2}{4}\right)^{-\frac{1}{6}}\right)\right] \right. \\ &\quad \left. +0.071 \left[1 - 0.6364 \left(-\frac{\kappa_l^2 \rho^2}{4}\right)^{-\frac{1}{4}}\right] \right\}. \end{aligned} \tag{7.86}$$

When the distance ρ between two points on the receiving section satisfies the condition $\rho \ll l_0$, the small parameter approximation is used:

$$ {}_1F_1(a;b;x) \sim 1 - \frac{a}{c}x. \tag{7.87} $$

Applying the approximation formula to Eq. (7.84) yields the approximation as

$$ \begin{aligned} D_n(\vec{\rho}) &= 1.685C_n^2\kappa_l^{-2/3}\left\{\left[\frac{2}{9}\left(\frac{\kappa_l^2\rho^2}{4}\right)\right] + \left[2.470\left(1-(1-\frac{1}{4}\frac{\kappa_l^2\rho^2}{4})\right)\right]\right. \\ &\quad \left. -0.071\left[1-\left(1-\frac{3}{8}\frac{\kappa_l^2\rho^2}{4}\right)\right]\right\} \\ &= 0.8131\frac{\kappa_l^2\rho^2}{4}. \end{aligned} \tag{7.88} $$

The obtained refractive-index structure function is introduced into the refractive-index structure function and phase-difference structure function relationship in Eq. (7.80) to obtain the expression for the unequal-halo error structure function as follows:

$$ \begin{aligned} D_{\Delta\phi}(\vec{r}) &= (k_1-k_2)^2 D_n(\vec{r}) \\ &= \begin{cases} 1.685(k_1-k_2)^2C_n^2\kappa_l^{-2/3}\left\{[0.9420\left(\frac{\kappa_l^2\rho^2}{4}\right)^{\frac{1}{3}} - 1] + \left[2.470\left(1-0.7639\left(-\frac{\kappa_l^2\rho^2}{4}\right)^{-\frac{1}{6}}\right)\right]\right. \\ \qquad \left. +0.071\left[1-0.6364\left(-\frac{\kappa_l^2\rho^2}{4}\right)^{-\frac{1}{4}}\right]\right\} (l_0 \ll \rho \ll L_0) \\ 1.685(k_1-k_2)^2C_n^2\kappa_l^{-2/3}\left\{\left[\frac{2}{9}\left(\frac{\kappa_l^2\rho^2}{4}\right)\right] + \left[2.470\left(1-\left(1-\frac{1}{4}\frac{\kappa_l^2\rho^2}{4}\right)\right)\right]\right. \\ \qquad \left. -0.071\left[1-\left(1-\frac{3}{8}\frac{\kappa_l^2\rho^2}{4}\right)\right]\right\} (\rho \ll l_0) \end{cases}. \end{aligned} \tag{7.89} $$

It can be seen from the above equation that the distribution of the wavefront-dispersion unequal-halo error of Gaussian beams of different wavelengths over the entire receiving aperture will be affected by the wavelengths of the two beams, the inner and outer turbulent scales, the atmospheric refraction up to the structure constant, and the distance between the two points.

However, the above equation does not explicitly contain the outer turbulent-scale parameter, so we introduce Eq. (7.72) from Sect. 7.5.1 into Eq. (7.81) to obtain the expression for the difference-structure function of the wavefront aberrations with turbulent extra scale explicitly included:

$$ \begin{aligned} D_n(R) &= 1.685C_n^2\kappa_m^{-2/3}\{{}_1F_1\left(-\frac{1}{3};\frac{3}{2};-\frac{\kappa_m^2R^2}{4}\right) - 1 \\ &\quad + 1.050C_n^2\kappa_0^{-2/3}\left[1 - {}_0F_1\left(-;\frac{2}{3};-\frac{\kappa_0^2R^2}{4}\right)\right], \quad \kappa_0 \ll \kappa_m, \end{aligned} \tag{7.90} $$

where $\kappa_m = 5.92/l_0$. According to the discussion in Sect. 7.5.1, $\kappa_0 = 8\pi/L_0$. The condition for the structure function to hold is $\kappa_0 \ll \kappa_m$, so the relationship between the outer and inner turbulence scales is required to be $L_0 \gg l_0$. Substituting this into the difference structure function and the refractive-index structure function relation to obtain:

$$D_{\Delta\phi}(\vec{r}) = 1.685C_n^2(k_1 + k_2)^2\kappa_m^{-2/3}\{{}_1F_1\left(-\frac{1}{3}; \frac{3}{2}; -\frac{\kappa_m^2\rho^2}{4}\right) - 1$$
$$+ 1.050C_n^2(k_1 + k_2)^2\kappa_0^{-2/3}\left[1 -{}_1 F_1\left(-\frac{1}{4}; \frac{2}{3}; -\frac{\kappa_0^2\rho^2}{4}\right)\right], \quad \kappa_0 \ll \kappa_m. \tag{7.91}$$

The above theoretical analysis shows that differences exist between different beam-wavefront aberrations. Moreover, the situation in the strong-turbulence region is more complex compared with the weak-turbulence region, so the analysis conditions and calculation process are more complicated. According to the calculation results, we can show that the differences between beam-wavefront aberrations are related to the internal and external turbulence scales, beam wavelength, and other factors. Hence, in the next section, the influence of each factor on the differences will be shown more intuitively in the form of simulation diagrams.

7.4.3 Simulation Analysis

The simulations were conducted under atmospheric conditions with outer turbulent scale $L_0 = 20$ m, inner turbulent scale $l_0 = 0.1$ m, and atmospheric refractive-index structure constant $C_n^2 = 1 \times 10^{-10}$ m$^{-2/3}$ [56]. Gaussian beams with wavelengths of 1550 nm, 850 nm, 632.8 nm, and 530 nm were transmitted at the same transmission distance $L = 50$ km [64]. Figure 7.15 shows the simulation results of the wavefront-phase distortion after an atmospheric-channel transmission.

Comparing the wavefront phase aberrations of beams of different wavelengths in three dimensions (3D), as the wavelength gradually increases from *a* to *d*, the degree of wavefront-phase undulation gradually decreases. However, the aberrations of short-wave beams are still stronger than those with long waves. Compared with the wavefront aberrations in the weak-turbulence region, those in the strong-turbulence region have a larger variation range, and the break between phases is severe, making it difficult to measure a more continuous wavefront phase.

As shown in the grayscale plots of Fig. 7.15a-(2), b-(2), c-(2), and d-(2), the red area is the main energy-concentration area of the spot, the yellow area is the broken area, and there is no spot-energy distribution. The longitudinal comparison shows that the complete spot is torn into several scattered spots, and some energy is distributed intermittently between them.

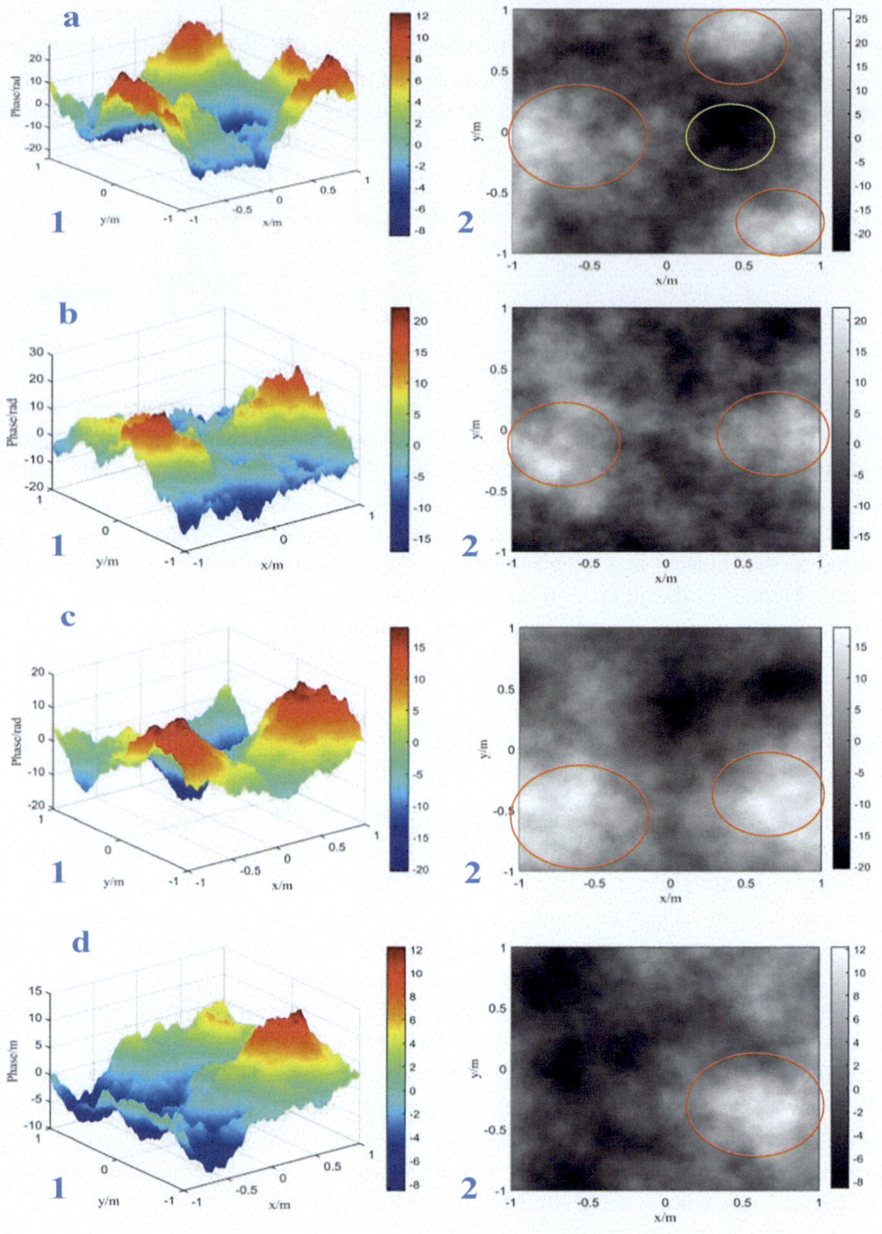

Fig. 7.15 Wavefront phase distortion of Gaussian beams with different wavelengths. (1) Three-dimensional diagram of a wavefront phase distortion; (2) grayscale image of a transmission cross-section spot. **a** 530 nm; **b** 632.8 nm; **c** 850 nm; **d** 1550 nm

According to the above simulation illustration, a wavefront sensor in the strong-turbulence region cannot measure the continuous phase distribution, and therefore cannot correct the difference between the wavefront aberrations using the relationship between the phases. It also cannot use the geometric-optics analysis method for the different wavelengths of the wavefront-phase analysis. Therefore, the statistical optics method, also known as the phase structure function method, is chosen.

The coefficient average, also known as the ensemble average, can be considered equal to the time average of the coefficient average and one random process time under the precondition of satisfying the generalized random smoothing process. According to the definitions of coefficient average and root mean square. By squaring the rms value of the wavefront aberration error, the root mean square (RMS) of the error between beams of different wavelengths can be obtained as:

$$RMS = \sqrt{D_{\Delta\phi}(r)}. \tag{7.92}$$

After the Zernike polynomial is fit, the relationship between the Strehl Ratio (SR) and RMS can be approximated as

$$SR \approx \exp(-RMS^2). \tag{7.93}$$

We obtained the trend of the RMS variation in the wavefront-distortion error for different factors. The simulation used the following settings: atmospheric conditions with outer turbulent scales of $L_0 = 20$ m, 30 m, 50 m, 60 m, and 80 m, inner turbulent scale $l_0 = 0.1$ m [90], and atmospheric refractive-index structure constant $C2\ n = 1 \times 10^{-10}\ \text{m}^{-2/3}$ V. The 1550-nm signal beam and 632.8-nm beacon beam were co-optically transmitted in the atmospheric channel for $L = 50$ km. Figure 7.16 shows that the RMS of the wavefront phase-distortion difference between different points on the received cross-section is affected by the outer turbulent scale.

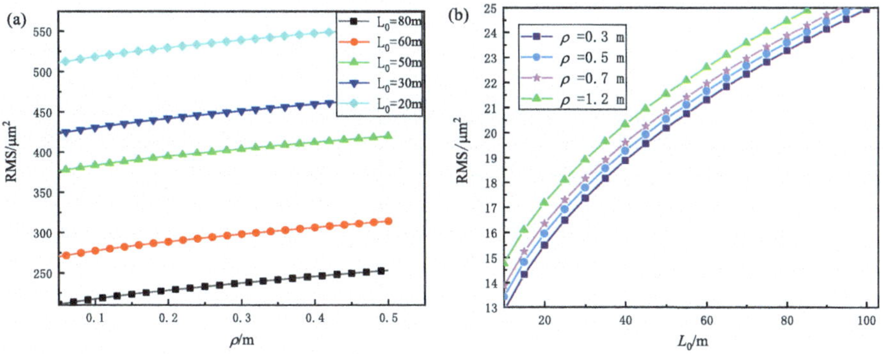

Fig. 7.16 Effect of the external turbulence scale on the wavefront difference of Gaussian beams with different wavelengths propagating through strong atmospheric turbulence. **a** Effect of the external turbulence scale on the error. **b** Effect of the distance between two points

Observing the transverse variation of Fig. 7.16a, it can be found that the influence of the distance between two points on the beam-wavefront difference is not too great in the receiver cross section, with a slight increasing trend and little overall change. This phenomenon is also reflected in the trend of the longitudinal values of Fig. 7.16b, thus indicating that the beam-wavefront difference is distributed more uniformly throughout the receiver cross section, and the difference values between different locations are not very different.

Longitudinally, it can be found that the beam-wavefront difference on the receiver cross section gradually increases with the increase of the outer turbulent scale, for which we simulate the wavefront difference corresponding to the outer turbulent scale [64] from 0 to 80 m and obtain the curve shown in Fig. 7.16a. With the increase of the outer turbulent scale, the wavefront difference increases; however, the slope of the curve gradually decreases, indicating that with the increase of the outer turbulent scale, its modulation of the beam gradually decreases. Finally, the influence on the wavefront difference will tend to stabilize and converge to a stable value.

Next, we simulated inner turbulent scales of $l_0 = 0.01$ m, 0.03 m, 0.05 m, 0.10 m, 0.15 m, 0.20 m, 0.25 m, 0.30 m, 0.35 m, and 0.40 m, and outer turbulent scale $L_0 = 40$ m [64]. The atmospheric refractive-index structure constant $C_n^2 = 1 \times 10^{-10}$ $\text{m}^{-2/3}$. Figure 7.17 shows that the RMS of the difference in wavefront-phase distortion between different points on the received cross section is affected by the inner turbulent scale after the 1550-nm signal beam and 632.8-nm beacon beam are transmitted in the atmospheric channel in a common optical path for $L = 50$ km.

When the distance between two points is smaller than the inner turbulent scale, the influence of different values of the inner turbulent scale on the difference of wavefront aberration is obvious. As the distance between two points increases and gradually approaches the value of the inner turbulent scale, the influence of different values of inner turbulent scale on the difference begins to converge and the curves gradually begin to fit a curve.

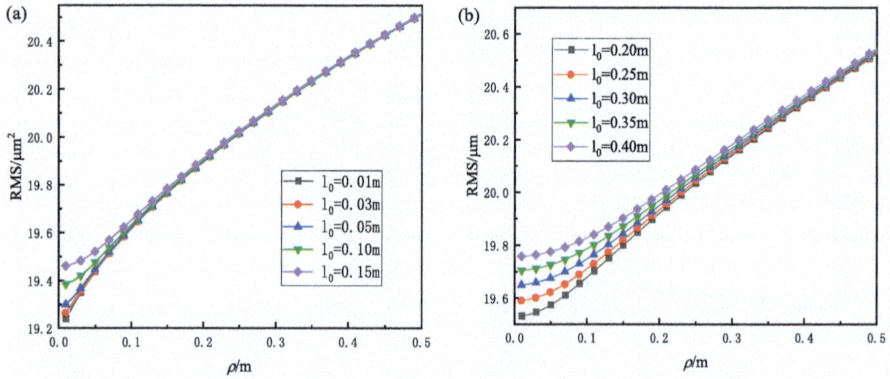

Fig. 7.17 Effect of the internal turbulent scale on the wavefront difference after Gaussian beams of different wavelengths propagate through strong atmospheric turbulence. **a** $l_0 = 0.01$ m–0.15 m. **b** $l_0 = 0.20$ m–0.40 m

When the distance between the two points exceeds the inner turbulent scale, the influence of different values is almost the same, and the slope no longer changes. This is because the diffraction of the beam inside the small-scale turbulence is more severe. When the distance exceeds the inner turbulence scale, the beam is refracted, and the influence of the small-scale turbulence on the beam decreases. Thus, the influence of the different sizes of the inner turbulence scale converges, and the difference can no longer be seen.

The following parameters were set: inner turbulent scale $l_0 = 0.10$ m, outer turbulent scale $L_0 = 50$ m, and the atmospheric refractive-index structure constant $C2\ n = 1 \times 10^{-10}\ \mathrm{m}^{-2/3}$. A 1550-nm signal beam and a beacon beam with wavelengths of 530 nm, 632.8 nm, 850 nm, 950 nm, and 1280 nm were co-opted in the atmospheric channel. After transmitting for $L = 50$ km in the atmospheric channel, the RMS of the wavefront phase-distortion difference between different points on the received cross section is affected by the wavelength difference, as shown in Fig. 7.18.

From Fig. 7.18a, we can see that in the receiving-aperture cross section, the variation in the distance between two points has little effect on the difference in wavefront aberrations between beams of different wavelengths, which is always maintained at a stable level. In the vertical comparison from top to bottom, the difference between the beacon-beam wavelength and signal-beam wavelength gradually decreases, and the corresponding difference of wavefront aberrations also gradually decreases.

When the wavelength of the beacon beam is 1280 nm, the difference between the signal beam and the wavefront aberrations of the beacon beam tends to be negligible. As shown in Fig. 7.18b, the mean-square value of the wavefront phase difference decreases as the wavelength difference between the signal and beacon beams decreases, and the wavefront phase difference is zero when the signal and beacon beams are at the same wavelength.

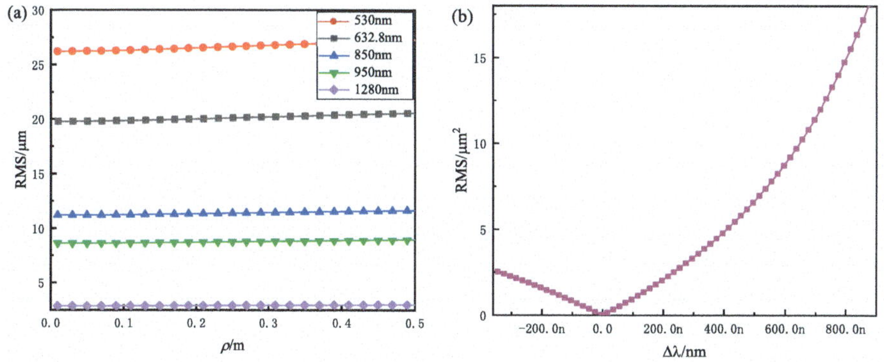

Fig. 7.18 Effect of the internal turbulent scale on the wavefront difference after Gaussian beams of different wavelengths propagate through strong atmospheric turbulence. **a** Phase difference in the receiving-aperture distribution under different wavelengths. **b** Phase differences at various wavelengths

The above simulations and results show that the wavelength is the main factor that determines the variation in the mean square of the difference between the wavefront aberrations. Only when the wavelengths of the signal and beacon beams are close to each other can the accuracy of the wavefront correction be further ensured by adaptive-optics.

In addition, comparing the variation curve of the mean-square difference of the wavefront aberration with the wavelength difference, we can see that the aberration difference is related to the wavelength difference and is unrelated to the wavelengths of the signal beam or beacon beam. This indicates that when choosing the wavelength of the beacon beam, we should consider its influence in practical applications, and the wavelength difference between it and the signal beam should be considered as an important factor.

7.4.4 No Wavefront Sensing Measurement Error Compensation Correction

7.4.4.1 No Wavefront Sensor Adaptive-Optics System

In the strong-turbulence region, the phase and logarithmic amplitude fluctuations produced by the beam transmission are much stronger than those in the weak-turbulence region. Thus, the optical wavefront phase plane is torn by atmospheric turbulence into a series of small scattered spots with discontinuous phase planes, a large number of phase singularities, and severe beam-intensity scintillation. At this time, the wavefront sensor cannot measure the complete and accurate wavefront phase distribution, and the measurement error is large, resulting in a significant decrease in the effect of the conjugate-phase correction [65].

The no wavefront sensor correction system uses a charge-coupled device (CCD) camera to measure the spot energy, Strehl ratio (SR), and other beam parameters of the laser beam as the optimization index of the control program, and converts them into the control signal of the deformation mirror [66]. It does not need to measure the wavefront phase and is free from the influence of beam-intensity scintillation, successfully solving the above problem.

Two general types of no wavefront sensor correction are used in space laser communication systems, as shown in Fig. 7.19a. In one, a no wavefront sensor correction system is applied at the receiver side, where a photodetector, such as a CCD camera, measures the performance index of the received beam as feedback information. The control algorithm uses the feedback information as an optimization parameter within the program and generates the control voltage of the deformation device to complete the phase-correction process [67].

Figure 7.19b shows a no wavefront sensor correction system applied to the transmitter, where the signal beam first passes through the photoelectric sensor and wavefront sensor without the correction voltage to reach the receiver. The receiver returns

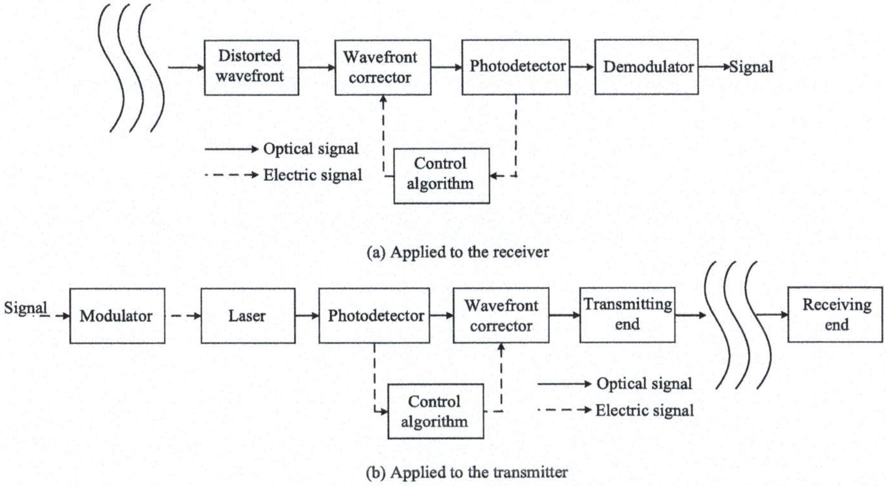

Fig. 7.19 Model of a no wavefront sensor AO system in a picture space laser communication system

part of the received signal to the photoelectric detector at the transmitter, in a time-division or wave-division manner. At this time, the detector reads the feedback signal and the control algorithm generates the control-voltage signal to correct the phase distortion. This correction method has a certain time delay, and the correction effect is lower than that with the correction method at the receiving end.

7.4.4.2 No Wavefront Sensor Correction Algorithm and Zernike Polynomial Fitting

Commonly used algorithms for no wavefront sensor adaptive-optical correction include the stochastic parallel gradient-descent algorithm (SPGD) [68], genetic algorithm (GA) [69], simulated-annealing algorithm (SA) [70], and mode method [71]. The genetic algorithm continuously selects, crosses, and compiles the coefficients of Zernike polynomials according to the law of biological-species evolution, and then determines the best-fitting coefficient according to the evaluation function.

The SA algorithm is developed based on the solid-annealing algorithm. After the solid is heated, it aggravates the disordered motion of the internal particles, and then returns to the ordered arrangement after cooling. On this basis, the simulated-annealing algorithm randomly determines the optimal solution for the objective function.

The SPGD algorithm randomly generates a random-disturbance voltage signal in accordance with the Bernoulli distribution, calculates the objective function after the forward and reverse voltages are applied, and finally iterates to obtain the desired voltage signal.

Regardless of the type of algorithm, the Zernike polynomial is mainly used to fit the wavefront distortion, control the deformable mirror, and optimize the control algorithm, according to the change in the beam performance index, to achieve the best correction effect.

The Zernike polynomials are a group of orthogonal polynomials in a unit circle. The coefficients of the polynomials, as a linear combination of basis functions, can accurately describe different types of distorted wavefronts. The wavefront aberration $\phi(\rho, \theta)$ is expressed as a linear combination of Zernike polynomials [72]:

$$\phi(\rho, \theta) = \sum_{i=1}^{N} a_i z_i(\rho, \theta), \tag{7.94}$$

where N is the total order of the aberration, a_i is the coefficient of the Zernike polynomial of order i, (ρ, θ) represents the polar coordinates, and the expression of the Zernike polynomial of order i in polar coordinates is $z_i(\rho, \theta)$.

$$Z_{evenj}(\rho, \theta) = \sqrt{n+1}\sqrt{2}R_n^m(\rho)\cos(m\theta), \quad m > 0 \tag{7.95}$$

$$Z_{oddj}(\rho, \theta) = \sqrt{n+1}\sqrt{2}R_n^m(\rho)\sin(m\theta), \quad m < 0 \tag{7.96}$$

$$Z_j(\rho, \theta) = \sqrt{n+1}R_n^0(\rho), \quad m = 0, \tag{7.97}$$

where

$$R_n^m(\rho) = \sum\nolimits_{s=0}^{(n-m)/2} \frac{(-1)^s(n-s)!}{s![(n+m)/2-s]![(n-m)/2-s]!}\rho^{n-2s}. \tag{7.98}$$

m and n are the angular and radial numbers of the Zernike polynomials in polar coordinates, respectively, and j is the order of the Zernike polynomial related to m and n.

The higher the order of the Zernike polynomial, the more complex the surface shape of the wavefront phase distortion and the higher the spatial frequency of the corresponding wavefront distortion. Therefore, when the Zernike polynomial is used to fit the wavefront phase distortion, the greater the order of the Zernike polynomial and the more types of aberrations considered, the higher the fitting accuracy.

The key to fitting the wavefront-distortion phase using Zernike polynomials is to determine the order of the Zernike polynomials and the coefficients of each order. According to reference [73], the coefficients of Zernike polynomials corresponding to wavefront phase disturbances caused by gas turbulence satisfy the Gaussian distribution under a zero mean and certain variance. The coefficients of a wavefront phase-distortion Zernike polynomial are as follows:

$$\mathbf{A} = \left[\, a_2, a_3, \ldots, a_N \,\right]^T. \tag{7.99}$$

The covariance matrix C is obtained by multiplying the coefficient vector A by its transpose vector:

$$C = E\left[A \cdot A^T\right] = \begin{bmatrix} E(a_1, a_1) & E(a_1, a_2) & \cdots & E(a_1, a_N) \\ E(a_2, a_1) & E(a_2, a_2) & \cdots & E(a_2, a_N) \\ \vdots & \vdots & \ddots & \vdots \\ E(a_N, a_1) & E(a_N, a_2) & \cdots & E(a_N, a_N) \end{bmatrix}, \tag{7.100}$$

where the elements $E(a_j, a_j)$ on the main diagonal of the matrix are the self-covariance matrix of the coefficients of the jth-order Zernike polynomial. $E(a_i, a_j)$ is the mutual-covariance matrix between the coefficients of the ith order- and jth-order Zernike polynomials [74]:

$$E(a_i, a_j) = \frac{2.2698(-1)^{(n+n\prime-2m)}\sqrt{(n+1)(n\prime+1)}\delta_z\Gamma[(n+n\prime-5/3)/2](D/r_0)^{5/3}}{\Gamma[(n+n\prime-17/3)/2]\Gamma[(n\prime+n-17/3)/2]\Gamma[(n+n\prime+23/3)/2]}, \tag{7.101}$$

where $\Gamma(\cdot)$ is the gamma function, m, n, m', and n' are the angular and radial numbers corresponding to Zernike polynomials, respectively; D is the diameter of the receiving section, and r_0 is the atmospheric-coherence length. δ_z is the Kronecker symbol, and its expression is

$$\begin{aligned} \delta_z &= \left(m = m'\right) \cap \left(\overline{\text{parity}(i, i')} \cup (m = 0)\right) \\ &= \begin{cases} 1, \ m = m' \neq 0 \\ 0, \ m \neq m' \text{or} m = 0 \end{cases} \end{aligned} \tag{7.102}$$

The covariance matrix C is an Emery matrix, and a YOU-Matrix exists that can transform matrix C into a diagonal matrix; therefore, the covariance matrix C is first decomposed into singular values:

$$C = XSX^{\mathrm{T}}, \tag{7.103}$$

where X is the YOU-Matrix of the covariance matrix C. The coefficient vector of another linear combination B of wavefront Zernike polynomials of the same dimension as A is $B = \left[b_2, b_3, \ldots, b_N\right]^T$. The coefficients of each Zernike polynomial of A can be found by the following relation:

$$A = X \cdot B. \tag{7.104}$$

7.4.4.3 No Wavefront Sensor Correction Measurement Error Compensation Coefficient Matrix

In space laser communication systems, no wavefront sensor adaptive-optics corrects the signal-beam wavefront, based on the beam-performance index of the measured beacon beam. The difference in beam index between the signal and beacon beams is caused by atmospheric-turbulence dispersion, absorption, and scattering differences of air particles at different wavelengths. There is a certain correction residual in the adaptive-optics system.

In the correction stage, the Zernike polynomial is used to separately fit the wavefronts of the signal and beacon beams. The wavefront Zernike-polynomial coefficients corresponding to the signal and beacon beams with wavelengths λ_1 and λ_2, respectively, are as follows:

$$a_1(\lambda_1), a_2(\lambda_1), a_3(\lambda_1), \ldots, a_j(\lambda_1) \tag{7.105}$$

$$a_1(\lambda_2), a_2(\lambda_2), a_3(\lambda_2), \ldots, a_j(\lambda_2). \tag{7.106}$$

The numerical relations between the Zernike polynomial coefficients of the same order can be established using the wavelength function [75]:

$$\begin{aligned} a_1(\lambda_1) &= f_1[a_1(\lambda_2), \lambda_1] \\ a_2(\lambda_1) &= f_2[a_2(\lambda_2), \lambda_1] \\ &\vdots \\ a_j(\lambda_1) &= f_j[a_j(\lambda_2), \lambda_1]. \end{aligned} \tag{7.107}$$

According to the actual laser-communication transmission process, the mode-analysis model is established to obtain the functional relationship between the polynomial coefficients. An infinitesimal region $\Delta\delta$ of the wavefront-phase plane is selected, and the signal and beacon beams are emitted in a common optical path, assuming that the wavefront-phase planes overlap completely. Thus, the corresponding paths $\Delta\delta$ in the transmission process of the atmospheric channel are identical, and the refractive-index changes are also identical.

According to the Rytov freezing hypothesis, we choose a certain moment t as the analysis moment. The superposition of the fluctuating atmospheric refractive index within the transmission distance of the beam at moment t can be equated to a lens with an inhomogeneous refractive-index distribution. It can be analyzed at this time using the commonly used material refractive-index dispersion formula in engineering. According to reference [66], the relationship between the Zernike polynomial coefficient and the wavelength is [75]

$$a_i = A_i + \frac{B_i}{\lambda} + \frac{C_i}{\lambda^{3.5}}, \tag{7.108}$$

where a_i is the coefficient of the ith-order Zernike polynomial, and A_i, B_i, and C_i are the coefficients of the Conrady–Zernike polynomial. The relationship between the two coefficients at different wavelengths is as follows:

$$a_i(\lambda_1) = \left(A_i + \frac{B_i}{\lambda_1} + \frac{C_i}{\lambda_1^{3.5}}/A_i + \frac{B_i}{\lambda_2} + \frac{C_i}{\lambda_2^{3.5}}\right)a_i(\lambda_2). \tag{7.109}$$

The scale-correction factor between the ith-order Zernike polynomial coefficients corresponding to the signal beam and fiducial beam is

$$Z_i = A_i + \frac{B_i}{\lambda_1} + \frac{C_i}{\lambda_1^{3.5}}/A_i + \frac{B_i}{\lambda_2} + \frac{C_i}{\lambda_2^{3.5}}. \tag{7.110}$$

A Zernike polynomial of order 15 was chosen as an example, and the Zernike-polynomial coefficient vector corresponding to the fiducial beam is

$$\mathbf{A}_{15}(\lambda_2) = \left[a_1, a_2, a_3, \ldots, a_{15}\right]^T. \tag{7.111}$$

By multiplying the coefficient vector of the fiducial beam by the corresponding coefficient-correction vector, the following result is obtained:

$$\begin{aligned}\mathbf{A}_{15'}(\lambda_2) &= \begin{bmatrix} z_1, 0, 0, \ldots, 0 \\ 0, z_2, 0, \ldots, 0 \\ \vdots \\ 0, 0, 0, \ldots, z_{15} \end{bmatrix} \left[a_1, a_2, a_3, \ldots, a_{15}\right]^T \\ &= \left[z_1a_1, z_2a_2, z_3a_3, \ldots, z_{15}a_{15}\right]^T.\end{aligned} \tag{7.112}$$

The covariance matrix is calculated by substituting the corrected coefficient vectors into Eq. (7.100), and the result is calculated as

$$\begin{aligned}C' &= E\left[A'_{15}(\lambda_2) \cdot A'^T_{15}(\lambda_2)\right] \\ &= \begin{bmatrix} E(z_1a_1, z_1a_1) & E(z_1a_1, z_2a_2) & \cdots & E(z_1a_1, z_Na_N) \\ E(z_2a_2, z_1a_1) & E(z_2a_2, z_2a_2) & \cdots & E(z_2a_2, z_Na_N) \\ \vdots & \vdots & \ddots & \vdots \\ E(z_Na_N, z_1a_1) & E(z_Na_N, z_2a_2) & \cdots & E(z_Na_N, z_Na_N) \end{bmatrix}.\end{aligned} \tag{7.113}$$

After performing the singular-value decomposition of matrix C′, we obtain

$$C\prime = X\prime SX\prime^{\mathrm{T}}. \tag{7.114}$$

The corresponding-signal optical-distortion wavefront Zernike-polynomial coefficients can then be determined using the following equation:

$$A' = X' \cdot B. \tag{7.115}$$

When the signal-beam wavelength is 1550 nm and the beacon-beam wavelength is 632.8 nm, the beams are emitted in the public optical path in a strong-turbulence region, and the wavefront phase aberration is obtained after transmission. The Strehl ratio (SR) is calculated by fitting the two wavefront aberrations using Zernike polynomials of orders 9 to 15. The Zernike dispersion relationship between the coefficients of the polynomials is shown in Fig. 7.20, which shows the principle of the distortion-difference correction without wavefront correction.

As shown in Fig. 7.20, the control program generates the coefficient matrix of the Zernike polynomial based on the SR value of the beacon beam; it also obtains the coefficient-dispersion relation based on the signal-beam wavelength, beacon-beam wavelength, and turbulence parameters. Then, it generates a correction matrix, corrects the polynomial-coefficient matrix of the generated beacon beam; and finally, generates the voltage control of the deformation mirror, based on the corrected Zernike-polynomial coefficient matrix signal.

After the correction at this point, the correction residual can be further reduced; in the same case, the number of iterations of the program can also be reduced, and thus, the running time of the program can be reduced.

In this section, the dispersion unequal-halo error between beams of different wavelengths in the strong-turbulence region is analyzed and studied as follows:

1. According to the strong-turbulence discussion, the statistical-optics method was determined to study the difference between the wavefront-phase distortion on the receiving cross section after the transmission of beams of different wavelengths through the strong-turbulence region. The structure function of the dispersion unequal-halo error of beams of different wavelengths was calculated, and the SR-value difference between the signal and beacon beams caused by the dispersion unequal-halo error was also calculated.
2. Analyzing the structure function shows that the chromatic aberration between beams of different wavelengths was influenced by the wavelength, the atmospheric refractive-index structure constant, inner and outer turbulent scales, and other factors.

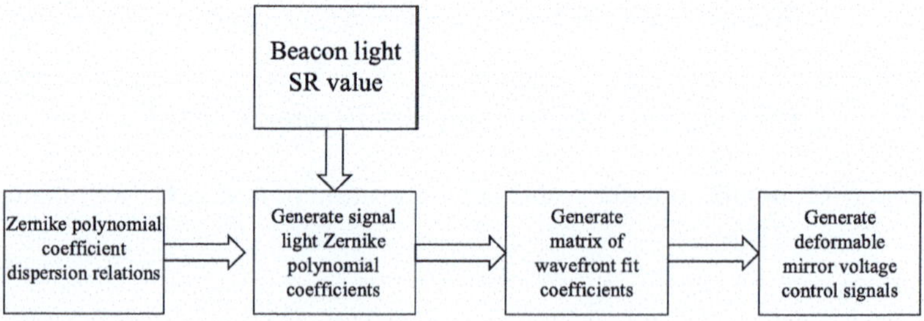

Fig. 7.20 Principle of chromatic-aberration correction without wavefront correction

3. The RMS of the wavefront phase-distortion difference of different wavelength beams was simulated numerically under different influencing factors. It was concluded from the simulation results that the increase of the outer turbulent scale led to a further expansion of the difference between wavefront aberrations of different wavelength beams. The difference stabilized after increasing the outer turbulent scale to a certain degree under certain receiving aperture conditions. When the distance between the two points was smaller than the inner scale, the influence of the difference between the inner turbulent scale and the wavefront aberration was more obvious.
4. The basic principle of no wavefront sensor correction was introduced, and a distortion-compensation method for no wavefront sensor correction was proposed. A Zernike polynomial was used to fit the wavefront phase, and the relationship between the Zernike-polynomial coefficients corresponding to beams of different wavelengths was determined.

7.5 Experimental Studies

This section introduces the external field experiments, including experimental design ideas, equipment, results, and analyses. Outdoor laser communication experiments were conducted using a space laser communication system and a new compensation-correction method to correct a distorted signal-beam wavefront. The new compensation-correction method had smaller correction residuals than the previous method. The system correction accuracy was significantly improved, and the correctness of the theoretical analysis was also illustrated.

7.5.1 Simulation Experiment of Wavefront-Distortion Difference-Compensation Correction for Dual-Wavelength Communication Systems

The correction matrix is calculated based on the quantitative relationship between the wavefront-aberration phases of Gaussian beams of different wavelengths, as described in Sect. 7.4.3. The measured wavefront-reconstruction matrix of the beacon beam is corrected to obtain the wavefront-reconstruction matrix of the signal beam to correct the signal beam. The aberrant wavefront can be expressed as follows:

$$W(x, y) = W^{'}(x, y) - W_0(x, y) = \sum_{j=0}^{j=\max} W_j Z_j(x, y), \tag{7.116}$$

where Z_j is the Zernike polynomial, W_j is its, and j is its order. $W^{`}(x, y)$ is the measured wavefront, $W_0(x, y)$ is the reference-plane wavefront, and the difference between them is the distorted wavefront. A Gaussian beam of wavelength λ_2 is used as the beacon beam, the signal beam of wavelength λ_1 is emitted in a common optical path, and a Shack–Hartmann wavefront sensor measures the beacon-beam wavefront to obtain the beacon-beam gradient matrix containing the aberrations:

$$\begin{bmatrix} b_{\lambda_2}(x,y)\big|_1 \\ b_{\lambda_2}(x,y)\big|_2 \\ \vdots \\ b_{\lambda_2}(x,y)\big|_k \\ c_{\lambda_2}(x,y)\big|_1 \\ c_{\lambda_2}(x,y)\big|_2 \\ \vdots \\ c_{\lambda_2}(x,y)\big|_k \end{bmatrix} = \begin{bmatrix} g_1(x,y)|_1 \; g_2(x,y)|_1 \; \cdots \; g_{j\max}(x,y)\big|_1 \\ g_1(x,y)|_2 \; g_2(x,y)|_2 \; \cdots \; g_{j\max}(x,y)\big|_2 \\ \vdots \\ g_1(x,y)|_k \; g_2(x,y)|_k \; \cdots \; g_{j\max}(x,y)\big|_k \\ h_1(x,y)|_1 \; h_2(x,y)|_1 \; \cdots \; h_{j\max}(x,y)\big|_1 \\ h_1(x,y)|_2 \; h_2(x,y)|_2 \; \cdots \; h_{j\max}(x,y)\big|_2 \\ \vdots \\ h_1(x,y)|_k \; h_2(x,y)|_k \; \cdots \; h_{j\max}(x,y)\big|_k \end{bmatrix} \cdot \begin{bmatrix} W_1 \\ W_2 \\ \vdots \\ W_{j\max} \end{bmatrix}, \tag{7.117}$$

where $\left.\frac{\Delta x(x,y)}{f}\right|_{\lambda_2}\Big|_k = b_{\lambda_2}(x,y)\big|_k$ and $\left.\frac{\Delta y(x,y)}{f}\right|_{\lambda_2}\Big|_k = c_{\lambda_2}(x,y)\big|_k$. According to Eq. 7.68, the offset gradient of the Gaussian-beam aberrations at two different wavelengths on the measurement surface of the sensor is calculated as follows:

$$\left.\frac{\Delta x(x,y)}{f}\right|_{\lambda_1}\Big|_k = \frac{1}{\gamma} \cdot \left.\frac{\Delta x(x,y)}{f}\right|_{\lambda_2}\Big|_k = b_{\lambda_1}(x,y)\big|_k \tag{7.118}$$

$$\left.\frac{\Delta y(x,y)}{f}\right|_{\lambda_1}\Big|_k = \frac{1}{\gamma} \cdot \left.\frac{\Delta y(x,y)}{f}\right|_{\lambda_2}\Big|_k = c_{\lambda_1}(x,y)\big|_k, \tag{7.119}$$

where $\frac{1}{\gamma} = \frac{\lambda_2}{\lambda_1} \cdot (\cos A\theta_2 + \frac{\sin A\theta_2}{\tan\theta_2})$. According to this relationship, a coefficient correction matrix can be obtained:

$$\gamma = \begin{bmatrix} \gamma_1 \, 0 \cdots 0_k \\ 0 \, \gamma_2 \cdots 0_k \\ \ddots \\ 0 \, 0 \cdots \gamma_k \end{bmatrix}. \tag{7.120}$$

The wavefront-reconstruction matrix of the beacon beam is multiplied by the coefficient-correction matrix to obtain the wavefront reconstruction matrix of the signal beam. The signal optical wavefront aberration is corrected in the AO control system by using the beacon optical wavefront reconstruction matrix and the beacon optical wavefront reconstruction matrix obtained after correction, respectively. The

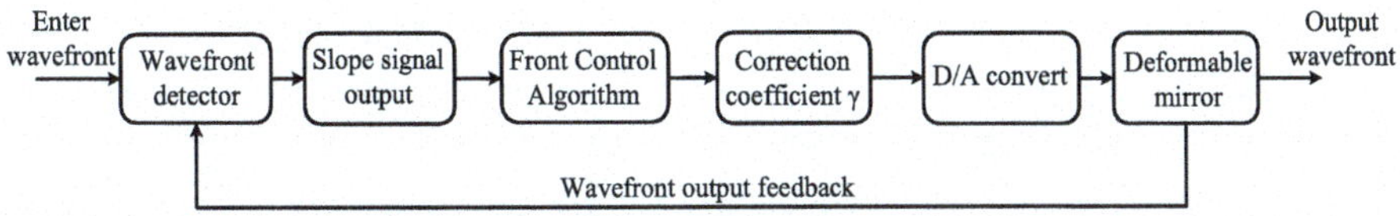

Fig. 7.21 Block diagram of the adaptive-optical control program with a beacon-beam correction coefficient

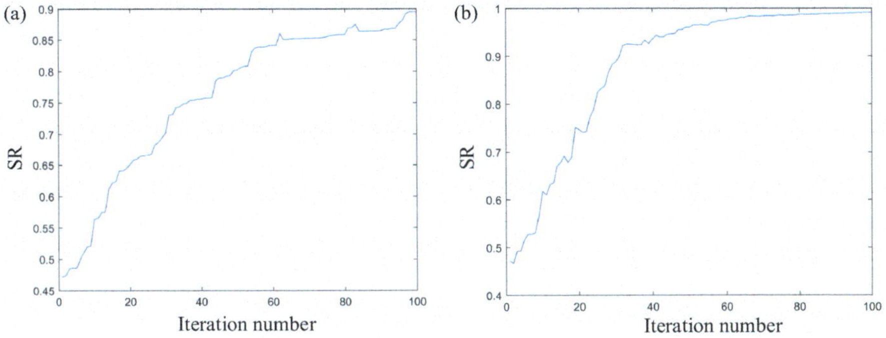

Fig. 7.22 Schematic diagram of the beacon beam and beacon-beam correction-matrix corrected signal. **a** Excluding the correction coefficient. **b** Including the wavefront correction coefficient

schematic of the closed-loop control system with the correction factor added is shown in Fig. 7.21.

Numerical simulation experiments of the correction procedure were conducted at a transmission distance $L = 10$ km, the signal-beam and beacon-beam wavelengths were 1550 nm and 632.8 nm, respectively, and the atmospheric refractive-index structure constant $C_n^2 = 1 \times 10^{-20}$ m$^{-2/3}$. After adding the correction coefficients to the calibration procedure, the signal-beam Strehl ratio was obtained in the simulation experiment, as shown in Fig. 7.22.

Figure 7.22 shows the change in the system Strehl Ratio (SR) during the correction process. (a) shows the SR curve when the beacon-beam wavefront aberration is directly corrected as the signal-beam wavefront aberration. (b) shows the SR curve when the corrected wavefront-phase matrix of the beacon-beam wavefront aberration is corrected as the signal-beam wavefront aberration using Eq. 7.68.

It can be seen from the figure that the SR of the corrected wavefront-reconstruction matrix converges to 1, and the overall SR improves by 10%, compared to the case where the wavefront-reconstruction matrix converges to 0.9 before the correction. The SR of the system converges quickly when the signal beam is corrected by the corrected wavefront-aberration matrix of the beacon beam, and stabilizes at 80 iterations. The jitter of the SR is significantly reduced, which verifies the correctness of the wavefront-phase relationship between the signal beam and the beacon beam derived from Eq. 7.68.

7.5.2 1.2-km Experimental Setup and Experimental Optical Path

To verify the correctness of the theoretical analysis, the group built a 1.2-km experimental optical path between the seventh floor of the North Science Building and the seventh floor of the fifth building of Xi'an University of Technology. The experimental schematic is shown in Fig. 7.23. The signal beam and beacon beam were transmitted through the same atmospheric channel to the seventh floor of the North Academic Building after the co-beamer, and then transmitted to the receiving antenna on the seventh floor of the fifth building after reflection by the plane reflector.

The combined beam was reflected by the deformation mirror and reflector and then incident to the beam splitter, which separated the signal beam and beacon beam. The wavefront sensor detected the wavefront aberrations of the beacon and signal beams, and corrected the signal-beam wavefront aberrations after adaptive proportional–integral–derivative (PID)-algorithm control. In the experiment, the signal-beam wavelength was 1550 nm and the beacon-beam wavelength was 650 nm. Figure 7.24 shows a photo of the experimental system, and Table 7.1 shows the corresponding experimental instrument parameters.

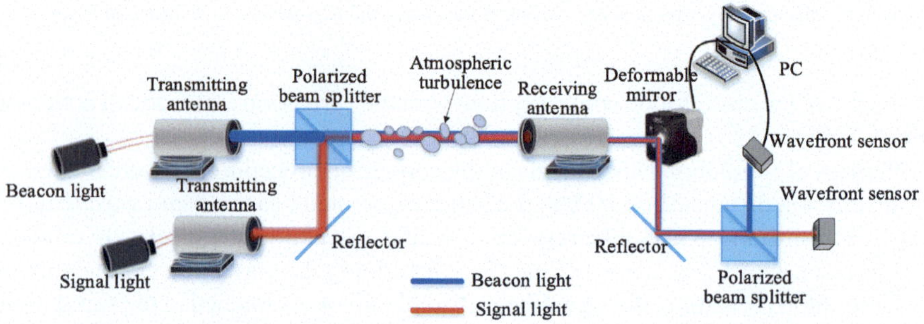

Fig. 7.23 Experimental structure and principle

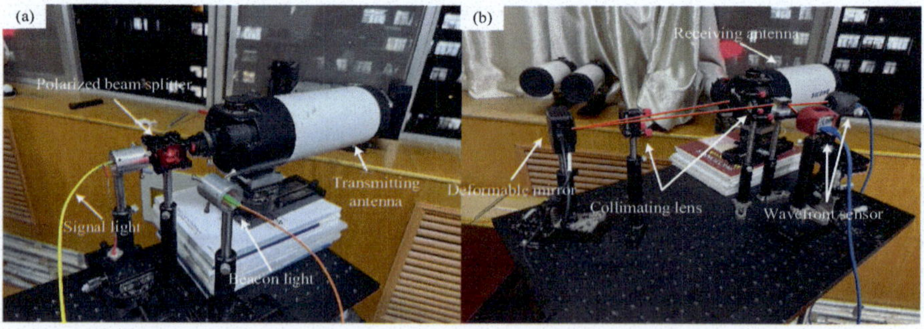

Fig. 7.24 Experimental optical path. **a** Transmitting end. **b** Receiving end

Table 7.1 Experimental-equipment parameters

Device name	Model	Parameters
Wavefront sensor	THORLABS-HASO4-NIR	Measuring wavelength range: 1500–1600 nm Microlens resolution: 32 × 40
Wavefront sensor	THORLABS-HASO4-FIRST	Measuring wavelength range: 400–1100 nm Microlens resolution: 32 × 40
Beacon beam laser	MW-GX-650	Wavelength: 650 nm, linewidth: 0–3 nm, Output power: 35 mW
Signal beam laser	KOHERA-BASIK -Module	Wavelength: 1550 nm, linewidth: 0.1 kHz, Output power: 0–200 mW
Deformable mirror	ALPAO-DM	Number of actuators: 69, effective aperture: 10.5 mm Frequency: 800 Hz
Optical antenna	Maksutov-Cassegrain	Diameter: 25 mm, focal length: 300 mm
Laser beam combiner	F810APC	Wavelength: 1550 nm/650 nm, focal length: 37.13 mm

7.5.3 10-km Adaptive-Optics Correction Experiment

Based on the previous 1.2-km experiment, a laser-communication link with a distance of 10 km was built, and adaptive-optical correction was performed at the receiving end. The transmitting antenna was at Xiaojiazhai on the White Deer Plain, and the receiving end was located in the Sixth Teaching Building of Xi'an University of Technology, as shown in Fig. 7.25, which illustrates the outfield experimental link. Figure 7.26a shows the laser and optical antenna at the transmitter end, and (b) shows the optical antenna and adaptive-optics correction equipment at the receiver end.

7.5.4 100-km Adaptive-Optics Correction Experiment

A 100-km experimental link was built over Qinghai Lake, with the transmitting end of the optical signal and its receiving end located at the Erlangjian scenic spot of Qinghai Lake and Quanji Township of Gangcha County, respectively. Figure 7.27 shows a diagram of the constructed outfield experimental-transmission link, and Fig. 7.28 shows photos of the transmitting and receiving ends.

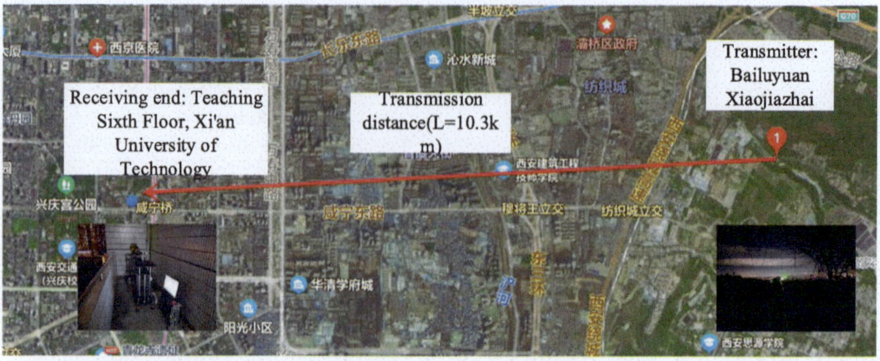

Fig. 7.25 10-km wireless-optical communication field experiment

Fig. 7.26 Photos of the 10.3-km wireless-optical coherent communication system. **a** Transmitter; **b** receiver

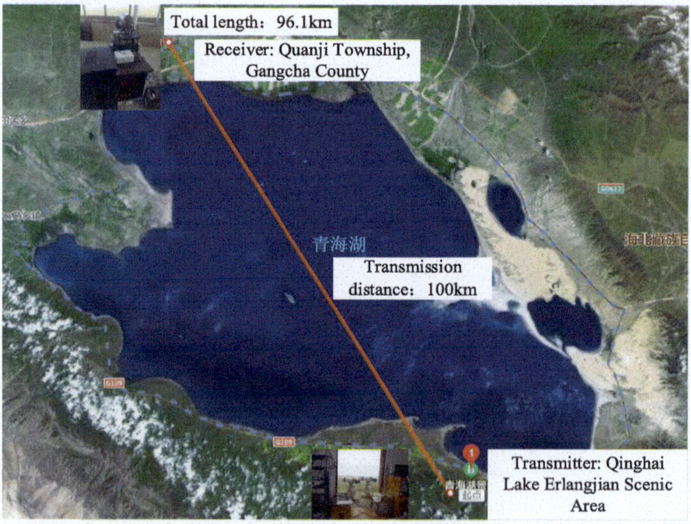

Fig. 7.27 100-km wireless-optical communication field experiment

Fig. 7.28 Photos of the 100-km wireless-optical coherent communication system. **a** Transmitter; **b** receiver

7.5.5 Experimental Results and Analysis

According to the wavefront-sensor parameters listed in Table 7.1, the size of the wavefront-slope matrix is 32 × 40. For the sake of illustration, some values of the partial matrix of the wavefront slope (5 × 9) are selected here for analysis. The wavefront-slope matrix for a 632.8-nm fiducial beam was measured as

$$\begin{bmatrix} 1.695649 & -232.678101 & -1.534530 & 1.373472 & 138.319489 \\ 1.534544 & -211.494125 & -1.695635 & 1.212480 & 121.542297 \\ 1.373486 & -190.090836 & -1.856775 & 1.051585 & 104.993004 \\ 1.212495 & -168.520920 & -1.534530 & 0.890840 & 88.660545 \\ 1.051600 & -146.797195 & -1.695635 & 0.730341 & 72.538742 \\ 0.890854 & -124.913033 & -1.212480 & 0.645228 & 53.086140 \\ 0.730355 & -102.856323 & -1.373472 & 0.483921 & 37.272217 \\ 0.645228 & -99.355789 & -156.190491 & -0.645228 & -0.730355 \\ 0.483921 & -77.025246 & -142.710129 & 172.641388 & 155.343063 \end{bmatrix}.$$

The wavefront-slope matrix of the 1550-nm signal beam was measured as

$$\begin{bmatrix} 1.035649 & -150.956345 & -0.956818 & 0.927685 & 100.935940 \\ 1.036523 & -140.412547 & -0.968546 & 0.689754 & 85.229863 \\ 0.879471 & -110.029023 & -1.125849 & 0.993528 & 65.498232 \\ 1.214295 & -98.659436 & -1.134262 & 0.995635 & 62.348757 \\ 0.706035 & -70.426181 & -0.749565 & 0.436325 & 40.962742 \\ 0.546754 & -70.336354 & -0.561424 & 0.413361 & 30.726241 \\ 0.420319 & -55.986219 & -0.956341 & 0.352981 & 22.954864 \\ 0.431875 & -65.952561 & -80.683419 & -0.356591 & -0.530355 \\ 0.687421 & -48.736929 & -73.289416 & 100.569631 & 122.598361 \end{bmatrix}.$$

The correction matrix was calculated for a 680-nm signal-beam wavelength, 1550-nm beacon-beam wavelength, and $\gamma = 0.53$. The correction matrix is multiplied by the beacon-beam wavefront slope matrix to obtain the corrected beacon-beam wavefront slope matrix:

$$\begin{bmatrix} 0.898693 & -123.478393 & -0.813300 & 0.727940 & 79.309359 \\ 0.813303 & -112.091886 & -0.898686 & 0.642614 & 61.417417 \\ 0.727947 & -100.748143 & -0.984090 & 1.051585 & 55.646292 \\ 0.642622 & -89.316087 & -0.813300 & 0.472145 & 46.990088 \\ 0.557348 & -77.802486 & -0.898686 & 0.387080 & 38.445533 \\ 0.472152 & -66.203907 & -0.642614 & 0.341970 & 28.135654 \\ 0.387088 & -54.513851 & -0.727940 & 0.256478 & 19.754275 \\ 0.341970 & -52.658568 & -82.783302 & -0.331370 & -0.387088 \\ 0.256478 & -40.823380 & -75.636368 & 91.499935 & 82.331823 \end{bmatrix}.$$

The number of correlations between the signal-beam wavefront-slope matrix and the beacon-beam wavefront-slope matrix before and after correction was calculated separately.

In subsequent experiments, the signal beam was corrected using the beacon-beam wavefront-slope matrix before and after the correction, and a change in the RMS value of the wavefront peak-to-valley (PV) was observed. The changes in the PV and RMS values during the correction process are shown in Fig. 7.29.

The stabilized wavefront PV values and wavefront-phase undulation RMS values of the signal beam before and after correction are shown in Figs. 7.30 and 7.31, respectively.

After the beacon-beam aberration data was directly corrected for the signal beam, the mean value of the wavefront PV of the signal beam was 2.08 μm and the variance was 0.90 μm^2; the mean value of the wavefront undulation RMS was 0.40 μm and the variance was 0.04 μm^2.

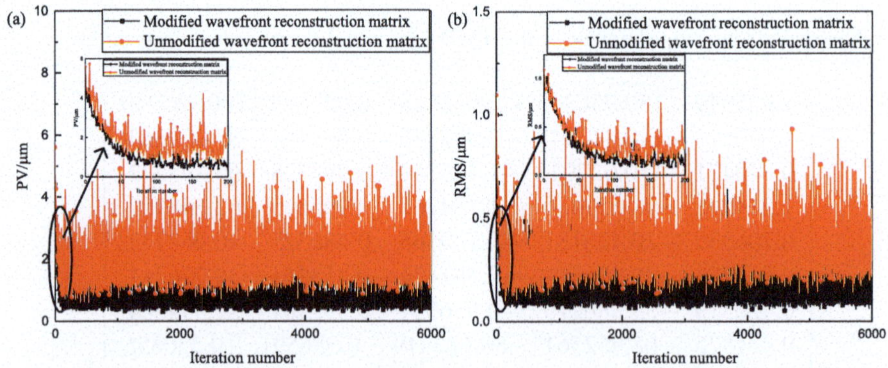

Fig. 7.29 Beacon-beam-corrected signal beam with correcting wavefront-reconstruction matrix. **a** Wavefront PV curve; **b** wavefront RMS curve

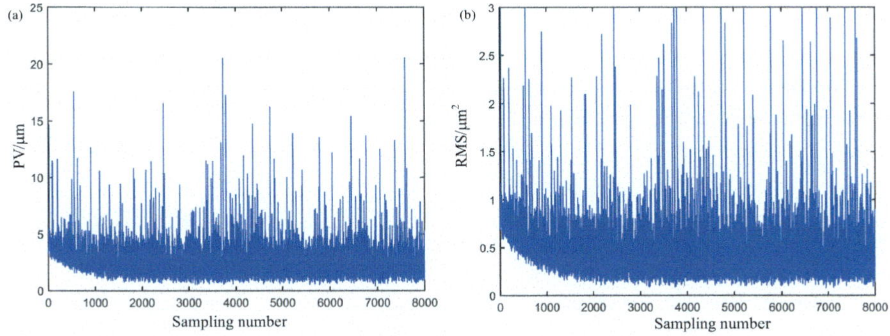

Fig. 7.30 Beacon-beam corrected signal beam. **a** Wavefront PV curve; **b** wavefront RMS curve

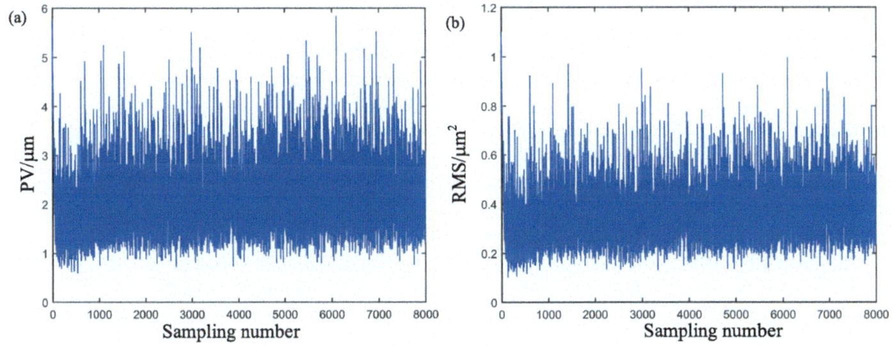

Fig. 7.31 Beacon-beam corrected signal beam with correcting wavefront-reconstruction matrix. **a** Wavefront PV curve; **b** wavefront RMS curve

Subsequently, the wavefront-reconstruction matrix of the beacon beam was multiplied by the coefficient-correction matrix to obtain the wavefront-reconstruction matrix of the signal beam. The AO system corrected the wavefront of the signal beam according to its wavefront-reconstruction matrix to obtain the wavefront PV values and the wavefront phase-undulation RMS values of the signal beam, as shown in Fig. 7.31.

After correcting the wavefront of the signal beam using the fiducial-beam corrected wavefront-reconstruction matrix, the mean PV value of the wavefront of the signal beam was 2.05 μm with a variance of 0.46 μm^2; the mean RMS value of the wavefront phase undulation was 0.35 μm and the variance of the wavefront phase undulation RMS was 0.01 μm^2.

The variance of the wavefront PV values decreased by 49.05%, and the variance of the wavefront-phase undulation RMS decreased by 35.58% in the signal beam reconstruction-matrix correction mode, compared to those before the wavefront reconstruction-matrix correction.

References

1. Landau LD (1963) Electrodynamics of continuous media, 2nd ed. People's Education Press, Beijing
2. Schliching H (1979) Boundary-layer theory. McGraw-Hill, New-York
3. Panchev S (1976) Random function and turbulence. Science Press, Beijing
4. Dubovikov MM, Tatarskii V (1987) The calculation of the asymptotic of the spectrum of locally isotropic turbulence in the viscous range. Zhurnal Eksperimental'noi i Teoreticheskoi Fiziki 93(1):1992–2001
5. Li YQ (2021) Numerical simulation and experimental study of propagation performance of super-continuum laser in turbulent atmosphere. University of Science And Technology of China, Hefei
6. RuiZhong R (2005) The propagation of light in turbulent atmosphere. An Hui Science & Technology Publishing House, Hefei
7. Fante RL (1985) Wave propagation in random media: a system approach. Prog Opt 22:341–398
8. Rytov SM, Kravtsov YA, Tatarskii VI (1989) Principles of Statistical radio physics 4-wave propagation through random media. Springer, Berlin
9. Andrews LC (1992) An analytical model for the refractive index power spectruni and its application to optical scintillations in the atmosphere. J Mod Opt 39(9):1849–1853
10. Tatarskii VI (1971) The effects of the turbulent atmosphere on wave propagation. Jerusalem: Israel Program Sci Transl 19(7):1–6
11. Andrews LC, Phillips RL (2005) Laser beam propagation through random media. SPIE Press, Washington
12. Zhou RenZhong (1996) Adaptive optics. Beijing Institute of Technology Press, Beijing
13. Ishimaru A (1978) Wave propagation and scattering in random media. IEEE Press, New York
14. Lukin VP (1979) Efficiency of some correction systems. Opt Lett 4(1):15–17
15. Sasiela (2007) Electromagnetic wave propagation in turbulence: evaluation and application of Mellin Transforms. SPIE Press, Washington
16. Fried DL (1971) Spectral and angular covariance of scintillation for propagation in a randomly inhomogeneous medium. Appl Opt 10(4):721–731
17. Wu J, Liu JB (2005) Optical transmission theory in the atmosphere. Beuing University of Posts and Telecommunications Press, Beijing, p 128
18. Rao R (2005) Modern atmospheric optics. Science Press, Beijing
19. Hogge CB, Butts RR (1982) Effects of using different wavelengths in wave-front sensing and correction. JOSA 72(5):606–609
20. Medecki H, Tejnil E, Goldberg KA et al (1996) Phase-shifting point diffraction interferometer. Opt Lett 21(19):1526–1528
21. Zhao Z (2021) Mathematical modeling and error analysis of adaptive optics system. Jiangnan University, Suzhou
22. Ziyun Zhao HuGu, Ma W et al (2021) Error budget and parameters optimization of adaptive optics system. Chin J Liquid Cryst Displays 36(05):663–672
23. Rao C, Jiang W (1996) Error and performance analysis of an infrared adaptive optics system at the 2.16 M telescope. J Astrophys 12(04):428–437

24. Rigaut FJ, Véran JP, Lai O (2002)Analytical model for Shack-Hartmann-based adaptive optics systems. In: Adaptive Optical System Technologies. International Society for Optics and Photonics, Waikoloa, Hi, 22 Aug 2002, vol 3353, pp 1038–1048.
25. Chipman RA (1987) Polarization aberrations. University of Arizona Press, Tucson
26. Greenwood DP (1979) Mutual coherence function of a wave front corrected by zonal adaptive optics. J Opt Soc Am 69(4):549–554
27. Hudgin R (1977) Wave-front compensation error due to finite corrector-element size. J Opt Soc Am 67(3):393–395
28. Parenti RR, Sasiela RJ (1994) Laser-guide-star systems for astronomical applications. J Opt Soc Am 11(1):288–309
29. Porter J, Queener H, Lin J et al (2006) Adaptive optics for vision science: Principles, practices, design, and applications. Wiley, Hoboken, pp 233–235
30. Fried DL (1982) Anisoplanatism in adaptive optics. J Opt Soc Am 72(1):52–61
31. Cao G (1994) Accuracy analysis of a Hartmann-Shack wavefront sensor operated with a faint object. Opt Eng 33(7):2331–2335
32. Huang C (2013) Research on key technologies of Shack Hartmann wavefront sensor. Zhejiang University, Hangzhou
33. Jiang WH, Xian H, Shen F (1998) detecting error of Shack-Hartmann wavefront sensor. Chinese J Quant Electron 2:218–227
34. Sasiela RJ (2012) Electromagnetic wave propagation in turbulence: evaluation and application of Mellin transforms. Springer Science & Business Media, Berlin
35. Breckinridge JB, Lam WST, Chipman RA (2015) Polarization aberrations in astronomy telescopes: the point spread function. Publ Astronom Soc Pacific 127(951):445–468
36. Stotts L (2017) Free space optical systems engineering design and analysis. Wiley-IEEE Press, Hoboken
37. Toselli T, Andrews L, Phillips R et al (2009) Free space optical system performance for a Gaussian beam propagating through non-Kolmogorov weak turbulence. IEEE Trans Antennas Propag 57(6):1783–1788
38. Gao C (2018) Research on propagation characteristics for Gaussian laser beam and estimation method of optical parameters through turbulent channel. University of Electronics Science and Technology of China, Chengdu
39. Cui L, Xue B (2015) Influence of asymmetry turbulence cells on the angle of arrival fluctuations of optical waves in anisotropic non-Kolmogorov turbulence. J Opt Soc Am A 32(9):1691–1699
40. Cui L (2013) Analysis of angle-of-arrival fluctuations for optical waves' propagation through weak anisotropic non-Kolmogorov turbulence. Opt Commun 305(5):36–41
41. Du W, Yu S, Tan L et al (2009) Angle-of-arrival fluctuations for wave propagation through non-Kolmogorov turbulence. Opt Commun 282(5):705–708
42. Devaney N, Goncharov AV, Dainty JC (2008) Chromatic effects of the atmosphere on astronomical adaptive optics. Appl Opt 47(8):1072–1081
43. Yang GQ (2007) Research on spreading and wander of laser beam in troposphere atmosphere turbulence. Xidian University, Xian
44. Li Z-M, Qi W-Z, Yu J-P (2010) Effect of angle-of-arrival fluctuation on angle precision of laser tracking system. Opt Optoelectron Technol 8(2):58–62 (in Chinese)
45. Qin ZQ (2010) The research of the impact of atmospheric turbulence on the space lingt-single mode fiber coupling efficiency. Harbin Institute of Technology, Harbin
46. Consortini A, Sun YY, Innocenti C et al (2003) Measuring inner scale of atmospheric turbulence by angle of arrival and scintillation. Opt Commun 216(13):19–23
47. Toselli I, Andrews LC, Phillips RL, Ferrero V (2007) Angle of arrival fluctuations for free space laser beam propagation through non-kolmogorov turbulence. Atmos Propag 65(51):23–28
48. Wu X, Safari M, Haas H (2017) Access point selection for hybrid Li-Fi and Wi-Fi networks. IEEE Trans Commun 65(12):5375–5385

49. Conan R, Borgnino J, Ziad A, Martin F (2000) Analytical solution for the covariance and for the decorrelation time of the angle of arrival of a wave front corrugated by atmospheric turbulence. J Opt Soc Am 17(10):1–22
50. Voelz D, Wijerathna E, Muschinski A et al (2018) Computer simulations of optical turbulence in the weak- and strong-scattering regime: angle-of-arrival fluctuations obtained from ray optics and wave optics. Opt Eng 57(10):1–9
51. Zhang YX (1987) Arrival angle of beam wave propagation in turbulent atmosphere. Chin J Quant Electron 12(1):70–75
52. Goodman JW, Qin KC (1992) Statistical optics. Science Press, Beijing
53. Gao C, Li X (2017) Effects of inner and outer scale on beam spreading for a Gaussian wave propagating through anisotropic non-Kolmogorov turbulence. Optica Appl 47(1):63–74.
54. Gao C, Li X (2018) An analytic expression for the beam wander of a Gaussian wave propagating through scale dependent anisotropic turbulence. Iran J Sci Technol Trans Sci 42(2):975–982
55. Yao Q (2002) Optics course, 3rd ed. Higher Education Press, Beijing
56. Andrews LC, Philips RL, Hopen CY (2001) Laser beam scintillation with applications. SPIE press, Bellingham
57. Hui T (2021) Study on beam wander under strong fluctuation condition. University of Science and Technology of China, Hefei
58. Zhengfang S (1990) The basis of applied atmospheric optics. China Meteerelogical Press, Beijing
59. Wang LL (2012) Study of the characteristic of beam wander and spreading in the atmosphere turbulence. Xidian University, Xian
60. Kolmogorov AN (1991) The local structure of turbulence in incompressible viscous fluid for very large Reynolds numbers. Proc R Soc Lond Ser A Math Phys Sci 434(1890):9–13
61. Collett E, Foley JT, Wolf E (1977) On an investigation of Tatarskii into the relationship between coherence theory and the theory of radiative transfer. J Opt Soc Am 4(4):465
62. Le NV (1971) The Von Karman integral method as applied to a turbulent boundary layer. J Aeronaut Sci 19(9):647–648
63. Primmerman CA, Price TR, Humphreys RA (1995) Atmospheric-compensation experiments in strong scintillation conditions. Appl Opt 34(12):2081–2088
64. Wang J (2020) Study on propagation and diffraction properties of partially coherent vortex beams in atmospheric turbulence. Xi'an University of Technology
65. Yang H, Cai D, Chen B, Li X, Jiang W (2008) Analysis of Adaptive optics techniques without a wave-front sensor and its application in atmospheric laser communications. Chin J Lasers 35(5)680–684 (in Chinese)
66. Primmermen CA, Price TR, Humphreys RA et al (1995) Atmospheric-compensation experiments in strong-scintillation conditions. Appl Opt 34(12):2081–2088
67. Wang Z (2009) The research of algorithm for wave-front distortion compensation in atmospheric optical communication. Huazhong University of Science and Technology, Wu Han
68. He X (2019) Study on sensor-less wave front correction algorithm in free space optical communication. Jinlin University, Chang Chun
69. Ke X, Zhang Y, Zhang Y, Lei S (2019) GPU Acceleration in wave-front sensorless adaptive wave-front correction system. Laser Optoelectron Progress 56(7):96–104 (in Chinese)
70. Ping Y, Wu M, Yuan L et al (2008) Adaptive optics genetic algorithm based on Zernike mode coefficients. Chin J Lasers 35(3):367–372
71. Li Z, Cao J, Zhao X et al (2015) Atmospheric compensation in free space optical communication with simulated annealing algorithm. Opt Commun 338(338):11–21
72. Spall JC (2002) Multivariate stochastic approximation using a simultaneous perturbation gradient approximation. IEEE Trans Autom Control 37(3):332–341
73. Roddier NA (1990) Atmospheric wavefront simulation using Zernike polynomials. Opt Eng 29(10):1174–1180

74. Zhang Q, Han S, Tang S, Luo C, Fu YG (2018) Study on Functional relationship between transmitted wavefront Zernike coefficients and wavelengths. Acta Optica Sinica 38(2):170–177 (in Chinese)
75. Wang W, Liu D, Xiang Y et al (2010) Applied optics. Huazhong University of Science & Technology Press, Wu Han, pp 167–177

Chapter 8
Multi-corrector Wavefront Correction Decoupling Control

Adaptive-optics is a type of optoelectromechanical integration technology. A deformable mirror, as the core device of an adaptive-optics system, corrects the wavefront by changing its own surface distribution to change the optical-path difference of the incident wavefront sensor. The travel amount and resolution of the deformable mirror directly determine the accuracy and extent of the wavefront correction. A single deformable mirror usually cannot correct the results of a strong-turbulence environment, high-power laser output, or large-aperture optical-system reception. These significantly distort the wavefronts of light waves, and the actuator service life is shortened when working near full stroke for a long time. Therefore, a fast steering mirror or multi-stage deformable mirror is introduced. A fast steering mirror (FSM) or large-travel deformable mirror is used to correct the low-order components of the wavefront, while a small-travel deformable mirror is used to correct the high-order components, thereby correcting the wavefront in a optical wireless communication system.

8.1 Problem

With the continuous expansion of applications in the field of adaptive-optics technology, more stringent application requirements for aberration-correction accuracy and correction amplitude follow. Achieving the high-precision correction of large strokes and eliminating high spatial-frequency wavefront aberrations are common problems that must urgently be solved in many application fields.

Owing to the limitations of the material, process, system-integration complexity, and other technical problems of different types of wavefront correctors, adaptive-optics systems with a single wavefront corrector often have difficulty correcting large-travel and high-spatial-frequency aberrations. Therefore, using multiple wavefront correctors with different performances to achieve the synchronous correction

X. Ke and C. Ke, *Performance-Improvement Limits in Optical Wireless Coherent Communication*, Optical Wireless Communication Theory and Technology,
https://doi.org/10.1007/978-981-96-2000-5_8

of large-amplitude, low-spatial-frequency aberrations and small-amplitude, high-spatial-frequency aberrations has become one of the most effective ways to solve the above problems.

To compensate for the large-travel and high-spatial-frequency aberrations beyond the correction ability of existing wavefront correctors, domestic and foreign scholars have proposed an adaptive-optics solution that combines multiple types of wavefront correctors. The most typical is the Woofer–Tweeter dual deformable-mirror adaptive-optics system.

A Woofer is a low-order deformable mirror with a large travel and low driver density; it is used to correct large-stroke low-spatial-frequency aberration components in wavefront distortions. A Tweeter is a high-order deformable mirror with a small stroke and high-driver density used to correct small-stroke high-spatial-frequency aberration components in wavefront distortions.

In theory, large-travel and high-spatial-frequency aberrations can be sufficiently compensated by selecting appropriate combinations of Woofer and Tweeter deformable mirrors. When the amplitude and spatial-frequency composition of the wavefront aberration to be corrected are more complex, wavefront correctors with different travel and spatial frequencies must be added, based on the distribution of the aberration travel and spatial frequency. Adaptive-optics systems with multiple wavefront correctors have become an effective method to reduce the pressure of developing wavefront correctors with high spatial frequencies and large travel. Moreover, in many application fields, they can compensate for aberrations better than a single wavefront corrector.

The introduction of multi-level correctors inevitably causes coupling issues between correctors. If the coupling effect between them is too large, the system will not operate stably, and the system correction performance will also be affected. Therefore, eliminating coupling has become a top priority in research on multi-wavefront-corrector adaptive-optics systems [1].

At present, a typical multi-wavefront-corrector adaptive-optics system has dual deformable mirrors. It uses their different travel and spatial-frequency characteristics to improve its correction accuracy and travel for wavefront aberrations.

However, because both deformable mirrors can correct aberrations, phase compensation in opposite directions, commonly known as the coupling effect [2], may occur. This causes the deformable mirrors to cancel each other out, leading to a significant decrease in the aberration-correction ability of the system and even to a system collapse.

Moreover, a Hartmann wavefront sensor cannot measure this coupling and can only measure the corrected residual aberration, wasting the travel of the deformable mirror and affecting the stability of the system. Therefore, when using dual deformable mirrors to synchronously correct aberrations, it is necessary to eliminate the potential coupling between them. This will enable them to work together efficiently, give full play to the advantages of a dual-deformable-mirror adaptive-optics correction system, and further improve its performance.

8.2 Fast Steering Mirror and Deformable Mirrors

A fast control mirror, also known as a fast steering mirror, uses a mirror surface to accurately control the direction of a beam between the light source and receiver. A fast-reflection mirror uses piezoelectric ceramics or voice-coil motors to quickly and slightly deflect the light beam.

A fast-reflection mirror is an important part of an optoelectronic precision-tracking system used to accurately control the direction of a beam. A fast-reflection mirror has a rapid response speed and high control accuracy. It can be used to correct tilt errors in the optical path and stabilize the beam direction, and is used in fast-tracking systems. A fast-reflection mirror controls the rapid and high-frequency rotation of the reflecting-mirror surface under the action of a driving element, achieving high-speed and accurate pointing, stability, and beam tracking [3].

Piezoelectric-ceramic drivers are the ideal driving method for fast-reflection mirrors. A fast steering mirror driven by piezoelectric ceramics consists of a mirror, flexible hinge, piezoelectric-ceramic driver, base, resistance-strain gauge sensor, and circuit structure. Piezoelectric ceramics used in fast-reflection mirrors require a high resolution at the nanoscale. Using piezoelectric ceramics can significantly improve the resonance frequency and sensitivity.

Piezoelectric ceramics utilize piezoelectricity. When certain substances are subjected to external forces, not only do their geometric dimensions change, but they also undergo internal polarization. The surface charge forms an electric field. When the external force disappears, it returns to the uncharged state. This effect is known as positive piezoelectricity.

In contrast, when an electric field is applied to a piezoelectric crystal, both polarization and deformation occur. This effect is known as inverse piezoelectricity. When a voltage is applied to piezoelectric ceramics, they experience a displacement that can reach the nanometer level. This displacement is approximately proportional to the applied voltage. In practical applications, stacked piezoelectric-ceramic actuators are often used to increase the displacement and output force.

Compared with traditional motor structures, piezoelectric ceramics do not require a transmission mechanism. They have high displacement-control accuracy, fast response speed, no mechanical-fit gap, and can achieve voltage-following displacement control. They have a strong output, low power consumption, and almost no power consumption while maintaining position. They are solid-state devices that can easily achieve closed-loop control in power supplies, sensors, microcomputers, etc.

A deformable mirror corrects the wavefront by changing the optical-path difference of the incident wavefront sensor by changing its own surface distribution. For a distorted wavefront, it can be seen as a filter that corrects the phase disturbance of the wavefront by filtering out the low-frequency components and retaining the high-frequency components.

Deformable mirrors have two main classification criteria: the mirror shape and driving method [4].

8.2.1 Mirror Shape

Deformable mirrors can be classified into two types, based on the shape of the mirror surface: discrete mirror-face and continuous mirror-face. A discrete mirror-face deformable mirror is composed of multiple independent submirrors. Each submirror is driven by a driver and forms the desired surface shape through translational or oblique motion. This type of deformable mirror has simple control and is easy to assemble; however, the gaps between individual mirror surfaces can lose light energy, affect the detection accuracy, and result in a larger size.

A continuous mirror-face deformable mirror is driven by magnetic force, piezoelectric force, or other forces to cause displacement and change its surface shape. Because of the absence of mirror gaps, this type of deformable mirror does not lose light energy, ensuring phase continuity; therefore, it has a higher correction accuracy than discrete mirror-face deformable mirror.

8.2.2 Driving Method

Depending on the driving method, deformable mirrors can be classified into electrostatic, piezoelectric, electromagnetic, and other types. Electrostatic driving deforms the mirror through the Coulomb force of the free charges between the plates. Owing to the electrostatic attraction, the shape variation of this deformable mirror is relatively small, which significantly limits its application range.

Piezoelectric deformable mirrors mainly include single piezoelectric deformable mirrors and piezoelectric microelectromechanical system (MEMS) deformable mirrors. The former has a simple structure, low price, small volume, and is easy to move; however, it can only accurately correction low-order aberrations. The latter has a short stroke and a hysteresis phenomenon. Therefore, it is necessary to use certain algorithms to overcome this impact to ensure the system-correction accuracy; this increases the algorithm complexity of the system and reduces its real-time performance.

The key components of an electromagnetic-driven deformable mirror are the electromagnetic coil and a permanent magnet, which generate a force between them through electrification that offsets the mirror surface. Because of its large stroke size and wide calibration range, it is widely used in optical communication systems [5].

8.3 Using Multiple Correctors to Correct Wavefront Distortions

8.3.1 Algorithm Principle

Because Zernike polynomials are orthogonal in the unit circle [6],

$$\int W(r) \cdot Z_i \cdot Z_{i'} d^2 r = \delta_{jj'} \begin{matrix} W(r) = \frac{1}{\pi} r \leq 1 \\ W(r) = 0 r > 1 \end{matrix}. \tag{8.1}$$

A fast steering mirror or low-resolution deformable mirror is used as a Woofer to correct low-order aberrations, and a small-path deformable mirror is used as a Tweeter to correct high-order aberrations; hence, there is no crosstalk between the two. Figure 8.1 shows a schematic diagram of the adaptive-optics closed-loop control algorithm, which uses multiple correctors based on the Zernike mode.

According to the principle of the push–pull method, for an adaptive-optics system composed of a Woofer and Tweeter, the Woofer corrects low-order aberrations and the Tweeter corrects high-order aberrations. They respectively solve the response matrix from the Woofer to the wavefront sensor and from the Tweeter to the wavefront sensor.

To solve any response matrix, the reflection state of the stationary plane in which the other corrector is in the initial position is required, which is recorded as $\mathbf{IM}_{P2z}$ and $\mathbf{IM}_{c2z}$, respectively [7]:

$$\mathbf{IM}_{P2z} = \mathbf{PZT}^{-1} \cdot \mathbf{Z} \tag{8.2}$$

$$\mathbf{IM}_{c2z} = \mathbf{cmd}^{-1} \cdot \mathbf{Z}. \tag{8.3}$$

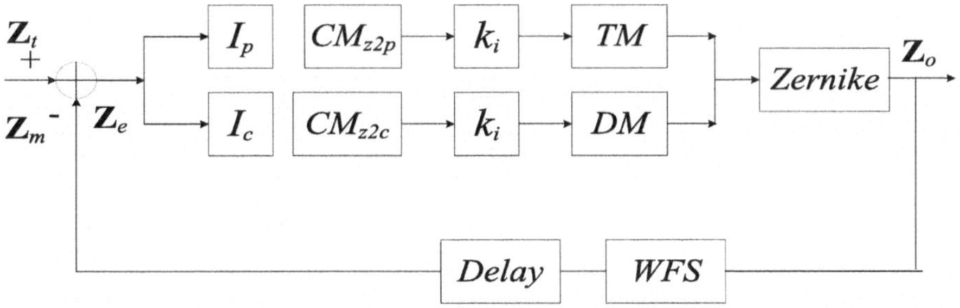

Fig. 8.1 Schematic of an adaptive-optics closed-loop control system with multiple correctors, based on the Zernike mode

In Eqs. (8.2) and (8.3), **PZT** and **cmd** are the command coefficients applied to the Woofer and Tweeter, respectively; $^{-1}$ is the inverse of the matrix; and **Z** is the Zernike coefficient collected by the wavefront sensor. The matrix sizes of $\mathbf{IM}_{P2z}$ and $\mathbf{IM}_{c2z}$ are 2 × 30 and 69 × 30, respectively.

After calculating the response matrix, the corresponding command matrices are calculated separately [8]:

$$\mathbf{CM}_{z2P} = (\mathbf{IM}_{P2z})^{-1} \tag{8.4}$$

$$\mathbf{CM}_{z2c} = (\mathbf{IM}_{c2z})^{-1}. \tag{8.5}$$

In Fig. 8.1, $\mathbf{Z}_t = 0$ represents 1 × 30 zero matrix, and $\mathbf{Z}_m$ is the Zernike coefficient collected by the wavefront sensor under the current-state 1 × 30 matrix. $\mathbf{Z}_e$ is the error Zernike coefficient matrix. The 30-order diagonal matrices, $\mathbf{I}_P$ and $\mathbf{I}_c$, separate the tilt component of the Zernike coefficient from the remaining components. After the corresponding command matrices $\mathbf{CM}_{z2P}$ and $\mathbf{CM}_{z2c}$ are converted, they are sent to the Woofer and Tweeter, respectively, via integration operations.

The wavefront sensor collects the incident-wavefront Zernike coefficient again and completes the next iteration after time-delay processing. The wavefront Zernike-coefficient matrix acquired by the wavefront sensor for the kth time is recorded as $\mathbf{Z}_I{}^k$ and the wavefront phase is $\mathbf{Phase}_I{}^k$. According to the principle of the integral control algorithm, the closed-loop iteration expression of the kth adaptive-optics system is [9]

$$\begin{aligned}\mathbf{Phase}_I^{k+1} &= \mathbf{Phase}_I^k + \left[\mathbf{PZT}^k + k_i \cdot (-\mathbf{Z}_I^k) \cdot \mathbf{I}_P \cdot \mathbf{CM}_{z2P}\right] \cdot \mathbf{IF}_{P2p} \\ &\quad + \left[\mathbf{cmd}^k + k_i \cdot (-\mathbf{Z}_I^k) \cdot \mathbf{I}_c \cdot \mathbf{CM}_{z2c}\right] \cdot \mathbf{IF}_{c2p},\end{aligned} \tag{8.6}$$

where $\mathbf{PZT}^k$ and $\mathbf{cmd}^k$ are the instruction matrices respectively applied to the Woofer and Tweeter for the kth time, k_i is the integral control coefficient, and $\mathbf{IF}_{P2p}$ and $\mathbf{IF}_{c2p}$ are the surface-influence functions of the Woofer and Tweeter, respectively. The 30th-order diagonal matrices $\mathbf{I}_P$ and $\mathbf{I}_c$ are

$$\mathbf{I}_P = \begin{bmatrix} 1 & & & & \\ & 1 & & & \\ & & 0 & & \\ & & & \ddots & \\ & & & & 0 \end{bmatrix} \quad \text{and} \quad \mathbf{I}_c = \begin{bmatrix} 0 & & & & \\ & 0 & & & \\ & & 1 & & \\ & & & \ddots & \\ & & & & 1 \end{bmatrix}. \tag{8.7}$$

When $\mathbf{I}_c = 0$, the Woofer is in a closed-loop state; when $\mathbf{I}_P = 0$, the Tweeter is in a closed-loop state. When $\mathbf{I}_c$ and $\mathbf{I}_P$ are those shown in Eq. (8.7), both the Woofer and Tweeter are in a closed-loop state.

Figure 8.2 shows an adaptive-optics system using multiple correctors to correct a wavefront distortion. One is a combination of a fast steering mirror and a 69-element

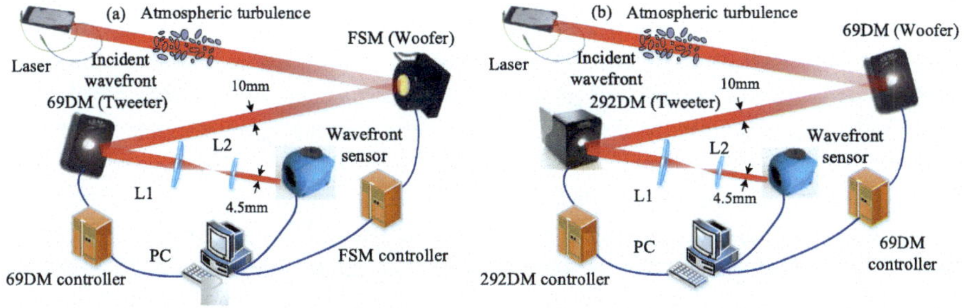

Fig. 8.2 Adaptive-optics system using multiple correctors to correct wavefront distortions. **a** Fast-steering mirror + 69-unit deformable mirror; **b** 69-unit deformable mirror + 292-unit deformable mirror

deformable mirror, and the other is a combination of a 69-element deformable mirror and a 292-element deformable mirror.

8.3.2 Working Principle of the Fast Steering Mirror and Deformable Mirror

The surface-profile distributions of the piezoelectric fast steering mirrors, small-element deformable mirrors, and multi-element deformable mirrors are shown in Fig. 8.3. The surface-profile diameter of the fast steering mirror is 25.4 mm (1 in.), and it is electrically connected through a four-point drive. It is driven by two pairs of independent piezoelectric ceramics. The two piezoelectric ceramics located on the $x(y)$ axis change their shape through the voltage to change the wavefront tilt in the $y(x)$ axis direction.

The fast steering mirror is driven by piezoelectric ceramics and uses a frictionless flexible-hinge differential parallel-guiding structure to tilt the mirror, achieving a precise angle adjustment at the rate of microradians. A fast steering mirror has a high

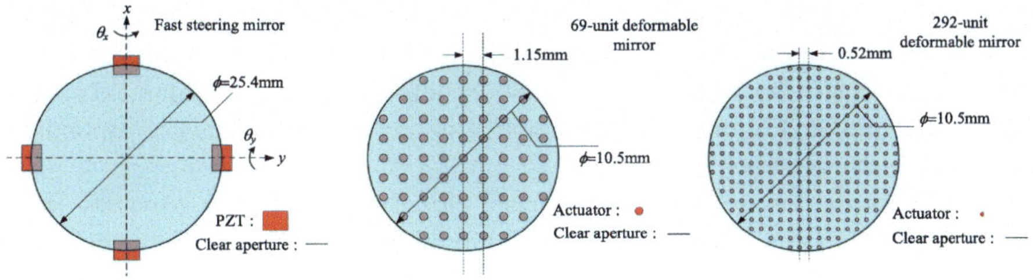

Fig. 8.3 Profile distribution of a wavefront corrector. **a** Fast-steering mirror; **b** 69-unit deformable mirror; **c** 292-unit deformable mirror

resonant frequency, fast scanning speed, high resolution, good stability, and frictionless movement. In the actual calculations, angle commands are sent to channels x and y, which are internally converted to generate the surface-angle tilt corresponding to the x and y axes, respectively, after applying the corresponding piezoelectric-ceramic voltage.

The 69-unit deformable mirror and 292-unit deformable mirror are electromagnetic continuous-surface deformable mirrors with 69 and 292 driving units, respectively. The luminous aperture is 10.5 mm, the actuator spacing is 1.15 mm and 0.52 mm, respectively, the maximum amount of tilt-front modulation is 60 μm, the nonlinear error is less than 3%, and the operating temperature is − 10 to 35 °C. The normalized voltage range applied to the drivers of the mirrors is between − 1 and 1 [10, 11].

The electromagnetic continuous deformable mirror consists of three parts: substrate, actuator, and reflector (mirror surface). The reflector uses thin continuous lenses, and the surface shape of the thin lenses changes with the actuator displacement. To ensure that the actuator motion is mostly reflected on the thin mirror surface, the substrate does not deform when the actuator moves. The reflector connected to the actuator deforms accordingly. This requires the substrate to be much harder than the reflector. Therefore, the substrate is typically made of harder metal materials.

The actuator material for separating the continuous-surface deformable mirror of the actuator is generally selected as a piezoelectric material (Pb (Zr, Ti) O_3) or electrostrictive material (Pb (Mg, Nb) O_3), and these two materials are used to make the actuator. When the driving voltage is applied to the deformable mirror, the actuator displacement will change, deforming the surface shape of the mirror connected to its corresponding position. The surface shape of the mirror on the adjacent actuator will also deform, which is similar to the Gaussian function [12].

8.3.3 Adaptive-Optics System with a Fast Steering Mirror + 69-Element Deformable Mirror

A numerical simulation of an adaptive-optics system using a combination of a fast steering mirror and 69-element deformable mirror (DM) was performed. The parameters were as follows: $k_i = 0.05$, Zernike coefficient order is 1–30, and atmospheric-turbulence intensity $D/r_0 = 0.1$. The 69-element deformable-mirror command matrix $\mathbf{CM}_{z2c}$ was provided by the ALPAO adaptive-optics company. The corresponding simulation curve is shown in Fig. 8.4.

When only the fast steering mirror has a closed loop, the peak-to-valley (PV) value of the wavefront decreases from 13.04 to 7.03 μm, and the root-mean-square (RMS) value of the wavefront decreases from 1.82 to 0.75 μm. When only the 69-unit deformable mirror forms a closed loop, the PV value of the wavefront decreases from 13.04 to 10.05 μm, and the RMS value of the wavefront decreases from 1.82 to 1.63 μm.

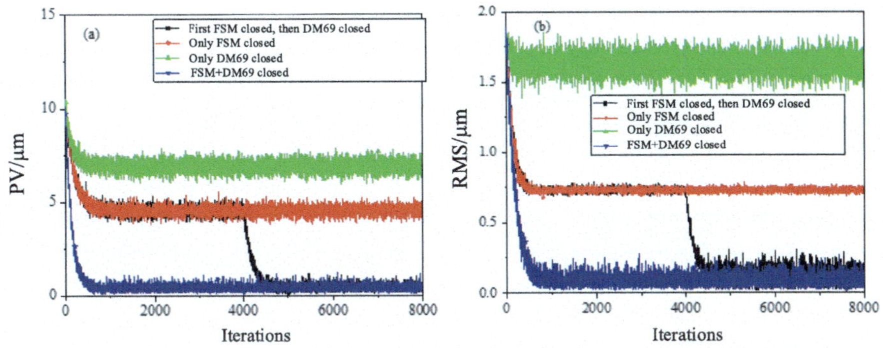

Fig. 8.4 Closed-loop curves of the adaptive-optics system with a fast-steering mirror + 69-unit deformable mirror. **a** Wavefront peak-to-valley value; **b** wavefront root-mean-square value

When the fast steering mirror and deformable mirror have closed loops simultaneously, the PV value of the wavefront decreases from 13.04 to 0.08 μm, and the RMS value of the wavefront decreases from 1.82 to 0.13 μm. This indicates that the proportion of the wavefront-tilt component is greater than that of the higher-order component, and the wavefront-correction effect using a combined fast steering mirror and deformable-mirror closed loop is better than that using single fast-reflector and single deformable-mirror closed loops.

Figure 8.5 shows the closed-loop voltage distribution of the adaptive-optics system using a combination of a fast steering mirror and 69-unit deformable mirror. Only the deformable mirror is an individually closed loop, and the total power of the deformable mirror is 0.7011. When the fast steering mirror and deformable mirror are closed loops simultaneously, the total power of the deformable mirror is 0.4362. Therefore, introducing a fast steering mirror reduces the output power of the 69-unit deformable mirror driver.

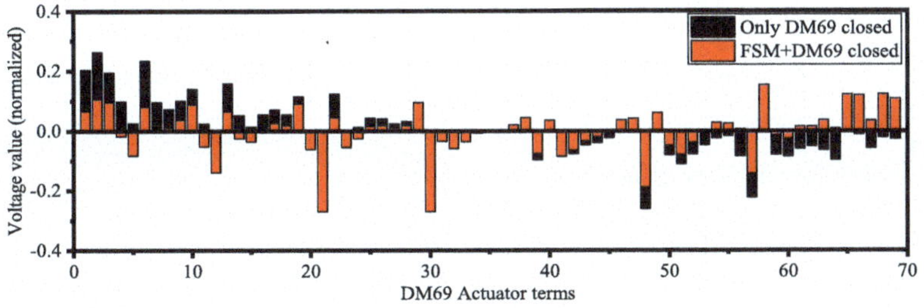

Fig. 8.5 Adaptive-optics system closed-loop voltage distribution when using a fast steering mirror and 69-unit deformable mirror

8.3.4 Adaptive-Optics System with 69-Element and 292-Element Deformable Mirrors

When using a combination of 69-unit and 292-unit deformable mirrors, the 69-unit deformable mirror serves as a Woofer for correcting low-order aberrations, and the 292-unit deformable mirror serves as a Tweeter for correcting high-order aberrations. The algorithm principle is the same as that of the adaptive-optics system composed of a fast steering mirror and a 69-element deformable mirror, except that the fast steering mirror is replaced by the 69-element deformable mirror, and the 69-element deformable mirror is replaced by the 292-element deformable mirror.

The adaptive-optics system with the 69-element and 292-element deformable-mirror combination was simulated, and the following simulation parameters were adopted: $k_i = 0.05$, Zernike coefficient order is 1–30, atmospheric turbulence intensity $D/r_0 = 0.1292$, and the element deformable-mirror command matrix $\mathbf{CM}_{z2c}$ was provided by ALPAO Company. Figure 8.6 shows the corresponding simulation curve.

When only the 69-unit deformable mirror is in the closed loop, the PV value of the wavefront decreases from 10.92 to 4.98 μm, and the RMS value of the wavefront decreases from 1.25 to 0.49 μm. When only the 292-unit deformable mirror is in the closed loop, the PV value of the wavefront decreases from 10.92 to 6.95 μm, and the RMS value of the wavefront decreases from 1.25 to 1.15 μm.

When both deformable mirrors are in a closed loop simultaneously, the PV value of the wavefront decreases from 10.92 to 0.82 μm, and the RMS value of the wavefront decreases from 1.25 to 0.13 μm. This shows that the wavefront-correction effect of the closed loop with 69-element and 292-element deformable mirrors is better than the single closed-loop correction effect. The correction effect does not differ significantly from that of the adaptive-optics system with the combined fast steering mirror and 69-element deformable mirror.

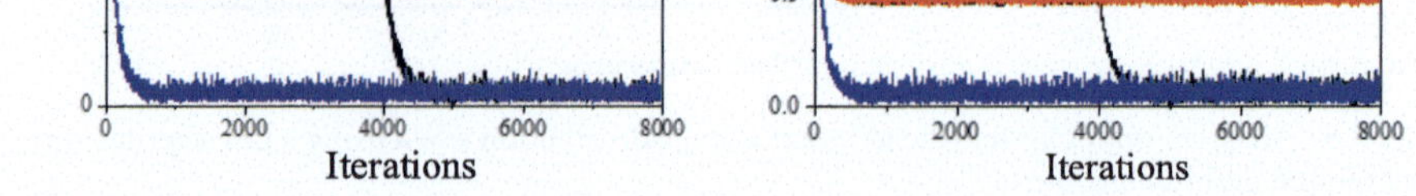

Fig. 8.6 Adaptive-optics system closed-loop curve with a 69-unit deformable mirror + 292-unit deformable mirror. **a** Wavefront PV value; **b** wavefront RMS value

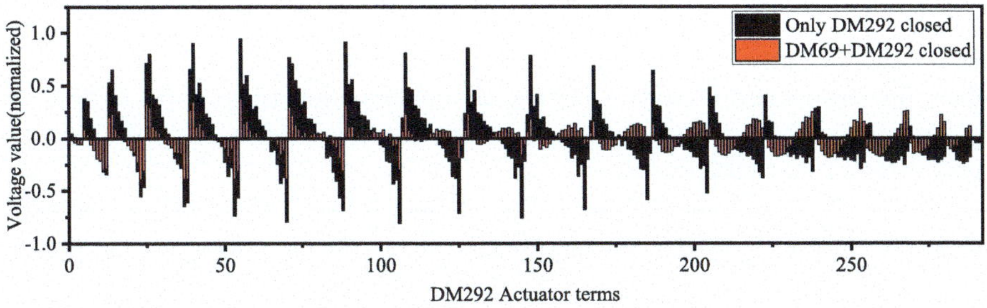

Fig. 8.7 Adaptive-optics system closed-loop voltage distribution using the 69-unit + 292-unit deformable mirrors

Figure 8.7 shows the closed-loop voltage distribution of the adaptive-optics system using the 69-element + 292-element deformable-mirror combination. When only the 292-units deformable mirror is an individually closed loop, the total power is 28.55. When 69 units of deformable mirrors and 292 units of deformable mirrors are simultaneously closed loops, the total power is 4.98. Therefore, the power of the 292-unit deformable-mirror driver was reduced by introducing 69 deformable-mirror units.

8.4 Correcting Wavefront Distortion Using a Multi-wavefront-Corrector Decoupling Algorithm

8.4.1 Algorithm Principle

The Hartmann sensor can measure the slope. In a dual-deformable-mirror correction system, its slope matrix can be represented as

$$g = g_w + g_t, \tag{8.8}$$

where g is the slope matrix directly measured by the Hartmann sensor, the dimension is $2m \times 1$, and m is the number of sub-apertures in the Hartmann sensor. g_w is the slope to be corrected by the lower-order deformable mirror and g_t is the slope to be corrected by the higher-order deformable mirror.

According to the wavefront-reconstruction principle of the Zernike mode method, the slope g can be expressed as

$$g = \mathrm{Z} \cdot Zernike. \tag{8.9}$$

We rewrite Eq. (8.9) as follows:

$$Zernike = \mathrm{Z}^{+} \cdot g, \tag{8.10}$$

where *Zernike* is the Zernike coefficient, and the dimension is 30 × 1. Z is the wavefront-reconstruction matrix, the dimensions are $2m \times 30$, and Z^{+} is the generalized inverse matrix of Z.

If a low-order deformable mirror corrects the low-order aberrations and assigns them aberrations, the expression is as follows:

$$Zernike_w = I_w \cdot Zernike = I_w \cdot \mathrm{Z}^{+} \cdot g, \tag{8.11}$$

where I_w is a 30×30 diagonal matrix. For example, if a low-order deformable mirror corrects first to second-order aberrations, the I_w expression is as follows:

$$I_w = \begin{bmatrix} 1 \\ 1 \\ 0 \\ \cdots \\ 0 \\ 0 \end{bmatrix}. \tag{8.12}$$

The slope of the low-order deformable-mirror correction can be expressed as

$$g_w = IM_w \cdot Zernike_w = IM_w \cdot I_w \cdot \mathrm{Z}^{+} \cdot g. \tag{8.13}$$

According to (8.13), the initial control voltage of the low-order deformable mirror can be obtained as

$$cmd_w = IM_w^{+} \cdot g_w = IM_w^{+} \cdot IM_w \cdot I_w \cdot \mathrm{Z}^{+} \cdot g, \tag{8.14}$$

where cmd_w is the control voltage of the low-order deformable mirror, IM_w is the slope-response matrix of the low-order deformable mirror, and $IM_w{}^{+}$ is the generalized inverse matrix of IM_w.

Subsequently, an incremental proportional–integral (PI) controller is used to solve the final control voltage of the low-order deformable mirror, which is sent to the low-order deformable mirror for operation. To accurately calculate the control voltage of the low-order deformable mirror, its control-voltage expression is

$$\begin{aligned} CMD_w(k+1) = {} & CMD_w(k) + k_{i\_w} \cdot cmd_w(k) \\ & + k_{p\_w} \cdot (cmd_w(k) - cmd_w(k-1)). \end{aligned} \tag{8.15}$$

where k_{i_w} and k_{p_w} are the integral and proportional parameters of the incremental PI controller, respectively.

In addition, the slope expression for the high-order deformable-mirror correction is

$$g_t = g - g_w = (I - IM_w \cdot I_w \cdot Z^+)g, \tag{8.16}$$

where the control voltage of the high-order deformable mirror can be obtained using the direct-slope method, and the expression is

$$cmd_t = IM_t^+ \cdot g_t = IM_t^+ \cdot (I - IM_w \cdot I_w \cdot Z^+)g, \tag{8.17}$$

where cmd_t is the control voltage of the high-order deformable mirror, IM_t is the slope-response matrix of the high-order deformable mirror, and $IM_t{}^+$ is the generalized inverse matrix of IM_t.

Owing to the fitting error of the deformable mirror for different orders of Zernike modes, coupling occurs when the high- and low-order deformable mirrors work simultaneously. Therefore, constraint matrix C was constructed to reset the control voltage of the high-order deformable mirror and suppress the coupling between the two deformable mirrors.

If the control voltage of the high-order deformable mirror does not include the coupling component with the low-order deformable mirror,

$$IM_{tw} V_t = 0. \tag{8.18}$$

At this point, the coupling effect between the two deformable mirrors is minimal.

In the equation, IM_{tw} is the projection-coefficient matrix. Based on the correlation between the influence function of each actuator of the high-order deformable mirror and the corresponding coefficients of each Zernike polynomial, the projection-coefficient matrix IM_{tw} of the high-order deformable mirror is constructed. The coefficient matrix is expressed as follows:

$$IM_{tw}(i,j) = k \frac{\iint f_i(x,y) Z_j(x,y) ds}{\iint Z_j(x,y) Z_j(x,y) ds}, \tag{8.19}$$

where $f_i(x, y)$ is the influence function of the ith actuator of the high-order deformable mirror and $Z_j(x, y)$ is the Zernike coefficient corresponding to the jth-order Zernike polynomial. k is the coupling-suppression factor used to adjust the degree of coupling suppression of the high-order deformable mirrors.

Equation (8.18) is derived, and the control-voltage expression of the decoupled high-order deformable mirror is as follows:

$$\begin{bmatrix} I \\ IM_{tw} \end{bmatrix} CMD_t' \approx \begin{bmatrix} CMD_t \\ 0 \end{bmatrix} \tag{8.20}$$

$$CMD_t' = \begin{bmatrix} I \\ IM_{tw} \end{bmatrix}^+ \begin{bmatrix} CMD_t \\ 0 \end{bmatrix} = C\begin{bmatrix} CMD_t \\ 0 \end{bmatrix}. \tag{8.21}$$

Then, Eq. (8.21) can also be expressed as

$$CMD_t(k+1) = C \times (CMD_t(k) + k_{i\_t} \cdot cmd_t(k) + k_{p\_t} \cdot (cmd_t(k) - cmd_t(k-1))). \tag{8.22}$$

At this point, the coupling effect between the high-order and low-order deformable mirrors is eliminated, and the final control voltage of the high-order deformable mirror does not contain the voltage component coupled with the low-order deformable mirror.

In the equation, k_{i_t} is the integral parameter of the incremental PI controller and k_{p_t} is the proportional parameter of the incremental PI controller. The simulation parameter settings for k_{i_w} and k_{p_w} are identical.

$$C = \begin{bmatrix} I \\ k \times IM_{tw} \end{bmatrix}^+ \tag{8.23}$$

From Eqs. (8.15) and (8.22), the final control voltages of the low-order and high-order deformable mirrors can be obtained, and the corrected wavefront phase can be obtained. The corrected wavefront phase is denoted as $phase^k$, and the iterative phase after the kth correction is expressed as

$$phase^{k+1} = phase^{k+1} + (CMD_w^k \cdot IF_w) + (CMD_t^k \cdot IF_t), \tag{8.24}$$

where IF_w is the shape-influence function of the low-order deformable mirror and IF_t is the shape-influence function of the high-order deformable mirror.

In practical applications, the multiple push–pull averaging method is used to measure, calculate, and save IM_w, IM_t, IM_w^+, and IM_t^+ in advance. They can then be called directly during the calibration process to simplify the program and significantly reduce its running time.

The dual deformable-mirror adaptive-optics correction system is shown in Fig. 8.8. In the adaptive-optics system, the deflection mirror can be regarded as a special deformable mirror that only compensates for tilt. One system is a dual deformable-mirror adaptive-optics correction system consisting of a deflection mirror and a 69-element deformable mirror. The other consists of two 69-element deformable mirrors.

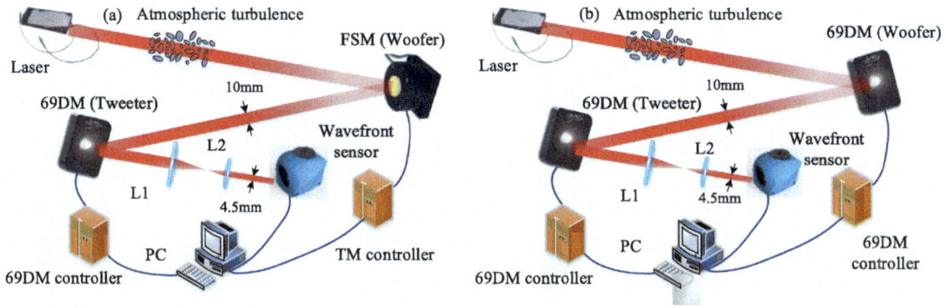

Fig. 8.8 Dual deformable-mirror adaptive-optics correction system. **a** Deflection mirror + 69-unit deformable mirror; **b** 69-unit + 69-unit deformable mirrors

8.4.2 Dual Deformable-Mirror Correction System Combining a Deflection Mirror and 69-Unit Deformable Mirror

Simulations were conducted using a dual deformable-mirror correction system composed of a polarizing mirror and 69-unit deformable mirror. Because of the inability of Hartmann sensing to measure the wavefront-aberration translation term, and because the translation term does not affect the degree of wavefront distortion, the Zernike coefficients of different atmospheric-turbulence-distorted wavefronts were simulated for 30 orders (except for the translation term, which corresponds to the tilt aberration in the x direction).

A multi-wavefront-corrector decoupling algorithm is used to correct distorted wavefronts under different turbulence intensities, with the polarizing mirror correcting tilt aberrations and the 69-unit deformable mirror correcting the remaining aberrations. D is the telescope aperture size and r_0 is the atmospheric coherence length. For weak turbulence, D/r_0 is approximately 2. For moderate turbulence, D/r_0 is approximately 10. For relatively strong turbulence, D/r_0 is approximately 20.

Figure 8.9 shows the wavefront phase diagram under weak turbulence conditions ($D/r_0 = 2$). The distorted wavefront under this turbulence intensity was corrected using a multi-wavefront-corrector decoupling algorithm. Figure 8.10 shows the PV and RMS values of the wavefront during the correction process.

Observing Figs. 8.9 and 8.10, it can be seen from the simulation diagram that the corrected-wavefront phase map is relatively flat after using the decoupling algorithm for correction. In the adaptive-optics system combining the deflection and deformable mirrors, the PV value of the wavefront decreases from 5.4224 to 0.0809 μm, and the RMS value of the wavefront decreases from 1.0887 to 0.0163 μm as the number of iterations increases. This indicates that, under weak turbulent interference, this algorithm can effectively correct the wavefront distortion.

When $D/r_0 = 10$, the multi-wavefront-corrector decoupling algorithm was used to correct the wavefront distortion under moderate turbulence intensity. The initial distorted-wavefront phase diagram and the corrected-wavefront phase diagram are

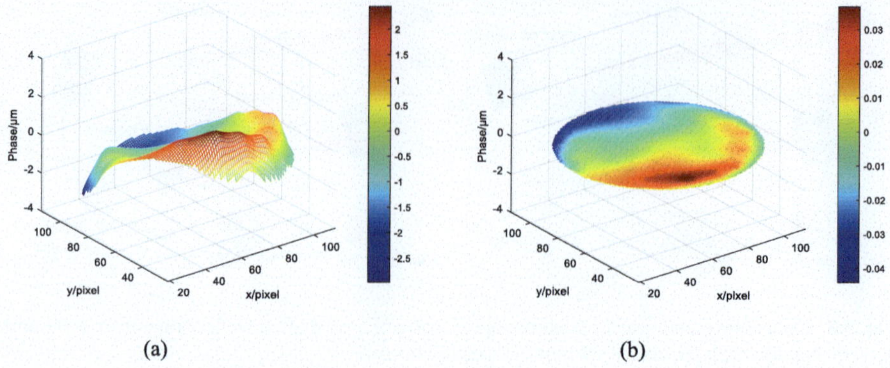

Fig. 8.9 Wavefront phase diagram under weak turbulence conditions. **a** Initial wavefront-distortion phase map; **b** decoupling algorithm–corrected wavefront-distortion phase map

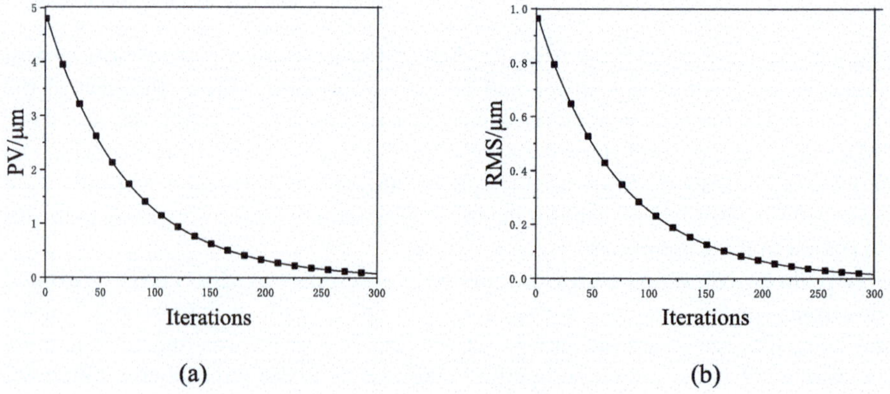

Fig. 8.10 Curves of PV and RMS values after decoupling-algorithm correction under weak turbulence conditions. **a** Wavefront PV value; **b** wavefront RMS value

shown in Fig. 8.11. The PV and RMS values of the wavefront during calibration are shown in Fig. 8.12.

The PV value of the wavefront in Fig. 8.11a is 23.3928 μm, and the RMS value of the wavefront is 5.4480 μm. Figure 8.11b shows a PV value of 0.1505 μm and an RMS value of 0.0335 μm. From the variation curve of Fig. 8.12, it can be seen that the PV and RMS values of the wavefront decrease with the increase of iteration times, indicating that the algorithm can still effectively correct the wavefront under medium-turbulence interference. The corrected PV and RMS values of the wavefront are relatively low, indicating a good wavefront-correction ability.

Figure 8.12 shows the wavefront phase diagram under strong turbulence ($D/r_0 = 20$) and the corrected wavefront using a multi-wavefront-corrector decoupling algorithm. Figure 8.13 shows the corrected residual wavefront PV and RMS values.

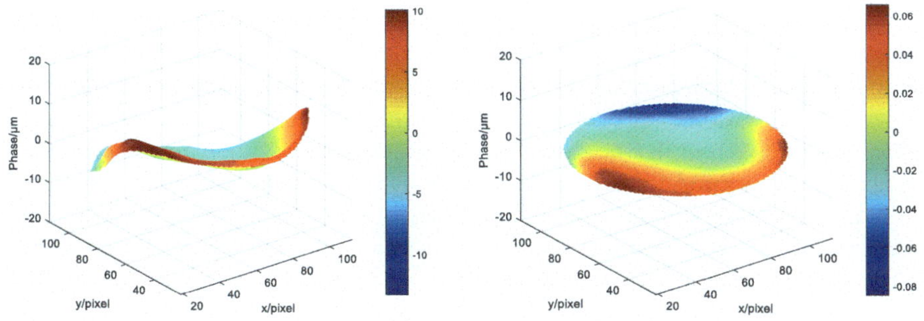

(a) (b)

Fig. 8.11 Wavefront phase diagram under moderate turbulence conditions. **a** Initial distorted-wavefront phase map; **b** decoupling algorithm–corrected wavefront phase map

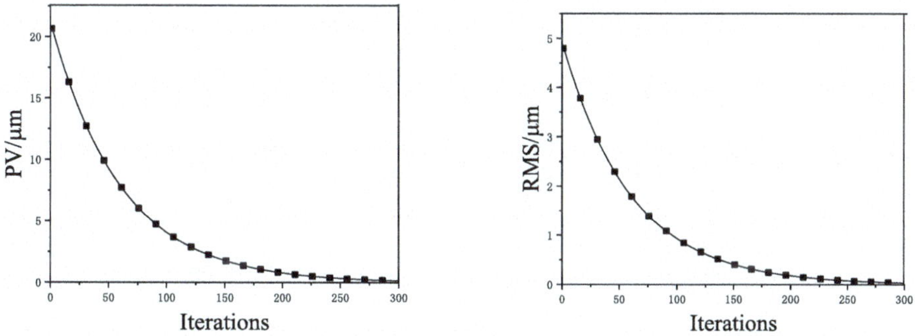

(a) (b)

Fig. 8.12 Curves of PV and RMS values after decoupling-algorithm correction under moderate turbulence. **a** Wavefront PV value; **b** wavefront RMS value

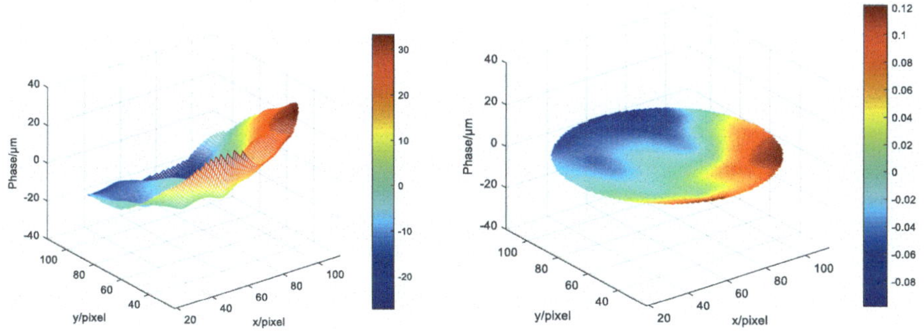

(a) (b)

Fig. 8.13 Wavefront phase diagram under strong turbulence conditions. **a** Initial distorted-wavefront phase map; **b** decoupling algorithm–corrected wavefront phase map

The simulation results in Fig. 8.13 show that, under strong turbulence conditions, the wavefront corrected using this decoupling algorithm is very flat. Observing the curve in Fig. 8.14, it can be seen that, as the number of iterations increases, the PV value of the wavefront decreases from 53.1531 to 0.22012 μm, and the RMS value of the wavefront decreases from 12.7908 to 0.05268 μm. This indicates that the algorithm has good correction ability under strong turbulence conditions.

In addition, the coupling-suppression ability of the algorithm was verified under different turbulence intensities. The coupling effect between the deflection mirror and deformable mirror is shown in Fig. 8.15.

As the number of iterations increases, the coupling coefficient of the system first increases and then stabilizes at a certain value. Under weak turbulence conditions, the coupling coefficient stabilizes at approximately 0.015. Under moderate turbulence

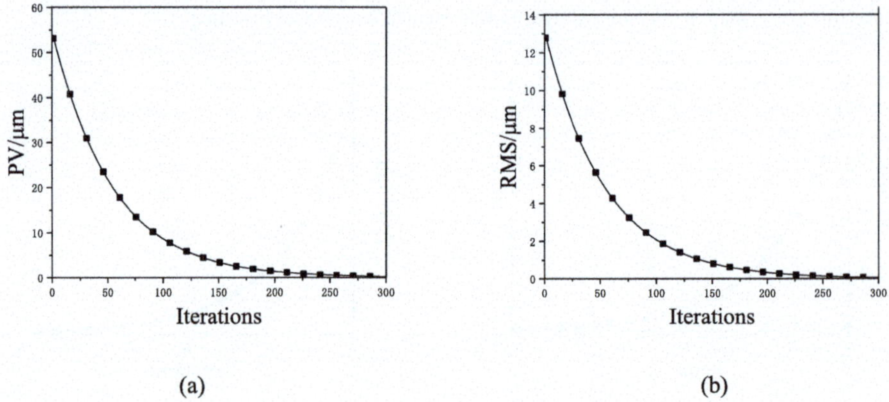

Fig. 8.14 Curves of PV and RMS values after decoupling-algorithm correction under strong turbulence. **a** Wavefront PV value; **b** wavefront RMS value

Fig. 8.15 Variation curve of the coupling coefficient during closed-loop correction of the deflection-mirror and deformable-mirror adaptive-optics system

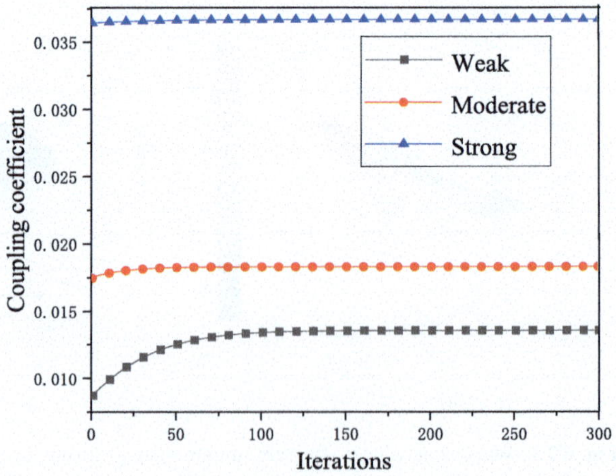

conditions, the coupling coefficient remains stable at approximately 0.020. Under strong turbulence conditions, the coupling coefficient remains stable at approximately 0.04. All values are less than 0.05, indicating that the algorithm has good coupling-suppression ability, and the coupling coefficient does not increase with the increase in iteration times.

8.4.3 Dual Deformable-Mirror Correction System Combining Two 69-Unit Deformable Mirrors

Tilt aberrations are usually corrected using a fast steering mirror or polarizing mirror. Therefore, a dual deformable-mirror correction system consisting of a combination of two 69-unit deformable mirrors was used to correct all aberrations, except for tilt aberrations. We set the values of the Zernike coefficients in the first and second order of Sect. 8.4.2 to 0, correct the third- to sixth-order aberrations using low-order deformable mirrors, and correct the remaining aberrations using high-order deformable mirrors.

Figure 8.16 shows the wavefront phase diagram with the tilt aberration removed under weak turbulence conditions ($D/r_0 = 2$) and the residual-wavefront phase diagram after wavefront-distortion correction by the dual deformable-mirror adaptive-optics system using the multi-wavefront corrector decoupling algorithm and the combination of two deformable mirrors. Figure 8.17 shows the PV and RMS values of the wavefront in the closed-loop correction process of the adaptive-optics system combining two 69-element deformable mirrors.

Observing the simulation diagram, it can be seen from Fig. 8.16b that the inclination of the wavefront at this time is very small and almost non-existent. Figure 8.17 shows the closed-loop correction of the adaptive-optics system using the decoupling

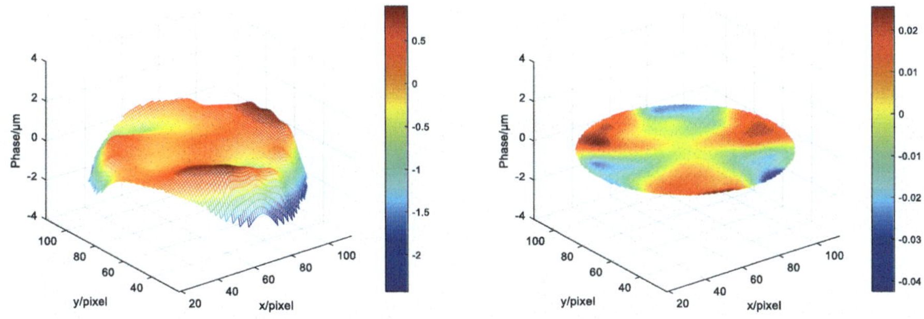

(a) (b)

Fig. 8.16 Wavefront phase diagram under weak turbulence conditions. **a** Distorted wavefront phase diagram after removing tilt aberrations; **b** distorted-wavefront phase diagram corrected by an adaptive-optical system combining two 69-unit deformable mirrors

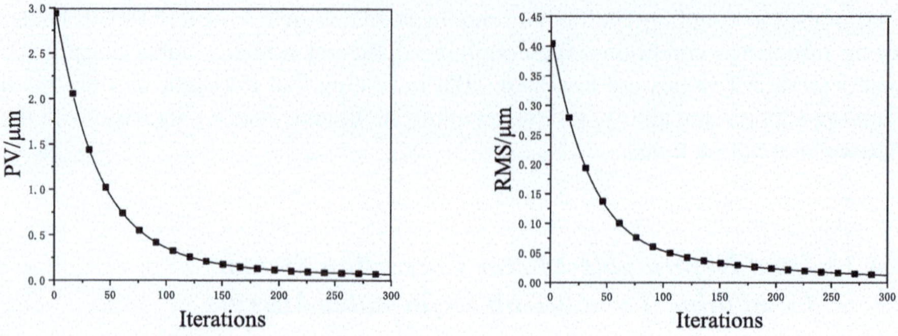

(a) (b)

Fig. 8.17 Changes in **a** PV and **b** RMS values for closed-loop correction of an adaptive-optical system with a combination of two 69-unit deformable mirrors under weak turbulence conditions

algorithm and the combination of two 69-element deformable mirrors. It can be seen that with the increase of the number of iterations, the PV values of the wavefront decrease from 2.94004 to 0.0681 μm, and the RMS values of the wavefront decrease from 0.4040 to 0.0124 μm. This shows that the decoupling algorithm is also suitable for the adaptive-optics system with two 69-element deformable mirrors under weak turbulence conditions.

Figure 8.18 shows the initial-wavefront phase diagram under moderate turbulence conditions ($D/r_0 = 10$) and the wavefront phase diagram corrected by the algorithm. The PV and RMS values of the wavefront corrected by the two 69-unit deformable mirrors closed loop are shown in Fig. 8.19.

The PV value of the wavefront in Fig. 8.18a is 23.3928 μm, and the RMS value of the wavefront is 5.4480 μm. Figure 8.18b shows a PV value of 0.1505 μm and an

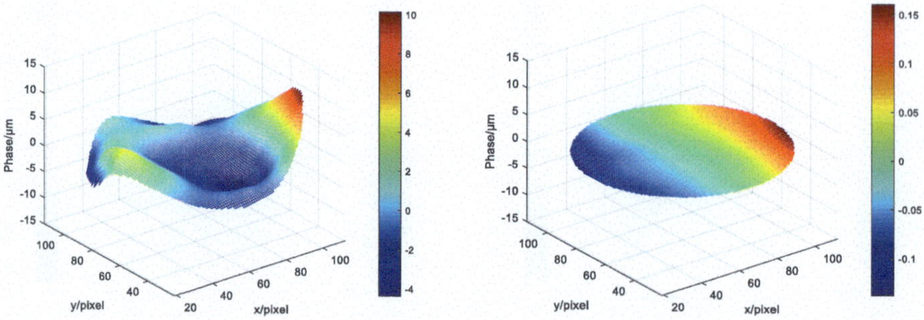

(a) (b)

Fig. 8.18 Wavefront phase diagram under moderate turbulence conditions. **a** After removing the tilt aberration; **b** wavefront-distortion phase diagram corrected by an adaptive-optical system combining two 69-unit deformable mirrors

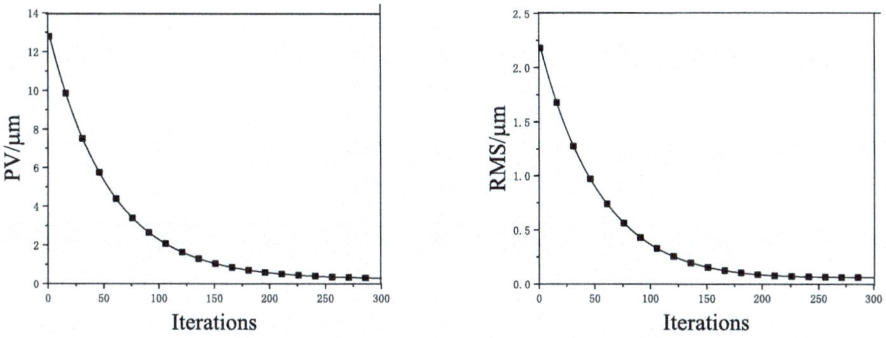

(a) (b)

Fig. 8.19 Changes in PV and RMS values for closed-loop correction of an adaptive-optical system with two 69-unit deformable mirrors under moderate turbulence conditions. **a** Wavefront PV value; **b** wavefront RMS value

RMS value of 0.0335 μm. From the variation curve of Fig. 8.19, it can be seen that the PV value and RMS value of the wavefront decrease with the increase of iteration times, indicating that the algorithm can still effectively correct the wavefront under medium turbulence interference. The corrected PV and RMS values of the wavefront are relatively low, indicating a certain wavefront-correction ability.

Figure 8.20 shows the wavefront phase diagram under strong turbulence conditions ($D/r_0 = 20$) and the correction of the wavefront distortion under strong turbulence intensity using a multi-wavefront-corrector decoupling algorithm. Figure 8.21 shows the corrected residual wavefront PV and RMS values.

Figure 8.20a shows the initial distorted-wavefront phase diagram, with a PV wavefront value of 26.3215 μm and an RMS wavefront value of 5.8760 μm. Figure 8.20b

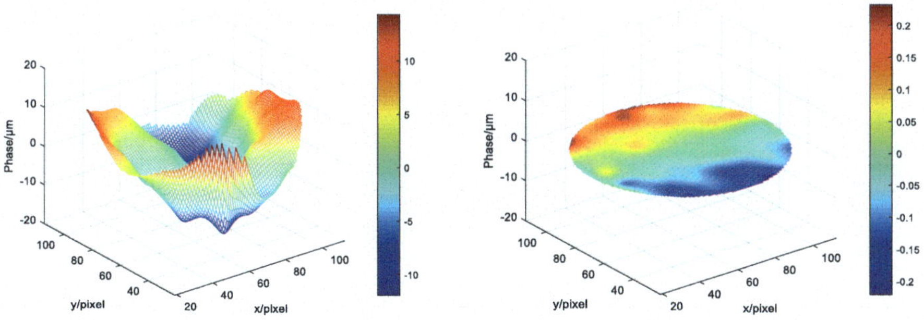

(a) (b)

Fig. 8.20 Wavefront phase diagram under strong turbulence conditions. **a** Distorted-wavefront phase diagram after removing tilt aberrations; **b** wavefront phase diagram after correction by an adaptive-optical system combining two 69-unit deformable mirrors

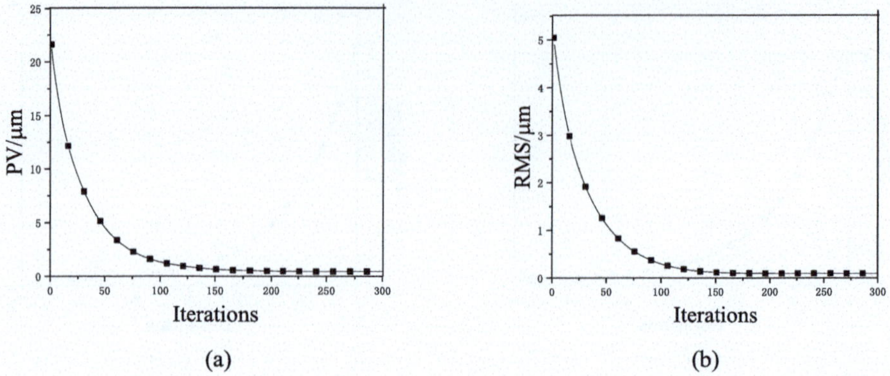

Fig. 8.21 Changes in PV and RMS values for closed-loop correction of an adaptive-optical system with a combination of two 69-unit deformable mirrors under strong turbulence conditions. **a** Wavefront PV value; **b** wavefront RMS value

shows the corrected wavefront phase diagram of the adaptive-optics system using a combination of two 69-element deformable mirrors. At this time, the PV value of the wavefront is 0.4534 μm, and the RMS value of the wavefront is 0.0929 μm.

From the variation curve of Fig. 8.21, it can be seen that both the PV and RMS values of the wavefront decrease with the increase of iteration times, indicating that the algorithm can still effectively correct distorted wavefronts under strong turbulence interference. The corrected residual wavefront is flat, demonstrating a certain wavefront-correction ability.

In addition, this study investigated the coupling effect between two sets of deformable mirrors and verified the coupling-suppression ability of the algorithm under different turbulence intensities. Figure 8.22 shows the variation curves of the coupling coefficients.

Fig. 8.22 Coupling-coefficient change curve of two 69-element deformable mirrors during a closed-loop correction of an adaptive-optics system

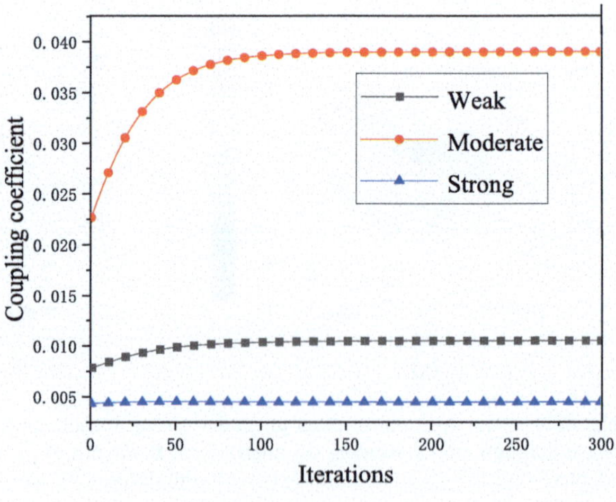

From the graph in Fig. 8.22, it can be seen that as the number of iterations increases, the coupling coefficient first increases and then stabilizes at a certain value under different turbulence conditions. When $D/r_0 = 2$, the coupling coefficient stabilizes at approximately 0.010. When $D/r_0 = 10$, the coupling coefficient remains stable at approximately 0.04. When $D/r_0 = 20$, the coupling coefficient remains stable at approximately 0.04. All of them are less than 0.05, indicating that the algorithm has good coupling-suppression ability under different turbulent environmental conditions. The coupling coefficient tends to stabilize, rather than increase with increasing iteration time.

8.5 Experimentation

Owing to the higher cost of 69-unit deformable mirrors and 292-unit deformable mirrors compared to fast steering mirrors, the actual system used a combination of fast steering mirrors and 69-unit deformable mirrors to verify the accuracy of the theoretical results. In this study, the experimental platform of a typical adaptive-optics dual deformable-mirror correction system (that is, a dual deformable-mirror correction system combining a fast steering mirror and 69-element deformable mirror) was built to correct the distorted wavefront. The experimental setup is shown in Fig. 8.23.

8.5.1 *Experimental Verification of Correcting Wavefront Distortion Using Multiple Wavefront Correctors*

The experimental results of correcting wavefront distortion using a multi-wavefront corrector are shown in Fig. 8.24. When only using the fast steering mirror closed

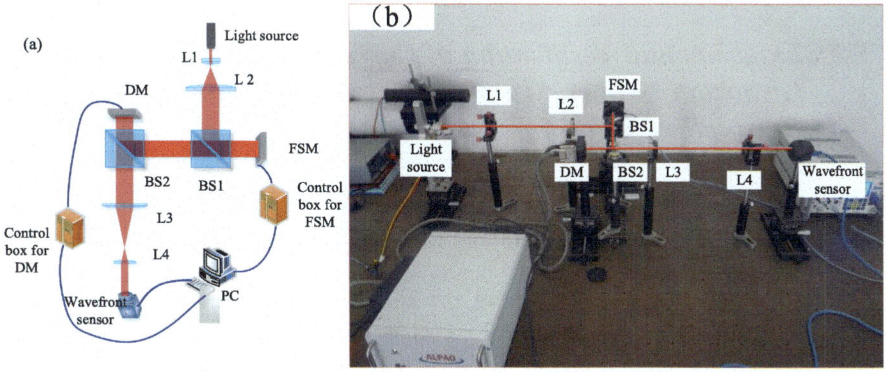

Fig. 8.23 Adaptive-optics wavefront correction system with a fast steering mirror + 69-unit deformable mirror. **a** Schematic optical-path diagram; **b** photograph of the system

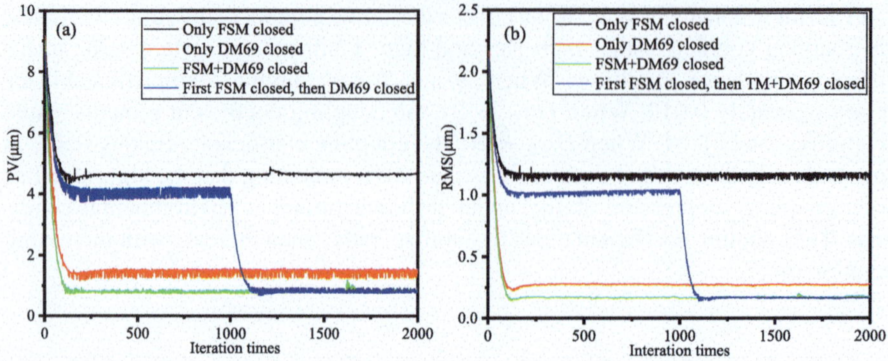

Fig. 8.24 Closed-loop curve using a fast steering mirror and 69-unit deformable mirror. **a** Wavefront PV value; **b** wavefront RMS value

loop, the PV value of the wavefront decreases from 9.21 μm to 4.63 μm, and the RMS value of the wavefront decreases from 2.41 to 1.27 μm. When using only the 69-unit deformable mirror closed loop, the PV value of the wavefront decreased from 9.21 to 1.38 μm, and the RMS value of the wavefront decreased from 2.41 to 0.26 μm.

When simultaneously using a fast steering mirror and 69 units in a closed loop, the PV value of the wavefront decreases from 9.21 to 0.84 μm, and the RMS value of the wavefront decreased from 2.41 to 0.26 μm. The simultaneous closed-loop correction of a fast steering mirror + 69-unit deformable mirror is better than that of a single fast steering mirror or a single 69-unit deformable mirror.

The effect of using a fast steering mirror + 69-unit deformable mirror for closed-loop correction is slightly inferior to that of using a deformable mirror alone to completely correct the wavefront distortion. This is because introducing a fast steering mirror increases the complexity of the optical system, and the adjustment errors introduced by each device gradually affects the final closed-loop correction.

8.5.2 Experimental Verification of Correcting Wavefront Distortion Using a Multi-wavefront-Corrector Decoupling Algorithm

A multi-wavefront-corrector decoupling algorithm was used for closed-loop correction, and the PV and RMS values of the wavefront during the correction process are shown in Fig. 8.25.

According to the ordinate values of the simulation shown in Fig. 8.25, as the number of iterations increases, the PV value of the wavefront decreases from 9.23 to 0.18 μm, and the RMS value of the wavefront decreases from 2.34 to 0.04 μm. According to the variation-curve trend, the convergence and stability of the corrected

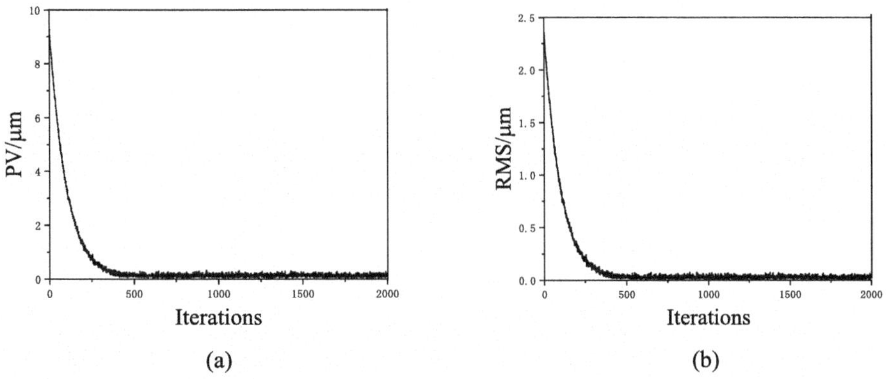

Fig. 8.25 Curves of PV and RMS values for closed-loop correction of wavefront distortion. **a** Wavefront PV value; **b** wavefront RMS value

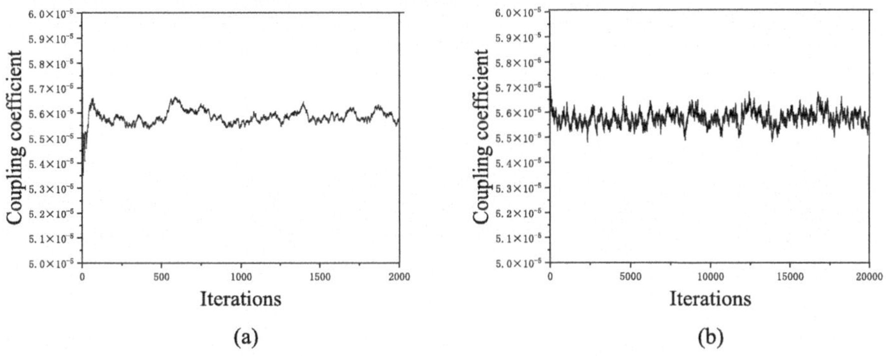

Fig. 8.26 Coupling coefficient during closed-loop correction process with different numbers of iterations. **a** 2000 times; **b** 20,000 times

residual wavefront PV and RMS curves are relatively good, indicating that the decoupling algorithm has a good correction ability.

Figure 8.26 shows the coupling coefficients of the system during the closed-loop correction.

Figure 8.26a shows the coupling coefficient when the number of iterations is 2000, and the system is subjected to long-term closed-loop correction. The coupling coefficients are shown in Fig. 8.26b after 20,000 times. It can be seen that the coupling coefficient tends to stabilize with an increase in the closed-loop time, and the increase in time does not increase the coupling effect of the system. This also no affects the stability of the double-deformable-mirror adaptive-optics correction system, indicating that the algorithm has a good coupling-suppression ability.

References

1. Liu W (2014) Research on decoupling control technology of multi wavefront correctors. Graduate School of Chinese Academy of Sciences (Institute of Optoelectronics Technology)
2. Liu W, Dong L, Yang P et al (2016) Zonal decoupling algorithm for dual deformable mirror adaptive optics system. Chin Opt Lett 14(2):1–5
3. Rao C, Zhu L, Zhang L et al (2018) Progress in solar adaptive optics technology. Optoelectronics 45(3):22–32
4. Wang Y, Cao Z, Li D et al (2018) Design of data acquisition and processing software for liquid crystal deformable mirror adaptive optics system. Opt Precis Eng 26(6):1507–1516
5. Han L, Wang Q, Xin TG et al (2010) Research on the performance of free-space optical communication system based on adaptive optics compensation. Appl Opt 31(2):301–304
6. Greenwood DP (1977) Bandwidth specification for adaptive optics systems. J Opt Soc Am 67(3):390–393
7. Li M, Cvijetic M (2015) Coherent free space optics communications over the maritime atmosphere with use of adaptive optics for beam wavefront correction. Appl Opt 54(6):1453
8. Doble N, Miller DT, Yoon G et al (2007) Requirements for discrete actuator and segmented wavefront correctors for aberration compensation in two large populations of human eyes. Appl Opt 46(20):4501–4514
9. Zou W, Qi X, Burns SA (2011) Woofer-tweeter adaptive optics scanning laser ophthalmoscopic imaging based on Lagrange-multiplier damped least-squares algorithm. Biomed Opt Express 2(7):1986–2004
10. Xu B, Wu J, Hou J et al (2006) Double-deformable-mirror adaptive optics system for phase compensation. Appl Opt 45(12):2638–2642
11. Li Z, Cao J, Zhao X et al (2014) Combinational-deformable-mirror adaptive optics system for atmospheric compensation in free space communication. Opt Commun 320:162–168
12. Cheng T, Liu W, Yang K et al (2018) Woofer tweeter adaptive optics system decoupling control algorithm based on Laplacian Eigenfunction. China Laser 45(09):295–304

Chapter 9
Effect of the Beam Mode on the Performance of Partially Coherent Optical-Coherence Detection

When a beam is transmitted through atmospheric turbulence, its characteristics change. When the signal light and local-oscillator light are coherently mixed in a wireless-optical coherent heterodyne-detection system, it causes a mismatch of the mode and polarization state of the two types of light, thus reducing the coherence efficiency and seriously affecting the performance of the entire communication system.

This chapter summarizes the research status of partially coherent optical-coherent heterodyne detection, and briefly introduces the basic theory of partially coherent Gaussian–Schell-model (GSM) beams. The variation in the heterodyne-detection efficiency of partially coherent beams with different factors, such as atmospheric turbulence, mismatch angle, detector scale, and optical-field parameters, is discussed. The influence of the beam mode on the performance of an optical heterodyne-detection system is analyzed both theoretically and experimentally, providing a theoretical and practical reference for the study of partially coherent optical-coherence detection.

9.1 Research Status

In the 1960s, People began to notice changes in the transmission of light in atmospheric turbulence, and the study of partially coherent optical heterodyne detection has not yet appeared.

In 1960, Chernov [1] used a geometric-optical approximation to determine the wave equation of a random continuum, but did not consider the finite size of the beam; thus, the application of this method was limited.

In 1961, Tatarskii [2] proposed a new method for studying beam propagation in turbulence. Based on Kolmogorov's modern turbulence theory, he used the Rytov approximation to represent the electric field in the form of exponents, and

X. Ke and C. Ke, *Performance Improvement Limits in Optical Wireless Coherent Communication*, Optical Wireless Communication Theory and Technology,
https://doi.org/10.1007/978-981-96-2000-5_9

expanded it into a series to solve it, thereby simplifying the phase- and amplitude-solving processes. This method successfully explained light transmission under weak turbulence and has become a classical theory for studying such.

After entering the 1970s, numerous experimental studies were conducted on various wave sources, including sound waves, lasers, and radio waves. Some new phenomena that appeared in the research process could not be explained by Tatarskii's classical theory. To solve this problem, theoretical methods applicable to strong turbulent environments were proposed, such as the Markov approximation [3] generalized Huygens–Fresnel principle [4], path-integration method [5], heuristic theory [6, 7], and phase-screen theory [8]. Research on the performance of heterodyne-detection systems for partially coherent beams was conducted gradually.

In 1970, Lahti et al. [9] calculated the average signal-to-noise ratio of a system based on a heterodyne-detection system of partially-coherent collimated beams and theoretically explained the aperture-saturation effect. They determined the spot size of the local-oscillator light corresponding to the optimal system signal-to-noise ratio and the coherence area, according to the parameters of the given signal light.

In 1979, Chiou [10] analyzed the influence of the temporal and spatial light-field coherence on the heterodyne-detection efficiency, based on the cross-spectrum function of the radiation light field. He provided the basic result that the heterodyne-detection efficiency varied with the spatial-coherence function.

In 1992, Tanaka et al. [11] theoretically analyzed the heterodyne efficiency of a heterodyne-detection system for partially coherent Gaussian signal and local-oscillator lights. It was found that the heterodyne efficiency decreased with the coherence of the signal light, and there was an optimal efficiency relationship between the beam parameters and the detector aperture.

In 2002, Ricklin et al. [12] derived the mathematical expressions of the average light intensity, phase curvature radius, and wavefront-coherence length of monochromatic Gaussian beams under atmospheric-turbulence conditions. They also discussed the effects of turbulence intensity and spatial coherence on wireless-optical communication links. In the same year, they found that partially coherent beams reduced the system bit-error rate caused by light-intensity flickering in the focal plane of the receiver [13].

In 2004, Salem and Dogariu [14] derived the expressions of the signal-to-noise ratio and heterodyne efficiency of a system when random electromagnetic beams were used for heterodyne detection, based on the unified theory of beam coherence–polarization, and analyzed the changes of the signal-to-noise ratio with the detector scale.

In 2008, Salem and Rolland [15] analyzed the influence of the beam-coherence and polarization characteristics on the heterodyne-detection performance during transmissions in free space based on random electromagnetic beams.

In 2010, Salem et al. [16] analyzed the effects of the spatial-angle mismatch, detector radius, and beam-overlap parameters on the mixing efficiency, based on a heterodyne-detection system of GSM beams. They concluded that, although partially coherent beams were more affected by the decrease in heterodyne efficiency than fully coherent beams, they were less affected by collimation errors than coherent beams.

In 2011, Pu et al. [17] used partially coherent beams with a uniform intensity distribution as the signal light and completely coherent Gaussian beams as the local-oscillator light to study the effects of the spatial-coherence characteristics and receiving-aperture size on the normalized carrier-to-noise ratio and field-of-view angle for remote LiDAR detection.

In 2012, Ren et al. [18] derived the closing expression of the mixing efficiency of a heterodyne-detection system in which both the signal light and local-oscillator light were partially coherent Gaussian–Schell beams. They discussed the effects of partially coherent light theory and atmospheric-coherence length on the performance of space optical-communication systems. The mixing efficiency could be maximized by adjusting the radius of the two overlapping beams using atmospheric-turbulence conditions and the detector radius.

In 2015, Li et al. [19] established a heterodyne-detection model in which both the signal light and local-oscillator light were partially coherent Gaussian–Schell beams, and discussed the effects of atmospheric turbulence, mismatch angle, detector scale, and optical-field parameters on the heterodyne efficiency of the system. The results showed that the turbulence and mismatch angle had a significant influence on the detection performance.

In 2016, Wang et al. [20] derived the expression for the mixing efficiency and signal-to-noise ratio when both the signal light and local-oscillator light were partially coherent GSM beams with missing supporting characters. They analyzed the influence of the polarization change on the performance of the differential detection system, and concluded that the mixing efficiency was closely related to the initial beam polarization state.

In 2020, Tan Zhenkun [21] based on the generalized Huygens–Fresnel principle and the incoherent mode-decomposition theory, Tan [21] derived the expression of the weight factor of a partially coherent GSM beam at the receiving end, and combined it with an actual heterodyne-detection system. It was experimentally verified on a 1.3-km external-field communication link.

In summary, research on partially coherent optical heterodyne-detection systems is still in the research stage. On the one hand, almost all beams emitted by lasers in practical applications are partially coherent beams, and atmospheric turbulence affects the intensity, quality, and directivity, as well as the spectral, polarization, and coherence characteristics, and equivalent curvature radius of the beams.

On the other hand, when a laser is transmitted in a strong turbulent environment and over a long distance, its coherence will be seriously damaged and it will degenerate into partially coherent light. Therefore, it is important to study the effect of partially coherent light transmitted through atmospheric turbulence on the performance of heterodyne-detection systems.

9.2 Partially Coherent Gaussian–Schell-Model Beams

Beam coherence includes spatial and temporal coherence. Generally, a partially coherent light source is generated by superimposing a time-varying random amplitude and phase distribution onto a fully coherent beam field. The added random amplitude and random phase distribution represent the temporal and spatial coherence of the optical field, respectively.

Assuming that the optical-field distribution of the initial fully coherent light source is $U_0(x, y, 0)$, the time-varying amplitude and phase distribution can be expressed as $\varphi(x, y; t) = \exp\left[i\xi(x, y; t)\right]$, and the optical-field distribution of the partially coherent beam at emission can be written as [22]

$$U_0(x, y; t) = U_0(x, y)\varphi(x, y; t), \tag{9.1}$$

where $\xi(x, y, t)$ is a random phase that varies with time and represents the partial coherence of the optical field of the light source.

The two-point cross-spectral density function of a partially coherent light source is defined as [23]

$$\begin{aligned} W_0(\boldsymbol{\rho}_1, \boldsymbol{\rho}_2) &= \left\langle U_0^*(\boldsymbol{\rho}_1)U_0(\boldsymbol{\rho}_2)\right\rangle \\ &= U_0^*(\boldsymbol{\rho}_1)U_0(\boldsymbol{\rho}_2)\left\langle \exp\left[i\xi_1(\boldsymbol{\rho}_1, t)\right]\exp\left[i\xi_2(\boldsymbol{\rho}_2, t)\right]\right\rangle, \end{aligned} \tag{9.2}$$

where $\boldsymbol{\rho}_1 = (x_1^2 + y_1^2)^{1/2}$, $\boldsymbol{\rho}_2 = (x_2^2 + y_2^2)^{1/2}$, and $\boldsymbol{\rho} = \left|\boldsymbol{\rho}_1 - \boldsymbol{\rho}_2\right|$ represents the two-dimensional vectors in the plane of the light source.

The cross-spectral density function of partially coherent Schell beams is expressed as

$$W_0(\boldsymbol{\rho}_1, \boldsymbol{\rho}_2) = \sqrt{I_0(\boldsymbol{\rho}_1)} \times \sqrt{I_0(\boldsymbol{\rho}_2)}\mu_0(\boldsymbol{\rho}_2 - \boldsymbol{\rho}_1), \tag{9.3}$$

where $I_0(\rho)$ is the average light intensity and $\mu_0(\rho_2 - \rho_1)$ represents the degree of coherence of the beam at $z = 0$. When the beam satisfies

$$I_0(\boldsymbol{\rho}) = A\exp\left(-\frac{2\rho^2}{w_0^2}\right) \quad \mu_0(\boldsymbol{\rho}) = \exp\left(-\frac{\rho^2}{2\sigma_g^2}\right) \tag{9.4}$$

it is called a partially coherent Gaussian–Schell-model (GSM) beam, where A is a constant, w_0 is the initial beam radius, and σ_g represents the degree of coherence of the beam. The larger the σ_g, the higher the degree of coherence. In contrast, the beam coherence is worse.

9.2.1 Space–Time Domain

A mutual-coherence function (MCF) is generally used to describe the coherence of partially coherent light fields in the spatial–time domain and is defined as [24]

$$\Gamma(\boldsymbol{\rho}_1, \boldsymbol{\rho}_2, \tau) = \langle U(\boldsymbol{\rho}_1, t+\tau) U^*(\boldsymbol{\rho}_2, t)\rangle, \tag{9.5}$$

where $U(\boldsymbol{\rho}_1, t+\tau)$ and $U^*(\boldsymbol{\rho}_2, t)$ are the complex amplitudes of the light field at point $(\boldsymbol{\rho}_1, t+\tau)$ and $(\boldsymbol{\rho}_1, t)$ in space and time, respectively. $\langle\cdot\rangle$ denotes the ensemble average. $\boldsymbol{\rho} = (x, y, z)$ is a spatial position vector, and z is a constant when the beam propagates along the z axis. Assuming that the radiation field has ergodic properties, and T is the measurement time, then

$$\langle U(\boldsymbol{\rho}_1, t+\tau) U^*(\boldsymbol{\rho}_2, t)\rangle = \lim \frac{1}{2T} \int_{-T}^{T} U(\boldsymbol{\rho}_1, t+\tau) U^*(\boldsymbol{\rho}_2, t) dt. \tag{9.6}$$

Let $\boldsymbol{\rho}_1 = \boldsymbol{\rho}_2 = \boldsymbol{\rho}$ and $\tau = 0$ in Eq. (9.6). Then, the average light intensity at point $\boldsymbol{\rho}$ can be expressed as

$$I(\boldsymbol{\rho}) = \langle U(\boldsymbol{\rho}, t) U^*(\boldsymbol{\rho}, t)\rangle = \Gamma(\boldsymbol{\rho}, \boldsymbol{\rho}, 0). \tag{9.7}$$

The MCF is normalized into a complex coherence degree and is expressed as

$$\gamma(\boldsymbol{\rho}_1, \boldsymbol{\rho}_2, \tau) = \frac{\Gamma(\boldsymbol{\rho}_1, \boldsymbol{\rho}_2, \tau)}{\sqrt{\Gamma(\boldsymbol{\rho}_1, \boldsymbol{\rho}_1, 0)}\sqrt{\Gamma(\boldsymbol{\rho}_2, \boldsymbol{\rho}_2, 0)}} = \frac{\Gamma(\boldsymbol{\rho}_1, \boldsymbol{\rho}_2, \tau)}{\sqrt{I(\boldsymbol{\rho}_1)}\sqrt{I(\boldsymbol{\rho}_2)}}. \tag{9.8}$$

According to the modulo of the complex coherence degree, $|\gamma(\boldsymbol{\rho}_1, \boldsymbol{\rho}_2, \tau)| (0 \le |\gamma(\boldsymbol{\rho}_1, \boldsymbol{\rho}_2, \tau)| \le 1)$ can determine the visibility of the interference fringes; thus, the beam coherence can be determined. When $|\gamma(\boldsymbol{\rho}_1, \boldsymbol{\rho}_2, \tau)|=1$, the beam is completely coherent; when $0 < |\gamma(\boldsymbol{\rho}_1, \boldsymbol{\rho}_2, \tau)| < 1$, it is partially coherent; and when $|\gamma(\boldsymbol{\rho}_1, \boldsymbol{\rho}_2, \tau)|=0$, it is completely incoherent.

In general, $\gamma(\boldsymbol{\rho}_1, \boldsymbol{\rho}_2, 0)$ and $\gamma(\boldsymbol{\rho}, \boldsymbol{\rho}, \tau)$ can be used to represent the spatial and temporal coherence of the beam, respectively; the latter is called the self-coherence function. It can be expressed as

$$\Gamma(\tau) = \Gamma(\boldsymbol{\rho}, \boldsymbol{\rho}, \tau) = \langle U^*(\boldsymbol{\rho}, t) U(\boldsymbol{\rho}, t+\tau)\rangle. \tag{9.9}$$

The average light intensity is $I(\boldsymbol{\rho}) = \Gamma(\boldsymbol{\rho}, \boldsymbol{\rho}, 0) = \Gamma(0)$; then, the autocoherence function is normalized: $\gamma(\tau) = \Gamma(\tau) / \Gamma(0)$, $\gamma(0) = 1$, $0 \le \gamma(\tau) \le 1$.

When the spectral width of the light source is less than the average frequency, the light field is quasi-monochromatic. The mutual-intensity function $J(\boldsymbol{\rho}_1, \boldsymbol{\rho}_2)$ can be used to describe the spatial coherence of the light field, instead of using the correlation

$\Gamma(\boldsymbol{\rho}_1, \boldsymbol{\rho}_2, 0)$ to describe the spatial coherence of the light field; that is

$$J(\boldsymbol{\rho}_1, \boldsymbol{\rho}_2) = \Gamma\left(\boldsymbol{\rho}_1, \boldsymbol{\rho}_2, 0\right) = \left\langle U^*\left(\boldsymbol{\rho}_1, t\right)U\left(\boldsymbol{\rho}_2, t\right)\right\rangle. \tag{9.10}$$

When $\boldsymbol{\rho}_1 = \boldsymbol{\rho}_2 = \boldsymbol{\rho}$, the average intensity can be expressed as

$$I(\boldsymbol{\rho}) = \Gamma(\boldsymbol{\rho}, \boldsymbol{\rho}, 0) = \left\langle U^*(\boldsymbol{\rho}_1, t)U(\boldsymbol{\rho}_2, t)\right\rangle. \tag{9.11}$$

After normalizing the mutual intensity, the complex coherence coefficient is expressed as

$$\gamma(\boldsymbol{\rho}_1, \boldsymbol{\rho}_2) = \frac{J(\boldsymbol{\rho}_1, \boldsymbol{\rho}_2)}{\sqrt{J(\boldsymbol{\rho}_1, \boldsymbol{\rho}_1)}\sqrt{J(\boldsymbol{\rho}_2, \boldsymbol{\rho}_2)}} = \frac{J(\boldsymbol{\rho}_1, \boldsymbol{\rho}_2)}{\sqrt{I(\boldsymbol{\rho}_1)}\sqrt{I(\boldsymbol{\rho}_2)}}, \quad 0 \leq \gamma(\boldsymbol{\rho}_1, \boldsymbol{\rho}_2) \leq 1. \tag{9.12}$$

9.2.2 *Space–Frequency Domain*

The cross-spectral density function (CSDF) is generally used to describe the coherence of light fields in the space–frequency domain and is defined as

$$W\left(\boldsymbol{\rho}_1, \boldsymbol{\rho}_2, \omega\right) = \left\langle \hat{U}^*(\boldsymbol{\rho}_1, \omega)\hat{U}\left(\boldsymbol{\rho}_2, \omega\right)\right\rangle, \tag{9.13}$$

where $\hat{U}(\boldsymbol{\rho}_{j=1,2}, \omega)$ is the Fourier transform of field variable $U(\boldsymbol{\rho}_{j=1,2}, t)$; that is,

$$\hat{U}(\boldsymbol{\rho}_j, \omega) = \int U(\boldsymbol{\rho}_j, t)\exp(2\pi i\omega t)dt. \tag{9.14}$$

Then, CSDF and MCF become Fourier transform pairs, and $W\left(\boldsymbol{\rho}_1, \boldsymbol{\rho}_2\right)$ is expressed as

$$\begin{gathered} W\left(\boldsymbol{\rho}_1, \boldsymbol{\rho}_2, \omega\right) = \int_{-\infty}^{\infty} \Gamma\left(\boldsymbol{\rho}_1, \boldsymbol{\rho}_2, \tau\right)\exp(2\pi i\omega\tau)d\tau \\ \Gamma\left(\boldsymbol{\rho}_1, \boldsymbol{\rho}_2, \tau\right) = \int_{0}^{\infty} W\left(\boldsymbol{\rho}_1, \boldsymbol{\rho}_2, \omega\right)\exp(-2\pi i\omega\tau)d\omega. \end{gathered} \tag{9.15}$$

When $\boldsymbol{\rho}_1 = \boldsymbol{\rho}_2 = \boldsymbol{\rho}$, the spectral-density function of the field point is $S(\omega) = W(\boldsymbol{\rho}, \boldsymbol{\rho}, \omega)$, which is the Fourier transform of the autocoherence function $\Gamma(\tau)$. Even when partially coherent light is transmitted in free space, the spectral density function changes; this is known as the Wolf effect [25].

The spectral-coherence degree, which describes the coherence of the light field in the spatial–frequency domain after normalizing the CSDF, can be expressed as

$$\mu(\boldsymbol{\rho}_1, \boldsymbol{\rho}_2, \omega) = \frac{W(\boldsymbol{\rho}_1, \boldsymbol{\rho}_2, \omega)}{\sqrt{W(\boldsymbol{\rho}_1, \boldsymbol{\rho}_1, \omega)}\sqrt{W(\boldsymbol{\rho}_2, \boldsymbol{\rho}_2, \omega)}} = \frac{W(\boldsymbol{\rho}_1, \boldsymbol{\rho}_2, \omega)}{\sqrt{S(\boldsymbol{\rho}_1, \omega)}\sqrt{S(\boldsymbol{\rho}_2, \omega)}}, \quad 0 \le \mu(\boldsymbol{\rho}_1, \boldsymbol{\rho}_2, \omega) \le 1 . \tag{9.16}$$

For a quasi-monochromatic field, $\hat{U}(\boldsymbol{\rho}, \omega) = U(\boldsymbol{\rho}) \exp(2\pi i\omega t)$, the corresponding CSDF is expressed as

$$W(\boldsymbol{\rho}_1, \boldsymbol{\rho}_2) = \langle \hat{U}^*(\boldsymbol{\rho}_1, \omega) U(\boldsymbol{\rho}_2, \omega) \rangle = \langle U^*(\boldsymbol{\rho}_1) U(\boldsymbol{\rho}_2) \rangle. \tag{9.17}$$

The cross-spectral density function $W(\boldsymbol{\rho}_1, \boldsymbol{\rho}_2)$ can be used to describe the coherence of the light field. In contrast to the mutual-intensity function $J(\boldsymbol{\rho}_1, \boldsymbol{\rho}_2)$ described in the previous section, the CSDF describes the coherence of the light field in the spatial–frequency domain, whereas the mutual-intensity function describes the coherence of the light field in the spatial–time domain.

9.3 Performance Analysis of a Partially Coherent–Beam Heterodyne-Detection System

9.3.1 Partially Coherent Optical Heterodyne-Detection System

Heterodyne detection is a type of coherent detection. In essence, the signal light and local-oscillator light mix on the detector surface (coherent superposition), and the detector responds to the light mixing and outputs the intermediate-frequency (IF) signal, as shown in Fig. 9.1.

Assuming that the polarization direction of the signal light, local-oscillator light, and wavefront are the same, the two beams coincide with each other. The transmission direction is parallel to the optical axis, the incidence is perpendicular to the detector surface, and the corresponding photoelectric fields can be expressed as [26]

$$E_s = A_s \cos(\omega_s t + \varphi_{s\_out}) \tag{9.18}$$

$$E_l = A_l \cos(\omega_l t + \varphi_l), \tag{9.19}$$

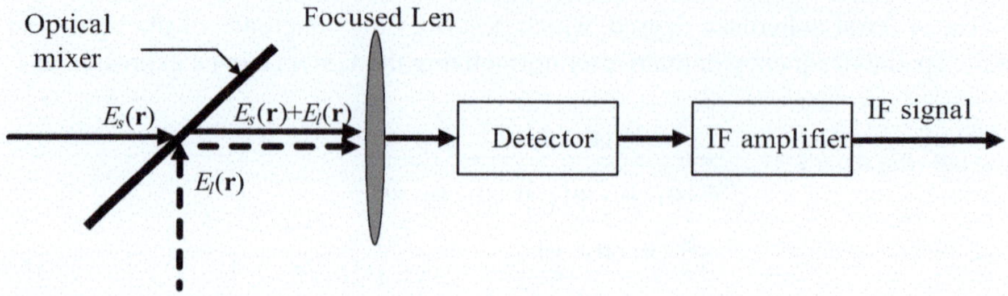

Fig. 9.1 Optical heterodyne-detection system

where A_s, ω_s, and φ_{s_out} and A_l, ω_l, and φ_l represent the amplitude, angular frequency, and phase of the signal light and local-oscillator light, respectively. If the signal is phase-modulated, then $\varphi_{s\_out} = \varphi_s + s(t)\pi$, where φ_s represents the initial phase of the signal light, and s(t) represents an electric-field modulated binary baseband signal, i.e., {0, 1}.

The photosensitive surface area of the optical mixer is S, and the total light intensity coherently synthesized by the two beams on the photosensitive surface is $|E_s + E_l|$. Because the detector current is proportional to the square of the total light intensity, the current on the detector's photosensitive surface can be expressed as [26]:

$$\begin{aligned} i_{\rm p} &= R\overline{[E_s(t) + E_l(t)]^2} \\ &= R\Big\{\overline{A_s^2 \cos^2(\omega_s t + \varphi_{s\_out})} + \overline{A_l^2 \cos^2(\omega_l t + \varphi_l)} \\ &\quad + \overline{A_s A_l \cos[(\omega_s + \omega_l)t + (\varphi_{s\_out} + \varphi_l)]} \\ &\quad + \overline{A_s A_l \cos[(\omega_s - \omega_l)t + (\varphi_{s\_out} - \varphi_l)]}\Big\}. \end{aligned} \tag{9.20}$$

In the formula, the horizontal line represents the time average over several optical-frequency periods, $R = e\eta/hv$ is the detector responsiveness, η is the quantum efficiency located on the entire photosensitive surface of the detector, e represents the electron charge, h is Planck's constant, and v is the carrier frequency.

The first and second terms in Eq. (9.20) are DC terms. The mean of the cosine squared is one-half. The third term is the sum-frequency term, and the detector cannot respond to this high frequency of light waves. The fourth term is the difference-frequency term, so that the angular-frequency difference $\Delta\omega = \omega_s - \omega_l$ and the initial phase difference $\Delta\varphi = \varphi_{s\_out} - \varphi_l$.

When $\Delta\omega$ is lower than the cutoff frequency of the detector, the photodetector has a photocurrent output with a frequency of $\Delta\omega/2\pi$; that is, the output photocurrent consists of two parts: the DC term and difference-frequency term:

$$i_{\rm p} = R\left[\frac{A_s^2}{2} + \frac{A_l^2}{2} + A_s A_l \cos(\Delta\omega t + \Delta\varphi)\right]. \tag{9.21}$$

If the frequency of the IF signal is limited to the difference passband range, the DC term in Eq. (9.21) is filtered out, and the IF current output through the bandpass filter with $\Delta\omega$ as the central frequency can be obtained as follows:

$$\begin{aligned} i_{IF} &= RA_s A_l \cos(\Delta\omega t + \Delta\varphi) \\ &= RA_s A_l \cos\left[(\omega_s - \omega_l)t + \left(\varphi_{s\_out} - \varphi_l\right)\right]. \end{aligned} \quad (9.22)$$

From Eq. (9.22), it can be seen that the amplitude $RA_s A_l$, frequency $\Delta\omega$, and phase $\Delta\varphi$ of the intermediate-frequency current all change with the signal light A_s, ω_s, and φ_l, which shows that the amplitude, frequency, and phase of the signal light, respectively, can be detected by coherent detection. In other words, coherent-detection technology can demodulate amplitude-modulated, frequency-modulated and phase-modulated optical signals. Equation (9.22) can also be expressed from the perspective of power as [27]

$$\begin{aligned} i_{IF} &= 2R\sqrt{P_s P_l}\cos(\Delta\omega t + \Delta\varphi) \\ &= 2R\sqrt{P_s P_l}\cos\left[(\omega_s - \omega_l)t + \left(\varphi_{s\_out} - \varphi_l\right)\right]. \end{aligned} \quad (9.23)$$

In Eq. (9.22), when the difference-frequency term $\Delta\omega \neq 0$, that is,$\omega_s \neq \omega_l$, the coherent-detection method at this time is called heterodyne detection. The expression of the intermediate-frequency current of the heterodyne-detection output is consistent with Eqs. (9.22) and (9.23). For heterodyne-detection technology, the polarization direction and wavefront curvature of the two beams must be strictly consistent before the signal light and the local-oscillator light are vertically incident light mixers. It is also necessary to ensure the monochromaticity of the two beams; in general, the narrower the spectral width, the better the monochromaticity.

9.3.2 Signal-to-Noise Ratio and Heterodyne Efficiency

The signal and local-oscillator beams are parallel to the optical axis and perpendicular to the detector surface. The instantaneous fields of both signals on the detector surface can be expressed as [19]

$$U_L(\boldsymbol{\rho}, t) = U_{LO}(\boldsymbol{\rho}, t)\exp(j\omega_{LO}t) \quad (9.24)$$

$$U_S(\boldsymbol{\rho}, t) = U_{SO}(\boldsymbol{\rho}, t)\exp(j\omega_{SO}t), \quad (9.25)$$

where the subscripts $_L$ and $_S$ denote the local-oscillator (LO) beam and the received signal beam, respectively. ω_L and ω_s are the central angular frequencies. Although the LO beam is typically coherent, we consider it to be partially coherent to obtain a general heterodyne-efficiency expression for heterodyne-detection systems that includes the coherence properties of the light source.

Based on the heterodyne-detection principle, the intermediate frequency (IF) signal-power output can be calculated according to the square-rate characteristic of the detector. We assume that the polarization direction and wavefront of the LO beam are consistent with the signal beam. $U(\boldsymbol{\rho}, t)$ is the total field at any point $\boldsymbol{\rho}$ on the detector surface, which can be expressed as

$$U(\boldsymbol{\rho}, t) = U_{LO}(\boldsymbol{\rho}, t)\exp(j\omega_{LO}t) + U_{SO}(\boldsymbol{\rho}, t)\exp(j\omega_{SO}t). \tag{9.26}$$

The coherent process of the signal light and the vibration light produces a mixing current, and the incremental value of this current at any point $\boldsymbol{\rho}$ on the detector surface can be given by [28]

$$di(\boldsymbol{\rho}, t) = \Re(\boldsymbol{\rho})\big[U(\boldsymbol{\rho}, t) \times U^{*}(\boldsymbol{\rho}, t)\big]d^{2}\rho, \tag{9.27}$$

where $\Re(\boldsymbol{\rho}) = e\eta(\boldsymbol{\rho})/hv$ is the detector responsivity at point $\boldsymbol{\rho}$, $\eta(\boldsymbol{\rho})$ is the quantum efficiency, e is the electronic charge, hv is the photon energy, and * denotes the complex conjugate. By substituting Eq. (9.26) into Eq. (9.27) and filtering the direct current (DC) of the photomixing current, the total IF current output from the detector can be expressed as

$$\begin{aligned} i_{IF}(t) = \iint \Re(\boldsymbol{\rho})\big\{&U_{LO}(\boldsymbol{\rho})U_{SO}^{*}(\boldsymbol{\rho})\exp\big[j(\omega_{LO} - \omega_{SO})t\big] + \\ &U_{SO}(\boldsymbol{\rho})U_{LO}^{*}(\boldsymbol{\rho})\exp\big[-j(\omega_{LO} - \omega_{SO})t\big]\big\}d^{2}\rho. \end{aligned} \tag{9.28}$$

Similarly, the power of the IF signal can be expressed as

$$\mathrm{p}_{IF} = \overline{i_{IF}(t)i_{IF}^{*}(t)}R_{L}, \tag{9.29}$$

where the bar denotes the time average over an interval of the IF signal and R_L is the load resistance of the detector. Equation (9.29) only provides a time average for the fluctuating fields. However, the signal beam is affected by atmospheric turbulence and other random factors, and the mixed total field in time and space fluctuates randomly because of the turbulence. Therefore, it is necessary to determine the ensemble average for all turbulence conditions, i.e.,

$$P_{IF} = \langle \mathrm{p}_{IF} \rangle. \tag{9.30}$$

By substituting Eqs. (9.28) and (9.29) into Eq. (9.30), the power of the IF signal can be expressed as

$$\begin{aligned} P_{IF} = \int\int\int\int \Re(\boldsymbol{\rho}_1)\Re(\boldsymbol{\rho}_2) \times \big[&\langle U_{LO}^{*}(\boldsymbol{\rho}_1)U_{SO}(\boldsymbol{\rho}_1)U_{SO}^{*}(\boldsymbol{\rho}_2)U_{LO}(\boldsymbol{\rho}_2)\rangle \\ &+ \langle U_{LO}(\boldsymbol{\rho}_1)U_{LO}^{*}(\boldsymbol{\rho}_2)U_{SO}(\boldsymbol{\rho}_2)U_{SO}^{*}(\boldsymbol{\rho}_1)\rangle\big]d^{2}\rho_1 d^{2}\rho_2. \end{aligned} \tag{9.31}$$

In a free-space coherent optical communication system, the signal beams propagate through the atmosphere over a long distance, and the LO beams at the receiving end are unaffected by atmospheric turbulence. As mentioned previously, the two fields are independent of each other. Equation (9.31) can be simplified to

$$P_{IF} = \int\int\int\int \Re(\boldsymbol{\rho}_1)\Re(\boldsymbol{\rho}_2) \times \mathrm{Re}\left[\Gamma_{LO}(\boldsymbol{\rho}_1, \boldsymbol{\rho}_2) \times \Gamma_S^*(\boldsymbol{\rho}_1, \boldsymbol{\rho}_2, L)\right] d^2\rho_1 d^2\rho_2, \tag{9.32}$$

where $\Gamma_{LO}(\boldsymbol{\rho}_1, \boldsymbol{\rho}_2)$ denotes the mutual-coherence functions of the LO beam and $\Gamma_{LO}(\boldsymbol{\rho}_1, \boldsymbol{\rho}_2, L)$ denotes the mutual-coherence functions of the signal beam at transmission distance L. The receiver noise is largely determined by the shot noise of the LO beam, which can be expressed with mutual-coherence functions as [19]

$$N = 2eB \iint \Re(\mathbf{r})\Gamma_{LO}(\mathbf{r}, \mathbf{r}) dr^2. \tag{9.33}$$

Based on the effective optical power expressed in Eq. (9.32) and the shot noise in Eq. (9.10), the signal-to-noise ratio (SNR) of the heterodyne-detection system of the partially coherent GSM beam can be defined as

$$SNR_{GSM} = \frac{\int\int\int\int \Re(\boldsymbol{\rho}_1)\Re(\boldsymbol{\rho}_2) \times \mathrm{Re}\left[\Gamma_{LO}(\boldsymbol{\rho}_1, \boldsymbol{\rho}_2) \cdot \Gamma_S^*(\boldsymbol{\rho}_1, \boldsymbol{\rho}_2, L)\right] d^2\rho_1 d^2\rho_2}{2\mathrm{eB} \iint \Re(\boldsymbol{\rho})\Gamma_{LO}(\boldsymbol{\rho}, \boldsymbol{\rho}) dr^2}. \tag{9.34}$$

Assuming that the detector responsivity $\Re$ does not vary with the position on the detector, it can be defined as $\Re = e\eta/hv$ at any position. η is the quantum efficiency of the photodetector, e is the electron charge, h is Planck's constant, and v is the carrier-wave frequency. Considering that these parameters do not affect the $\mathrm{SNR_{GSM}}$ variation, Eq. (9.34) can be further simplified as

$$SNR_{GSM'} = \frac{\int\int\int\int \mathrm{Re}\left[\Gamma_{LO}(\boldsymbol{\rho}_1, \boldsymbol{\rho}_2) \cdot \Gamma_S^*(\boldsymbol{\rho}_1, \boldsymbol{\rho}_2, L)\right] d^2r_1 d^2r_2}{\iint \Gamma_{LO}(\boldsymbol{\rho}, \boldsymbol{\rho}) dr^2}. \tag{9.35}$$

Comparing Eq. (9.34) with Eq. (9.35), it can be observed that $SNR_{GSM'} = 2eBSNR_{GSM}/\Re$.

Another common measure of heterodyne performance is the dimensionless heterodyne efficiency. The heterodyne efficiency of a partially coherent GSM beam can be defined as

$$\eta_{het\_GSM} = \frac{\int\int\int\int \Re(\boldsymbol{\rho}_1)\Re(\boldsymbol{\rho}_2) \times \mathrm{Re}\left[\Gamma_{LO}(\boldsymbol{\rho}_1, \boldsymbol{\rho}_2) \cdot \Gamma_S^*(\boldsymbol{\rho}_1, \boldsymbol{\rho}_2, L)\right] d^2\rho_1 d^2\rho_2}{\iint \Re(\boldsymbol{\rho})\Gamma_{LO}(\boldsymbol{\rho}, \boldsymbol{\rho}) dr^2 \iint \Re(\boldsymbol{\rho})\Gamma_{SO}(\boldsymbol{\rho}, \boldsymbol{\rho}) dr^2}. \tag{9.36}$$

Similarly, assuming that the responsivity $\Re$ is uniform across the detector surface, Eq. (9.36) can be further simplified as

$$\eta_{het\_GSM}{}' = \frac{\int\int\int\int \mathrm{Re}\left[\Gamma_{LO}(\rho_1, \rho_2) \cdot \Gamma_S^*(\rho_1, \rho_2, L)\right] d^2\rho_1 d^2\rho_2}{\iint \Gamma_{LO}(\rho, \rho) dr^2 \iint \Gamma_{SO}(\rho, \rho) dr^2}. \quad (9.37)$$

The signal and LO beams are partially coherent GSM beams at the source field, the cross-correlation function of which can be expressed as [16]

$$\Gamma_\alpha(\mathbf{r}_1, \mathbf{r}_2, 0) = I_\alpha \exp\left[-\frac{\left(\mathbf{r}_1^2 + \mathbf{r}_2^2\right)}{4\sigma_\alpha^2} - \frac{(\mathbf{r}_1 - \mathbf{r}_2)^2}{2\delta_\alpha^2}\right], \quad (9.38)$$

where $\mathbf{r}_1$ and $\mathbf{r}_2$ denote any two-point position vector at the source field $L = 0$. I_α is the intensity of the source. $I_\alpha = 1$ indicates the ideal intensity at the transmitter. σ_a is the waist radius; δ_a is the coherence length of the signal beam and indicates the partially coherent characteristics of the source. The subscript $a = L$ indicates the LO beam. S denotes the signal beam.

9.3.3 *Analysis of Factors Affecting the Optical Heterodyne-Detection Performance*

9.3.3.1 Mismatch Angle

Assuming that the focal plane of the receiving antenna coincides with the detector surface and the local-oscillator light is perpendicular to the detector surface, the angle θ between the optical axis of the signal light and the local-oscillator light is called the mismatch angle. It is assumed that the direction of the signal-light wave vector is at angle θ with optical axis z, where the local-oscillator light $\mathbf{K}_\mathrm{L}$ is vertically incident, the signal light $\mathbf{K}_\mathrm{S}$ is incident on a slanted path, and both beams are vectors. Suppose that the phase of the signal light at any point on the photosensitive surface of the detector can be expressed as $kr \cos\varphi \sin\theta$, according to the following integral formula:

$$\int_0^{2\pi} exp(ikrcos\varphi sin\theta) d\varphi = 2\pi J_0(krsin\theta). \quad (9.39)$$

Because the mismatch angle is small, $\sin\theta$ can be approximated as θ, and the heterodyne-efficiency expression with the mismatch angle can be obtained through simplification; that is [21],

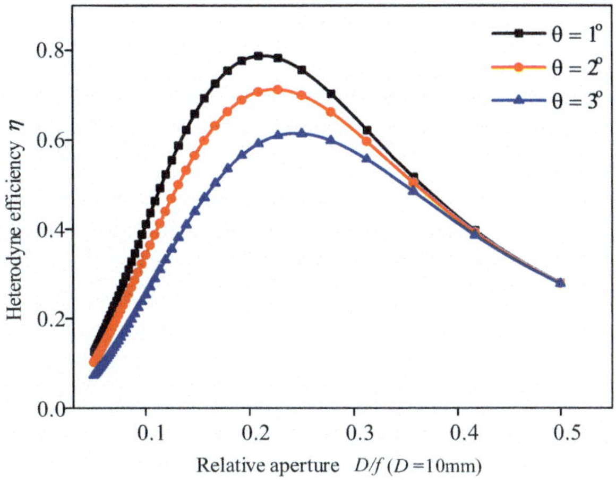

Fig. 9.2 Variation of the heterodyne efficiency with relative aperture [21]

$$\eta_{het\theta} = \frac{4}{\omega_0^2}\frac{\left|\int_0^\infty \exp(-\mathrm{i}kf)\exp\left(-\frac{\mathrm{i}kr^2}{2f}\right)\exp\left(-\frac{r^2}{\omega_0^2}\right)\mathrm{J}_1\left(\frac{\pi Dr}{\lambda f}\right)\mathrm{J}_0(kr\theta)\mathrm{d}r\right|^2}{\int_0^\infty\left[\mathrm{J}_1\left(\frac{\pi Dr}{\lambda f}\right)\right]^2\Big/ r\mathrm{d}r}. \tag{9.40}$$

When other conditions are fixed, the relationship between the heterodyne efficiency of different mismatch angles and the relative aperture is shown in Fig. 9.2 [21]. As shown in the figure, the heterodyne efficiency decreases with an increase in the mismatch angle, and there is an optimal aperture for maximizing the heterodyne efficiency.

9.3.3.2 Atmospheric Turbulence and the Detector Aperture

Atmospheric turbulence and the detector aperture have a great influence on the detection efficiency of a heterodyne system. The specific performance is as follows [29]. When the mismatch angle of the light field is 0, the higher the atmospheric-turbulence intensity, the lower the heterodyne efficiency. The heterodyne efficiency decreases faster when a mismatch angle is present. When the heterodyne efficiency drops to a certain extent, the detection efficiency of a heterodyne system in a strong turbulent region changes more slowly than that in a weak turbulent region, which may be due to the beam spreading caused by turbulence.

With the increase of the detector aperture, the detection efficiency will gradually decrease; this phenomenon is more obvious when there is a mismatch angle. When the detector scale is small, the allowable mismatch-angle range is larger. Therefore, it is necessary to select a suitable detector aperture to improve the influence of factors, such as the mismatch angle and turbulence, on the external differential efficiency in practice.

9.3.3.3 Optical-Field Parameters

When the beam radii of the signal light and local-oscillator light are different, the variation of the heterodyne-detection efficiency is different. Studies have shown that [19] when the beam radius of the signal light remains unchanged, the beam radius of the local-oscillator light can be adjusted to ensure a better match between the two mixing light fields, to maximize the detection efficiency.

When the detector scale is fixed and the beam radii of the two light fields are greater than or equal to the detector scale, the system can still work with high detection efficiency when the mismatch angle is small. In addition, the transverse-length variation of the signal light and local-oscillator light also affects the detection efficiency. When the transverse-coherence length of the signal light is small, the detection efficiency changes little with the increase of the transverse-coherence length of the local-oscillator light.

When the transverse-coherence length of the signal light is larger than the detector scale, the detection efficiency changes obviously with a change in the transverse-coherence length of the local-oscillator light. When the transverse-coherence length of the local-oscillator light increases to a certain value, the transverse-coherence length of the signal light has little effect on the detection efficiency.

9.4 Influence of the Beam Mode on the Performance of a Heterodyne-Detection System

9.4.1 Mode Decomposition

The one-dimensional intensity of partially coherent GSM beams can be expressed as the superposition of the one-dimensional intensity of Hermite–Gaussian (H-G) beams with independent modes [28]. According to the theory of incoherent mode decomposition, the M^2 factor of a partially coherent beam at the receiver is expressed as [21]

$$M^2 = \sum_{n=0}^{\infty} (2n + 1)\lambda_n, \tag{9.41}$$

where λ_n is the weighting factor and can be expressed as

$$\lambda_n = c_n / \left(\sum_{n=0}^{\infty} c_n \right), \tag{9.42}$$

where c_n is the mode coefficient and n is the order.

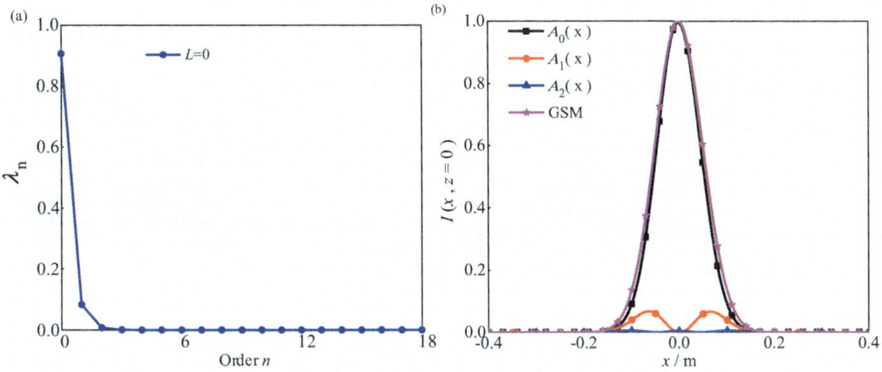

Fig. 9.3 Mode decomposition of the GSM beam at the source field [21]. **a** Variation of the weighting factor with the order. **b** Superposition of H-G beams of different orders

The mode decomposition of a partially coherent GSM beam at the source field is shown in Fig. 9.3 [21]. As can be seen from Fig. 9.3(a), the modes at the source field are mainly fundamental and first-order modes, accounting for 91% and 8.3%, respectively. Figure 9.3b shows the one-dimensional intensity distribution of a partially coherent GSM beam and the corresponding one-dimensional intensity distribution of each order H-G beam after decomposition.

$A_0(x)$ represents the zero-order H-G beam, that is, the fundamental-mode Gaussian beam;$A_1(x)$ represents the first-order H-G beam; $A_2(x)$ represents the second-order H-G beam; and the star line describes the intensity of the GSM beam after the superposition of the three modes. The figure shows that the light-intensity curve of $A_0(x)$ almost coincides with the star line, indicating that the basal film accounts for the largest proportion of the partially coherent GSM beam at the source field.

9.4.2 *Influence of the Beam Mode on the Heterodyne-Detection System Performance*

The GSM signal beam is transmitted to the receiving end along the uplink in atmospheric turbulence and coherently mixed with the GSM local-oscillator light. Figure 9.4 shows the influence of the light-source parameters, including the beam coherence length and the fundamental-mode content when the beam waist radius is different, on the external differential efficiency [21].

As shown in Fig. 9.4a, b, the higher the content of the base film, the greater the heterodyne efficiency. With an increase in the partial coherence length, the content of the fundamental mode increases, and the corresponding heterodyne efficiency increases. When the waist radius of the signal beam gradually increases, the fundamental-mode content first increases and then decreases.

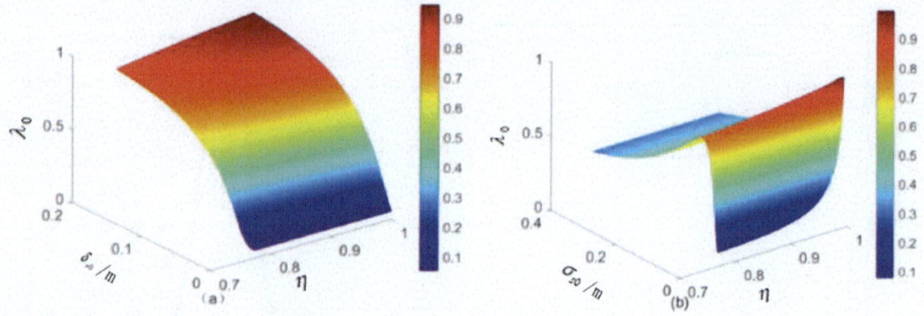

Fig. 9.4 Effect of the fundamental-mode content on the external differential efficiency [21]. **a** Coherence length, **b** waist radius

This is because when the waist radius of the signal beam increases to match that of the local-oscillator beam, the heterodyne efficiency is the largest, and the corresponding fundamental-mode content is also the largest. This indicates the existence of a signal-light beam waist radius that maximizes the heterodyne efficiency and optimizes the detection-system performance.

In addition to being affected by the light-source parameters, the normalized M^2 factor and heterodyne efficiency are also affected by the inner turbulence scale, outer turbulence scale, zenith angle, and turbulence intensity, as shown in Fig. 9.5 [21]. With a decrease in the inner scale and an increase in the outer scale, the zenith angle increases, the turbulence intensity increases, the normalized M^2 factor increases, and the beam quality deteriorates. This conclusion is consistent with that in reference [30].

In addition, the influence of the outer scale on the M^2 factor and the heterodyne efficiency is much smaller than that of the inner scale. In other words, with an increase in atmospheric-turbulence intensity, the beam diffraction becomes more serious, the corresponding mode order increases, the normalized M^2 factor increases, and the beam quality decreases, which eventually leads to a decrease in heterodyne efficiency.

9.4.3 Experimental Verification of a 1.3-km Heterodyne-Detection System

An external-field 1.3-km experimental communication link was established between the 11th floor of the Kasen Fujing Yayuan Community building, East Second Ring Road, Xi'an City, Shaanxi Province, and the sixth floor of Xi'an University of Technology. An experimental diagram of the heterodyne-detection communication system is shown in Fig. 9.6 [21] and the instrument used in the figure is described in Table 9.1.

At the transmitting terminal, the phase modulator modulates the high-definition (HD)-video digital information to the 1550-nm signal optical carrier, amplifies the signal optical power with the modulation information using an erbium-doped fiber amplifier (EDFA), and sends it out by the optical antenna.

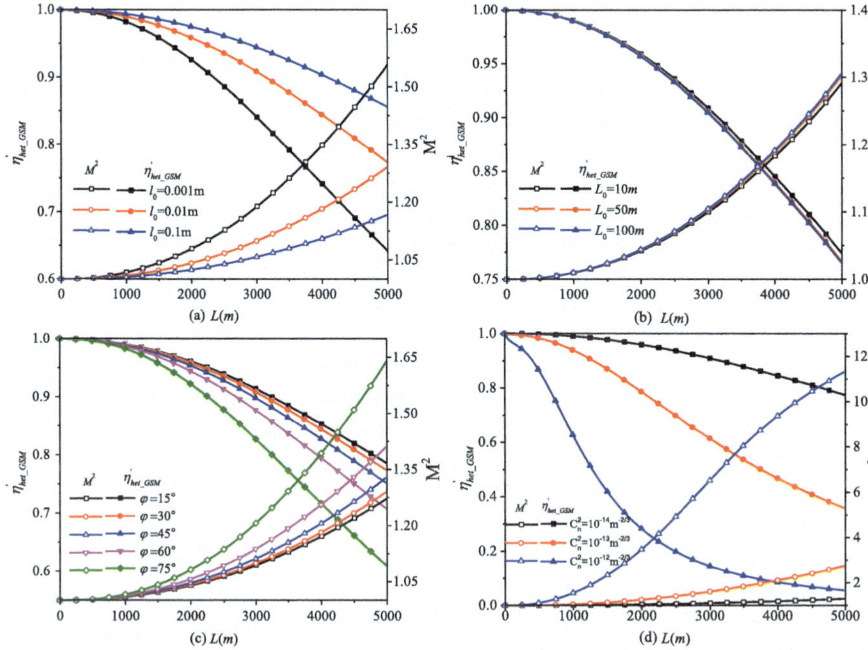

Fig. 9.5 Effect of atmospheric turbulence on the normalized M^2 factor and heterodyne efficiency [21]. **a** Different inner scales l_0, **b** different outer scales L_0, **c** different zenith angles φ, and **d** different atmospheric refractive-index structure constants C_n^2

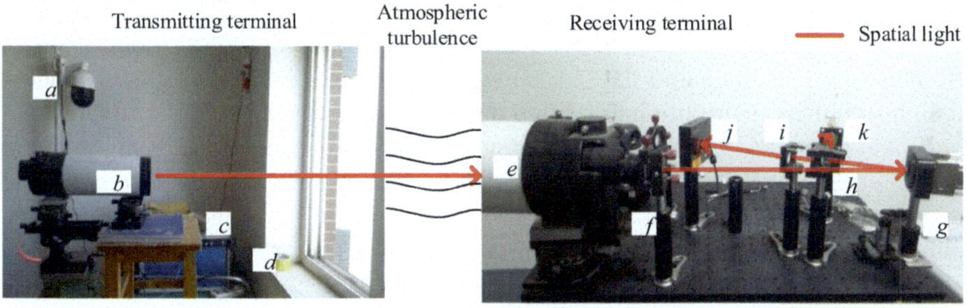

Fig. 9.6 Heterodyne-detection communication system with a 1.3-km transmission distance [21]

At the receiving terminal, a deformable mirror (DM) is used to correct the distorted wavefront, and the corrected beam passes through a dichroic-direction splitting mirror, which separates the beacon light and the signal light. The signal light is coupled to the single-mode fiber by a coupling lens for heterodyne detection.

Table 9.1 Experimental equipment for heterodyne detection

	Transmitting end
a	HD camera
b	Optical transmitting system
c	Erbium-doped fiber amplifier
d	Transmitting box
	Receiving end
e	Receiving optical system
f	Focusing lens with a 100-mm focal length
g	Deformable mirror
h	Dichroic splitter
i	Focusing lens with a 175-mm focal length
j	Spiricon beam analyzer
k	Focusing lens with a 125-mm focal length

In this experiment, a beam analyzer was placed at point *j* with a 632.8-nm measuring wavelength. Relevant experimental data, such as the light spot, Gaussian coincidence, and Gaussian roughness, were collected under different weather conditions.

Owing to the limitations of the measuring instrument, the proportion of the fundamental mode after the beam transmission can only be reflected indirectly by the Gaussian coincidence [21] and Gaussian roughness [21] in the experiment. A higher fundamental-mode content corresponds to a higher Gaussian fit, and a smaller fitting roughness indicates a smaller fitting roughness of the beam.

The measured spot-distribution results under different weather conditions at different times of the same day are shown in Fig. 9.7 [21]. Figure 9.7a, b show the light-intensity distribution measured during moderate rain and light rain, respectively, and (c) and (d) show the light-intensity distribution measured at night and at night after the rain stops, respectively. It can be seen that the spot distribution is the most dispersed and the light-intensity distribution is the most uneven in Fig. 9.7a. With a decrease in rainfall, the spot energy gradually increases, and the spot quality is best at night.

Table 9.2 [21] shows the two-dimensional fitting coincidence and two-dimensional fitting roughness of the light intensity at different times, corresponding to Fig. 9.7. It can also be seen from the table that with a decrease in rainfall, the fitting coincidence gradually increases, whereas the opposite fitting roughness gradually decreases. This indicates that, with a decrease in rainfall, the proportion of the fundamental mode in the beam gradually increases.

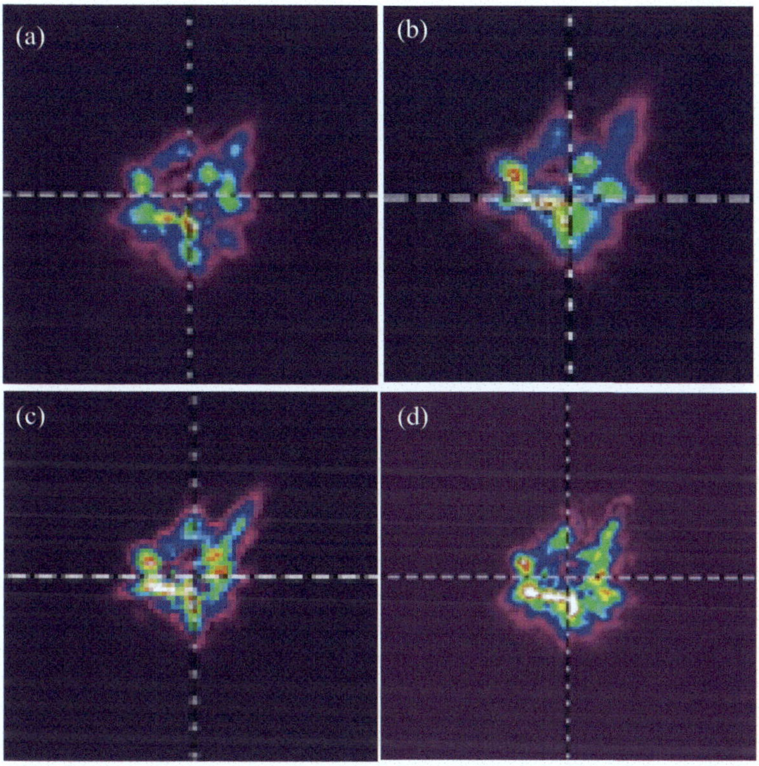

Fig. 9.7 Light spots at different time periods on May 2, 2018 [21]. **a** Moderate rain at 4:30 p.m. **b** Light rain at 4:30 p.m. **c** No rain at 8:30 p.m. **d** Cloudy at 10:30 p.m

Table 9.2 Coincidence and roughness results of the Gaussian fit for different time periods on May 2, 2018 [21]

Time of experiment	4:30 p.m (%)	4:30 p.m (%)	8:30 p.m (%)	10:30 p.m (%)
Coincidence of two-dimensional Gaussian fit	42	65	72	78
Roughness of two-dimensional Gaussian fit	99.7	85.1	81.4	71

9.5 Effect of the Beam-Polarization State on the Sensitivity of Partially Coherent Optical Heterodyne Detection

In Sect. 4.6, we discussed in detail the issue of heterodyne-detection-system sensitivity when a partially coherent EGSM beam travels in an oblique range under atmospheric turbulence. This section analyzes the change in the coherent-detection sensitivity when the signal light and local-oscillator light in the heterodyne-detection system are EGSM beams, in the case of a horizontal transmission, and their polarization states are different. Determining the best combination of polarization states of the two beams can produce the best mixing performance, which can provide an important theoretical basis for the design of a coherent detection system.

9.5.1 Detection Sensitivity of Partially Coherent EGSM Beams for Coherent Detection

When a beam is horizontally transmitted to the receiver, the wave-structure function in Eq. (4.78) can be expressed as [19]

$$\langle \exp[\psi^*(\boldsymbol{\rho_1}, \boldsymbol{r}_1, z_\beta) + \psi(\boldsymbol{\rho}_2, \boldsymbol{r}_2, z_\beta)] \rangle \cong \exp[-\frac{1}{2}(M\boldsymbol{r}_d^2 + \boldsymbol{r}_d\boldsymbol{\rho}_d + \boldsymbol{\rho}_d^2)], \quad (9.43)$$

where $\boldsymbol{\rho}_d = \boldsymbol{\rho}_1 - \boldsymbol{\rho}_2$, $\boldsymbol{r}_d = \boldsymbol{r}_1 - \boldsymbol{r}_2$, and M can be expressed as [19]

$$M = \frac{1}{3}\pi^2 k^2 z \int_0^\infty \kappa^3 \Phi_n(\kappa) d\kappa. \quad (9.44)$$

The non-Kolmogorov turbulence spectrum was used, that is, Eq. (4.81). The existing non-Kolmogorov turbulence power spectrum does not consider the equivalence of the refractive-index structure constants of non-Kolmogorov and Kolmogorov turbulence; hence, the relationship between the turbulence statistics and the power law of the power spectrum is very complex. To simplify the problem, C_n^2 is used as the physical quantity characterizing the atmospheric-turbulence intensity. The non-Kolmogorov turbulence refractive-index fluctuation power-spectrum model of a horizontal link is established considering the equivalence between $\tilde{C}_n^2$ and C_n^2 [31]:

$$\Phi_n(\kappa) = h(\alpha)\frac{\exp[-\kappa^2/\kappa_m^2]}{(\kappa^2 + \kappa_0^2)^{\alpha/2}}, \quad 0 \le \kappa < \infty, \quad 3 < \alpha < 4, \quad (9.45)$$

where

$$h(\alpha) = -\frac{\Gamma(\alpha)(k/z)^{\alpha/2-11/6}C_n^2}{8\pi^2\Gamma(1-0.5\alpha)[\Gamma(0.5\alpha)]^2\sin(0.25\pi\alpha)}. \quad (9.46)$$

Then, under non-Kolmogorov turbulence, M can be written as

$$M = -\frac{1}{3}\pi^2 k^2 z \frac{1}{2(\alpha-2)} \cdot \frac{\Gamma(\alpha)(k/z)^{\alpha/2-11/6}C_n^2}{8\pi^2\Gamma(1-0.5\alpha)[\Gamma(0.5\alpha)]^2\sin(0.25\pi\alpha)} \times \left[\kappa_m^{2-\alpha}\beta\exp\left(\frac{\kappa_0^2}{\kappa_m^2}\right)\Gamma\left(2-\frac{\alpha}{2}, \frac{\kappa_0^2}{\kappa_m^2}\right) - 2\kappa_0^{4-\alpha}\right], \quad (9.47)$$

Where $\beta = 2\kappa_0^2 - 2\kappa_m^2 + \alpha\kappa_m^2$.

Substituting Eq. (9.43) into Eq. (4.77) and using the integral operation yields

$$J_{\beta ij}(\rho_1, \rho_2, z_\beta) = \frac{I_{\beta ij}}{\Delta_{\beta ij}} \exp\left[-\left(\frac{1}{2\theta_{\beta ij}^2 \Delta_{\beta ij}} + (1 + \frac{2}{\Delta_{\beta ij}})M - \frac{M_2^2 z_\beta^2}{2w_\beta^2 k^2 \Delta_{\beta ij}}\right)(\rho_1 - \rho_2)^2\right]$$
$$\times \exp\left[-\frac{1}{8w_\beta^2 \Delta_{\beta ij}}(\rho_1 + \rho_2)^2\right] \times \exp[-\mathrm{i}kT_{\beta ij}(\rho_1^2 - \rho_2^2)], \tag{9.48}$$

where

$$\Delta_{\beta ij} = 1 + \left(\frac{z_\beta}{kw_\beta \theta_{\beta ij}}\right)^2 + \frac{2z_\beta^2 M}{k^2 w_\beta^2}$$
$$\frac{1}{\theta_{\beta ij}^2} = \frac{1}{4w_\beta^2} + \frac{1}{\delta_{\beta ij}^2},\ T_{\beta ij} = \frac{k^2 w_\beta^2 \Delta_{\beta ij} + M_2 z_\beta^2 - k^2 w_\beta^2}{k^2 z_\beta w_\beta^2 \Delta_{\beta ij}}. \tag{9.49}$$

By substituting Eq. (9.48) into Eq. (4.72), the power of the IF signal can be expressed as follows, after a polar-coordinate operation:

$$P_{\mathrm{IF}ij} = 2R^2 \int_0^{2\pi}\int_0^{2\pi}\int_0^{D/2}\int_0^{D/2} \mathrm{Re}[J_{\mathrm{o}ij}(\rho_1, \rho_2, z_\beta)$$
$$\times J_{\mathrm{s}ij}^*(\rho_1, \rho_2, z_\beta)]\rho_1\rho_2 \mathrm{d}\rho_1 \mathrm{d}\rho_2 \mathrm{d}\varphi_1 \mathrm{d}\varphi_2 \tag{9.50}$$

In the calculation, using the relation $W^2 = D^2 / 8$ [19], Eq. (9.50) can be expressed as

$$P_{\mathrm{IF}ij} = 2R^2 \frac{I_{\mathrm{o}ij} I_{\mathrm{s}ij}}{\Delta_{\mathrm{o}ij}\Delta_{\mathrm{s}ij}} \int_0^{2\pi}\int_0^{2\pi}\int_0^{\infty}\int_0^{\infty}$$
$$\mathrm{Re}\left\{\frac{I_{\mathrm{o}ij}}{\Delta_{\mathrm{o}ij}} \exp\left[-\left(\frac{1}{2\theta_{\beta ij}^2 \Delta_{\beta ij}} + (1 + \frac{2}{\Delta_{\beta ij}})M - \frac{M_2^2 z_\beta^2}{2w_\beta^2 k^2 \Delta_{\beta ij}}\right)(\rho_1 - \rho_2)^2\right]\right.$$
$$\times \exp\left[-\frac{1}{8w_0^2 \Delta_{\mathrm{o}ij}}(\rho_1 + \rho_2)^2\right] \times \exp[-\mathrm{i}kT_{\mathrm{o}ij}(\rho_1^2 - \rho_2^2)]$$
$$\times \frac{I_{\mathrm{s}ij}}{\Delta_{\mathrm{s}ij}} \exp\left[-\left(\frac{1}{2\theta_{\beta ij}^2 \Delta_{\beta ij}} + \left(1 + \frac{2}{\Delta_{\beta ij}}\right)M - \frac{M_2^2 z_\beta^2}{2w_\beta^2 k^2 \Delta_{\beta ij}}\right)(\rho_1 - \rho_2)^2\right]$$
$$\exp\left[-\frac{1}{8w_\mathrm{s}^2 \Delta_{\mathrm{s}ij}}(\rho_1 + \rho_2)^2\right]$$
$$\left.\times \exp[\mathrm{i}kT_{\mathrm{s}ij}(\rho_1^2 - \rho_2^2)] \exp\left[-\frac{\rho_1^2 + \rho_2^2}{W^2}\right]\right\}\rho_1\rho_2 \mathrm{d}\rho_1 \mathrm{d}\rho_2 \mathrm{d}\varphi_1 \mathrm{d}\varphi_2. \tag{9.51}$$

Using the relation $\boldsymbol{\rho}_1\boldsymbol{\rho}_2 = \rho_1\rho_2\cos(\varphi_1 - \varphi_2)$, Eq. (9.50) can be arranged as

$$\begin{aligned} P_{\mathrm{IF}ij} &= 2R^2\frac{I_{\mathrm{o}ij}I_{\mathrm{s}ij}}{\Delta_{\mathrm{o}ij}\Delta_{\mathrm{s}ij}}\int_0^{2\pi}\int_0^{2\pi}\int_0^{\infty}\int_0^{\infty}\mathrm{Re}\{\exp[-a_{ij}(\rho_1^2+\rho_2^2)] \\ &\times \exp[\mathrm{i}c_{ij}(\rho_1^2-\rho_2^2)] \\ &\times \exp[2b_{ij}\rho_1\rho_2\cos(\varphi_1-\varphi_2)]\}\rho_1\rho_2\mathrm{d}\rho_1\mathrm{d}\rho_2\mathrm{d}\varphi_1\mathrm{d}\varphi_2, \end{aligned} \tag{9.52}$$

where

$$\begin{aligned} a_{ij} &= \frac{1}{2}\left(\frac{1}{\theta_{\mathrm{o}ij}^2\Delta_{\mathrm{o}ij}}+\frac{1}{\theta_{\mathrm{s}ij}^2\Delta_{\mathrm{s}ij}}\right)+2M\left(1+\frac{1}{\Delta_{\mathrm{o}ij}}+\frac{1}{\Delta_{\mathrm{s}ij}}\right)-\frac{1}{2}\left(\frac{M_2^2z_o^2}{w_o^2k^2\Delta_{\mathrm{o}ij}}+\frac{M_2^2z_s^2}{w_s^2k^2\Delta_{\mathrm{s}ij}}\right) \\ &+\frac{1}{8}\left(\frac{1}{w_0^2\Delta_{\mathrm{o}ij}}+\frac{1}{w_s^2\Delta_{\mathrm{s}ij}}\right)+\frac{1}{W^2} \\ b_{ij} &= \frac{1}{2}\left(\frac{1}{\theta_{\mathrm{o}ij}^2\Delta_{\mathrm{o}ij}}+\frac{1}{\theta_{\mathrm{s}ij}^2\Delta_{\mathrm{s}ij}}\right)+2M\left(1+\frac{1}{\Delta_{\mathrm{o}ij}}+\frac{1}{\Delta_{\mathrm{s}ij}}\right)-\frac{1}{2}\left(\frac{M_2^2z_o^2}{w_o^2k^2\Delta_{\mathrm{o}ij}}+\frac{M_2^2z_s^2}{w_s^2k^2\Delta_{\mathrm{s}ij}}\right) \\ &-\frac{1}{8}\left(\frac{1}{w_o^2\Delta_{\mathrm{o}ij}}+\frac{1}{w_s^2\Delta_{\mathrm{s}ij}}\right) \\ c_{ij} &= k(T_{\mathrm{s}ij}-T_{\mathrm{o}ij}). \end{aligned} \tag{9.53}$$

After simplification, it can be expressed as

$$P_{\mathrm{IF}ij} = \frac{2\pi^2R^2I_{\mathrm{o}ij}I_{\mathrm{s}ij}}{\Delta_{\mathrm{o}ij}\Delta_{\mathrm{s}ij}(a_{ij}^2+c_{ij}^2-b_{ij}^2)}, \quad (i,j=x,y). \tag{9.54}$$

The local optical power can be expressed as

$$\begin{aligned} P_{\mathrm{o}ii} &= R\iint \boldsymbol{J}_{\mathrm{o}ii}(\boldsymbol{\rho},\boldsymbol{\rho})\mathrm{d}^2\boldsymbol{\rho} = R\int_0^{2\pi}\int_0^{\infty}\boldsymbol{J}_{\mathrm{o}ii}(\boldsymbol{\rho},\boldsymbol{\rho})\exp\left(-\frac{\rho^2}{W^2}\right)\rho\mathrm{d}\rho\mathrm{d}\varphi \\ &= \frac{2R\pi I_{\mathrm{o}ii}W^2w_o^2}{W^2+2w_o^2\Delta_{\mathrm{o}ii}}, \quad (i=x,y). \end{aligned} \tag{9.55}$$

Similarly, the signal optical power can be written as

$$\begin{aligned} P_{\mathrm{s}ii} &= R\iint \boldsymbol{J}_{\mathrm{s}ii}(\boldsymbol{\rho},\boldsymbol{\rho})\mathrm{d}^2\boldsymbol{\rho} = R\int_0^{2\pi}\int_0^{\infty}\boldsymbol{J}_{\mathrm{s}ii}(\boldsymbol{\rho},\boldsymbol{\rho})\exp\left(-\frac{\rho^2}{W^2}\right)\rho\mathrm{d}\rho\mathrm{d}\varphi \\ &= \frac{2R\pi \mathrm{I}_{\mathrm{s}ii}\mathrm{W}^2\mathrm{w}_s^2}{\mathrm{W}^2+2\mathrm{w}_s^2\Delta_{\mathrm{s}ii}}, \quad (i=x,y). \end{aligned} \tag{9.56}$$

After substituting Eqs. (9.55) and (9.56) into Eqs. (4.101) and (4.102), the sensitivity expressions of the partially coherent EGSM detection system for horizontal transmission can be obtained.

9.5.2 *Influence of the Polarization State on the Exodyne-Detection Sensitivity*

To maintain fixed beam-polarization characteristics in the atmospheric channel [32, 33], the beam spatial lengths of the signal light and local-oscillator light were set as $\delta_{\mathrm{o}xy} = \delta_{\mathrm{o}yx} = 0.2$ mm, $\delta_{\mathrm{o}xx} = \delta_{\mathrm{o}yy} = 0.2$ mm, $\delta_{\mathrm{s}xx} = \delta_{\mathrm{s}yy} = 0.4$ mm, and $\delta_{\mathrm{s}xy} = \delta_{\mathrm{s}yx} = 0.4$ mm.

First, the numerical values of the global-variable simulation parameters are given. Unless otherwise specified, the numerical values of the parameters are listed in Table 9.3. Here, we mainly analyze the detection-sensitivity values under different polarization-state combinations of signal light and local-oscillator light for a given bit-error rate, that is, BER $= 10^{-9}$.

According to Eq. (4.91), when the values of $I_{\mathrm{s}xx} = I_{\mathrm{s}yy} = 1$, $I_{\mathrm{o}xx} = I_{\mathrm{o}yy} = 1$, $I_{\mathrm{s}xy} = I_{\mathrm{s}yx} = 1$, and $I_{\mathrm{o}yx} = I_{\mathrm{o}xy} = 0$, 0.5, and 1, the corresponding polarization degree P is 0, 0.5, and 1, respectively. At this time, the signal light is fully polarized.

When $I_{\mathrm{s}xx} = I_{\mathrm{s}yy} = 1$, $I_{\mathrm{o}xx} = I_{\mathrm{o}yy} = 1$, $I_{\mathrm{s}xy} = I_{\mathrm{s}yx} = 0.5$, and $I_{\mathrm{o}yx} = I_{\mathrm{o}xy} = 0$, 0.5, and 1, the corresponding polarization degree P is 0, 0.5, and 1, respectively, and the signal light is partially polarized.

When $I_{\mathrm{s}xx} = I_{\mathrm{s}yy} = 1$, $I_{\mathrm{o}xx} = I_{\mathrm{o}yy} = 1$, $I_{\mathrm{s}xy} = I_{\mathrm{s}yx} = 0$, and $\mathrm{I}_{\mathrm{oyx}} = \mathrm{I}_{\mathrm{oxy}} = 0$, 0.5, and 1, the corresponding polarization degree P can be obtained as 0, 0.5, and 1, respectively. At this point, the signal light is unpolarized.

Figure 9.8a, b show the change curves of the system detection sensitivity with the power-spectrum power law when the signal light is fully and partially polarized, respectively, after mixing it with local light in different polarization states.

It can be seen from the figure that, with a horizontal transmission path, the detection sensitivity of the system first increases and then gradually tends to a constant value, with the increase of the power-spectrum power law. When the local-oscillator light is fully polarized, the detection sensitivity of the system reaches the maximum, regardless of whether the signal light is fully or partially polarized. When the local-oscillator light is unpolarized, the detection sensitivity of the system reaches a minimum, regardless of whether the signal light is fully or partially polarized. When

Table 9.3 Values of simulation parameters

Parameter name	Parameter value	Parameter name	Parameter value
Wavelength λ	1550 nm	Atmospheric structure constant C_n^2	5×10^{-14} m$^{-2/3}$
Detector diameter D	2 mm	Local-oscillator beam waist radius w_o	3 mm
Planck's constant h	6.623×10^{-34} J·s	Signal beam waist radius w_s	3 mm
Quantum efficiency η	0.8	Carrier frequency v	1.9×10^{14} Hz
Propagation distance z_s	2000 m	Detector bandwidth B	200 MHz

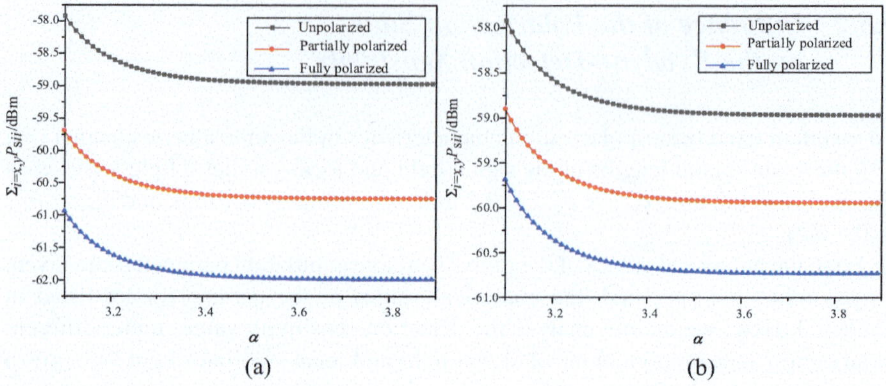

Fig. 9.8 Detection sensitivity versus the power-spectrum power law. **a** Signal light is fully polarized. **b** Signal light is partially polarized

the local light is partially polarized, the detection sensitivity of the system is between the maximum and minimum values.

While comparing Fig. 9.9a, b, it can be found that when the local-oscillator light is fully polarized, the detection sensitivity of the system when the signal light is fully polarized is 1.3 dBm higher than that when the signal light is partially polarized. When the local-oscillator light is partially polarized, the detection sensitivity of the system when the signal light is fully polarized is 1.4 dBm higher than that when the signal light is partially polarized.

Figure 9.9 shows the detection-sensitivity change of the system after the non-polarized EGSM signal beam is mixed with a local-oscillator light of different polarization states in coherent detection. The unpolarized signal light has no significant effect on the detection sensitivity of the system. This is because, for an unpolarized signal light, the interference effects between different polarization states will statistically cancel each other, so that it does not have a significant impact on the detection sensitivity of the system. Therefore, for an unpolarized signal light, choosing a specific polarization state for the local-oscillator light cannot significantly enhance or weaken the interference effect, so it has no obvious effect on the detection sensitivity of the system.

Figure 9.10 shows the detection-sensitivity change curve of the system when both the signal and local light are fully polarized. Figure 9.10a shows the detection-sensitivity changes of the system under different local-oscillator beam waist radii when the signal beam waist radius is 1 mm. It can be found that when the waist radius of the local-oscillator beam is larger, the detection sensitivity of the system is smaller. When the waist radii of the local-oscillator beam and the signal beam are equal, the detection sensitivity of the system is the smallest.

Figure 9.10b shows the detection-sensitivity curve of the system changing with the diameter of the photosensitive area of the detector under different turbulence

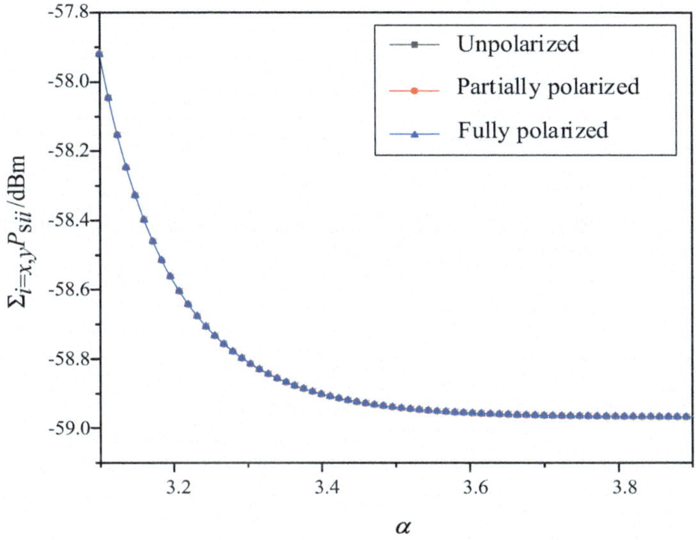

Fig. 9.9 Detection sensitivity versus the power-spectrum power law

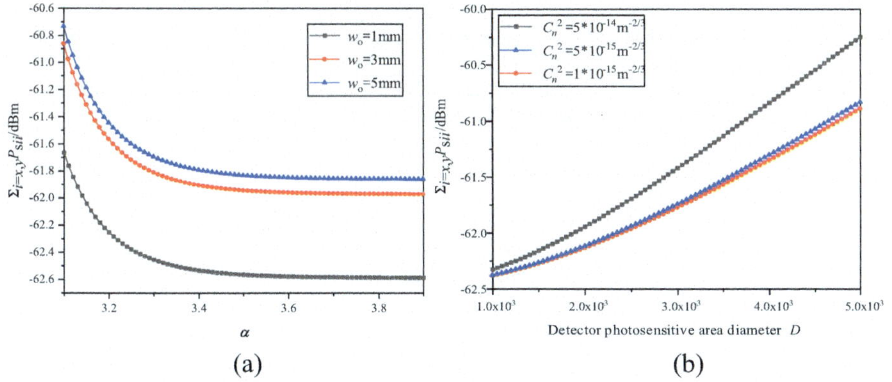

Fig. 9.10 Detection-sensitivity curve under different conditions. **a** Different local-oscillator beam waist radii. **b** Different turbulence intensities

intensities. It is obvious from the figure that, under a certain atmospheric refractive-index structure constant, the detection sensitivity of the system decreases with the increase of the detector diameter; the larger the atmospheric refractive-index structure constant, the more obvious this phenomenon.

This is due to the spatial-coherence limitation of the signal light; when the detector size is small, the signal light and the local-oscillator light match better, and the detection sensitivity of a system with a smaller detector size is greater.

References

1. Chernov LA (1960) Wave propagation in a random medium. McGraw-Hill, New York
2. Tatarskii VI (1961) Wave propagation in a turbulent medium. Dover Publication, New York
3. Klyatskin VI, Kon AI (1972) On the displacement of spatially-bounded light beams in a turbulent medium in the Markovian-random-process approximation. Radiophys Quantum Electron 15(9):1056–1061
4. Lutomirski RF, Yura HT (1971) Wave structure function and mutual coherence function of an optical wave in a turbulent atmosphere. J Opt Soc Am 61(4):482–487
5. Wu ZS (1986) Intensity correlation and scintillation in the saturated regime for wave propagation through random media. Chin J Radio Sci 01(01):76–88
6. Hill RJ (1982) Theory of saturation of optical scintillation by strong turbulence plane-wave variance and covariance and spherical-wave covariance. J Opt Soc Am 72(2):212–222
7. Hill RJ, Clifford SF (1981) Theory of saturation of optical scintillation by strong turbulence for arbitrary refractive-index spectra. J Opt Soc Am 71(6):675–686
8. Flatte SM, Gerber JS (2000) Irradiance-variance behavior by numerical simulation for plane-wave and spherical-wave optical propagation through strong turbulence. J Opt Soc Am A 17(6):1092–1097
9. Lahti JN, Nagel CM (1970) Mixing partially coherent fields with Gaussian irradiance profiles optimization criteria. Appl Opt 9(1):115–124
10. Chiou W (1979) Optical heterodyne detection of partially coherent radiation. In: 23rd Annual technical symposium, vol 194, pp 165–186
11. Tanaka T, Taguchi M et al (1992) Heterodyne efficiency for a partially coherent optical signal. Appl Opt 31:5391–5394
12. Ricklin JC, Davidson FM (2002) Atmospheric turbulence effects on a partially coherent Gaussian beam: implications for free-space laser communication. J Opt Soc Am A 19(9):1794–1802
13. Ricklin JC, Davidson FM, Weyrauch T et al (2002) Free-space laser communication using a partially coherent laser source. In: Proceedings of SPIE—The International Society for Optical Engineering, vol 4538, pp 13–23
14. Salem M, Dogariu A (2004) Optical heterodyne detection of random electromagnetic beams. J Mod Opt 51(15):2305–2313
15. Salem M, Rolland JP (2008) Effects of coherence and polarization changes on the heterodyne detection of stochastic beams propagating in free space. Opt Commun 281(20):5083–5091
16. Salem M, Rolland JP (2010) Heterodyne efficiency of a detection system for partially coherent beams. J Opt Soc Am A 27(5):1111–1119
17. Pu LL, Zhou Y, Sun JF et al (2011) Receiving characteristics of coherent ladar under partially coherent condition. Acta Optica Sinica 31(12):260–268
18. Ren Y, Dang A et al (2012) Heterodyne efficiency of a coherent free-space optical communication model through atmospheric turbulence. Appl Opt 51(30):7246–7254
19. Li CQ, Wang TF, Zhang HY et al (2015) The performance of heterodyne detection system for partially coherent beams in turbulent atmosphere. Opt Commun 356:620–627
20. Wang Y, Li CQ, Wang TF et al (2016) The effects of polarization changes of stochastic electromagnetic beams on heterodyne detection in turbulence. Laser Phys Lett 13(11):116006
21. Tan ZK (2019) Effects of factors on the performance of heterodyne detection and experimental investigation in wireless optical communication. Xi'an University of Technology
22. Xiao XF, Voelz D (2006) Wave optics simulation approach for partial spatially coherent beams. Opt Express 14(16):6986–6992
23. Ricklin JC, Davidson FM (2002) Atmospheric turbulence effects on a partially coherent Gaussian beam: implications for free-space laser communication. J Opt Soc Am A Opt Image Sci Vis 19(9):1794–1802
24. Wang L (2006) Propagation properties and spectral changes of partially coherent beams. Southwest Jiaotong University, Chengdu

25. Zhang XX, Dan YB, Zhang B (2012) Spreading of partially coherent flat-topped beams propagating along a slant path in turbulent atmosphere. Acta Optica Sinica 32(12):1201001-1–1201001-17
26. Liu C, Chen SQ, Li XY et al (2014) Performance evaluation of adaptive optics for atmospheric coherent laser communications.Optics Express 22(13):15554–15563
27. Wang QZ, Hu Y, Lin CJ (1993) Photoelectric detection technique. Electronic Industry Press, Chengdu
28. Osche GR (2002) Optical detection theory. Wiley, New York
29. Li CQ (2016) Investigating the effect of optics factors on the performance of laser heterodyne detection system. University of Chinese Academy of Sciences
30. Tanaka K, Ohta N (1987) Effects of tilt and offset of signal field on heterodyne efficiency. Appl Opt 26(4):627–632
31. Chao ZHAI, Li-ying TAN, Si-yuan YU et al (2015) Fiber coupling efficiency for a Gaussian-beam wave propagating through non-Kolmogorov turbulence. Opt Express 23(12):15242–15255
32. Jiang GY,Chen M,Xiao Y.Effect of the turbulent atmosphere on DOP of light beams. Ship Electron Eng 40(06):180–186
33. Du XY, Zhu YB, Zhao DM (2009) Propagation properties of completely unpolarized beams and completely polarized beams. Acta Optica Sinica 29:79–83

The manufacturer's authorised representative in the EU is Springer Nature Customer Service Centre GmbH, Europaplatz 3, 69115 Heidelberg, Germany. If you have any concerns regarding our products, please contact ProductSafety@springernature.com

Printed and bound by CPI Group (UK) Ltd, Croydon, CR0 4YY
01/10/2026
02225926-0016